# Secondary Surveillance Radar

# Secondary Surveillance Radar

Michael C. Stevens
*Chief Scientist, Aviation Systems Division, Cossor Electronics Ltd*

Artech House
Boston and London

**British Library Cataloguing in Publication Data**

Stevens, Michael C.
Secondary surveillance radar.
1. Secondary radar
I. Title
621.3848

ISBN 0-89006-292-7

**Library of Congress Cataloging-in-Publication Data**

Stevens, Michael C., 1936–
Secondary surveillance radar.

Bibliography: p.
Includes index.
1. Secondary surveillance radar. I Title.
TL696.R25S84 1988 629.136′6 88–3301

ISBN 0–89006–292–7



**ARTECH HOUSE, INC.**
**685 Canton Street**
**Norwood, MA 02062**



**International Standard Book Number: 0-89006-292-7**
**Library of Congress Catalog Card Number: 88-3301**

**10 9 8 7 6 5 4 3 2 1**

# Contents

# Acknowledgments

The assistance of the following people and organizations with the preparation of this book is gratefully acknowledged: Donald Aaron, Michel Biot, Kevin Bond, John Boyachoc, David Clifford, Nicholas Comyn, Claude Eymann, Michael Gee, Clinton MacNeil, Vincent Orlando, Philip Platt, Melvin Rees, Terence Roblett, John Shaw, David Spalding, Peter Stanley, James Thurgood, Richard Trim, Brian Watson and Brian Wyndham; Allied Bendix Aerospace, Civil Aviation Authority, Eurocontrol, Hazeltine Corporation, Lincoln Laboratory, Royal Signals and Radar Establishment, Sivers IMA, Télécommunications Radioélectroniques et Téléphoniques, and especially Cossor Electronics Ltd.

## Chapter 1

# Introduction

Civil aviation is concerned with the safe and expeditious movement of people and freight from one point on the globe's surface to some other desired point. The responsibility for this safe passage is that of the pilots of the aircraft involved. In this responsibility they are assisted by ground-based systems and personnel. Safety is maintained by assuring adequate separation of aircraft, and to this end appropriate procedures and recommended practices have been devised by the International Civil Aviation Organization (ICAO). Many countries provide an air traffic control (ATC) service in at least part of their airspace. Pilots are required to file flight plans before their departure so that their intentions are understood and their flight can be integrated with other air traffic.

The degree of separation required for aircraft is influenced by the quality and accuracy of the position data available. In sparsely populated areas external aids to the pilot can be few and far between and separations of as much as 30 min of travel can be set for aircraft on the same course. These procedural separations are very generous since they are dependent on the aircraft's own navigation accuracy and allowance must be made for possible errors in navigation. However, they are inappropriate in areas of dense traffic, and the use of ground radar is required in order to handle demands on airspace without prejudicing safety.

The word "radar", which is an acronym of RAdio Detection And Ranging is usually taken to mean what is more accurately called primary radar. In primary radar a high power pulse of radio frequency energy is transmitted from a rotating antenna and if this strikes the body of an aircraft then a small amount of that energy is reflected back towards the rotating antenna. The skill of the primary radar design engineer is directed to the detection of this small amount of reflected energy. When an aircraft has been detected in this way its range can be determined by measuring the elapsed time between the transmission and subsequent reception of the reflection since the speed of a radio wave is known. The aircraft's bearing is indicated by the direction of the antenna at the time that the aircraft is detected.

Like any system, primary radar has its good and bad points. The major good point is that, for its operation, it does not require anything special of the aircraft other than that it presents a reasonable radar-reflecting surface. The bad points are that primary radar also obtains returns from rain, from the ground and even from birds, and these extraneous returns make it difficult to distinguish between real aircraft and the background clutter. Many special techniques have been developed to overcome these problems with some considerable success. However, primary radar is unable to distinguish one aircraft from another similar aircraft and in most cases cannot determine height to sufficient accuracy.

Secondary radar overcomes these difficulties. Secondary radar, more precisely secondary surveillance radar (SSR), requires each aircraft to carry on board a special transmitter–receiver equipment known as a transponder. An SSR ground station transmits pulses of radio frequency energy from a rotating antenna in the same way as for primary radar except at a very much lower power level. Only sufficient power to overcome the one-way path loss to the transponder is required compared with the two-way path loss for primary radar. SSR transmits pulses of a few hundred watts in power compared with a few hundreds or thousands of kilowatts in the case of primary radar. Upon reception of the ground-station signal, the transponder transmits a reply on a different frequency. The use of a separate frequency prevents the reply from being confused with possible returns from rain, birds or terrain. The aircraft range and bearing are determined from the time delay and antenna direction in the same way as for primary radar.

The main advantage of SSR is that it can obtain extra information from the aircraft. The availability of the on-board SSR equipment allows a rudimentary data link to be added to the radar function. The ground SSR station codes its transmission by emitting pairs of pulses of different spacing. At the same time as determining the aircraft's range and bearing the ground station can also ask the aircraft "Who are you?" or "What is your altitude?". The response of the airborne transponder is in the form of a multiple-pulse reply carrying as data either its identity number or its flight level. The ability of SSR to obtain this additional information about the aircraft is of immense value to the air traffic controller on the ground. As a result SSR is becoming the major sensor for ATC.

Other advantages of SSR include the ability to obtain long-range cover (over 200 nautical miles if required) at lower cost, as SSR is only a fraction of the cost of an equivalent primary radar. The latter fact is producing a trend to the use of SSR only at *en route* sites away from major airports.

The SSR ground equipment comprises a rotating antenna, a tower with antenna-turning equipment, a transmitter–receiver, which is usually called an interrogator, and a reply-signal processor known as a plot extractor or digitizer. The ground station transmits to the aircraft on a frequency of 1030 MHz and receives the aircraft's response on a frequency of 1090 MHz. Figure 1.1 shows a typical SSR antenna, in this case mounted on a colocated primary radar antenna with which it shares the tower and turning gear. The SSR interrogator and a plot extractor are normally installed in an equipment room, and examples are shown in Fig. 1.2. The plot extractor converts the reply data into a target report for each aircraft and sends these reports via land lines to ATC centers. The target report data are displayed to the air traffic controller in the form of a map-like presentation. The display shows the position of each aircraft, with either its SSR identity number or its associated

**Fig. 1.1** SSR antenna (Cossor type CRS512) mounted on a primary radar antenna (TI type ASR8) at Geneva airport.

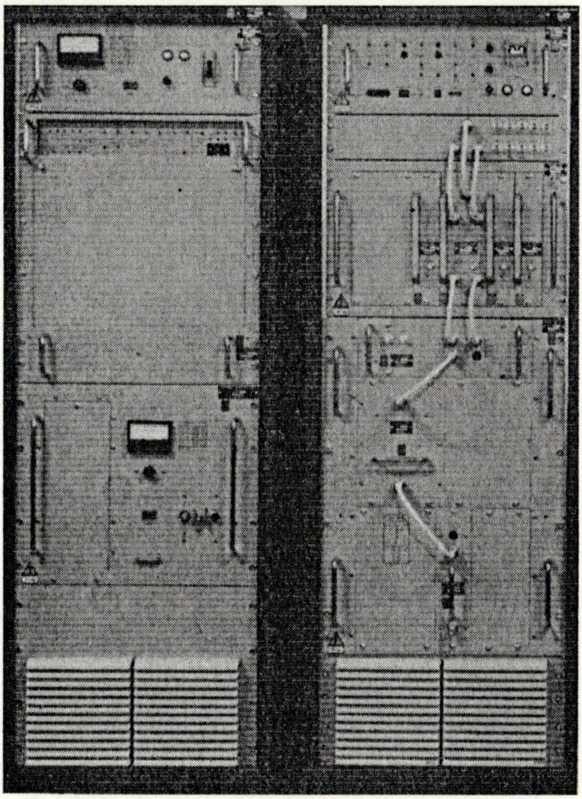

**Fig. 1.2** Typical interrogator (right) and plot extractor (left) (Cossor types 9610 and 9620 respectively).

flight number, together with its height. An example of such an ATC display is shown in Fig. 1.3.

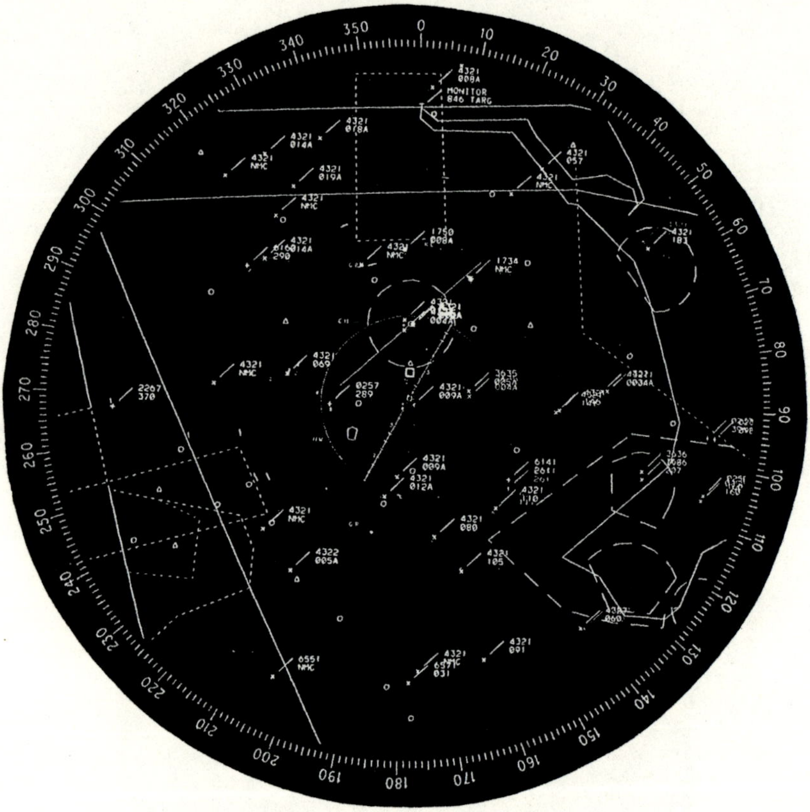


**Fig. 1.3** Modern ATC raster scan display (Cossor type 9691).

The aircraft equipment comprises an antenna, which is usually mounted on the under surface of the aircraft's fuselage, a transmitter–receiver known as a transponder which detects the signal transmitted from the ground station and transmits a reply back to it, a control unit by which the pilot can set the SSR identity number as allocated by ATC and an encoding barometer for indicating aircraft height. An example of a transponder and control unit is shown in Fig. 1.4.

A considerable advance in the performance of SSR has been obtained in recent years through the application of monopulse direction-finding techniques. Monopulse requires the use of a ground antenna with an additional beam and a second receiver channel in order to measure the direction of arrival of each aircraft reply pulse. This technique improves the bearing measurement accuracy and also increases the ability to decode the aircraft replies accurately. In this book emphasis is placed on monopulse processing when describing ground receivers and plot extractors because of the importance of this development.

The airborne element—the transponder—is described in sufficient detail for the design and operation of the ground equipment to be understood.

SSR has attracted its share of jargon. A further complication is that the jargon differs in different parts of the world; for example, in the USA SSR is known as the Air Traffic Control Radar Beacon System (ATCRBS, pronoun-

(a)

(b)

**Fig. 1.4** (a) Airborne transponder and (b) control unit (Cossor type SSR2700).

ced at-crabs). Where an ICAO term exists, that is usually used here. A glossary is given at the end of the book to assist the reader.

Many sources of information were used in the compilation of this book. Principal amongst these are *Annex 10* of ICAO [1.1]. The discussion of monopulse plot extractors (Chapter 7, Sections 7.1–7.4) borrows heavily from the work of the Lincoln Laboratory of MIT [1.2, 1.3]. Much other material is taken from design specifications, published papers and unpublished work of Cossor Electronics Ltd.

## References

1.1 Aeronautical Telecommunications, *Annex 10 to the Convention on International Civil Aviation*, Vol. 1, International Civil Aviation Organization, Montreal, April 1985.

1.2 Gertz, J. L. The ATCRBS mode of DABS, *Lincoln Laboratory ATC-65*, *Rep. FAA-RD-76-39*, Lincoln Laboratory, MIT, Lexington, MA, 31 January 1977.

1.3 Nelson, R. G. and Nuckols, J. H. A hardware implementation of the ATCRBS reply processor used in DABS, *Lincoln Laboratory ATC-78*, *Rep. FAA-RD-77-92*, Lincoln Laboratory, MIT, Lexington, MA, 19 September 1977.

## Chapter 2

# History

The exact date when radar was invented is a matter of debate and of word definition. Thomas Edison filed a patent in 1885 on the use of radio for the prevention of collision at sea. Such an application was also considered, independently, by Edouard Branley, Nikola Tesla and Christian Hulsmeyer, with the latter demonstrating equipment using a frequency of 600 MHz in 1904. In the 1920s and early 1930s workers in many different countries used radio waves for echo sounding off the ionosphere, for altimetry and for conducting investigations into the perturbations in radio links caused by reflecting objects. The use of radio for the specific purpose of the detection of objects was proposed in many different countries including Germany, Italy, the USA, France, the UK, Russia, Japan and Holland [2.1].

Radar had to happen and its origin was perhaps more of a discovery than an invention. In the buildup to the Second World War many countries again recognized its military importance although for obvious reasons did not give it widespread publicity. The UK was more fortunate than most in having people in positions of influence who had sufficient vision and determination to brush aside the doubts, bureaucratic obstacles and conservatism to force through the development of radar.

The role that radar played during the war is well known. Development was rapid, as it was with other electronic navigation aids which were to form the basis of civil systems when hostilities had ceased. In wartime the ability to detect aircraft by the use of radar was valuable, but with it went the need to discriminate between friend and foe. Equipment was developed, to be carried by friendly aircraft, which would detect the Allies ground radar transmissions and transmit a signal to indicate that the airplane was a friend. This system, known as Identification Friend or Foe (IFF), was the mother of civil secondary surveillance radar (SSR).

The IFF Mk I, later followed by Mks II and III, was developed by F. C. Williams and Dr Vivian Bowden (later Lord Bowden) [2.2, 2.3]. Mks I and II were designed to respond to the ground radar transmission, while Mk III used a separate signal transmitted in the frequency band 157–187 MHz.

In 1943 Dr Bowden moved to Washington to head the British contingent of the Combined Research Group located at the Naval Research Laboratories. Considerable resources were made available to the Group, including those of the Hazeltine Corporation who had assisted in adapting the UK Mk III sets for mass production in the USA. The Mk V IFF, which was also known as the United Nations Beacon (UNB), was produced; this operated in the 950–1150 MHz band, the same as used by today's SSR.

Further US development after the war gave rise to the IFF Mk X (X initially meant "undefined" but was later taken to be the Roman numeral 10) which operated on 12 channels, at a nominal 17 MHz spacing, from 952 to 1139 MHz. The basic Mk X IFF had very simple interrogation and reply signals which did not permit the individual identification of replying aircraft. This omission was made good by the development of Mk X (SIF), where SIF stands for selective identification facility. The interrogations were confined to 1030 MHz and the replies to 1090 MHz. Replies consisted of pulse trains up to 15 pulses in length which allowed for individual aircraft identification. Mk X (SIF), with further refinements, is still in use today.

After the war thought was given to the civilian uses of the wartime developments, and allocations of the frequency bands for civil aviation were made at the 1947 International Telecommunications Union (ITU) Conference. Secondary radar was given the three bands 3246–3266 MHz, 5440–5460 MHz and 9300–9320 MHz. The bands 960–1215 MHz and 1600–1660 MHz were allocated to worldwide aeronautical navigation, with a possible future use of the band 1300–1600 MHz for a worldwide integrated air navigation/traffic control system.

Ken Harris [2.3] of the Telecommunication Research Establishment (TRE), later called the Royal Radar Establishment (RRE) and then the Royal Signal and Radar Establishment (RSRE), had been convinced of the need for SSR and, after moving to Cossor in 1950, worked with the support of the Ministry of Civil Aviation to produce a demonstration secondary radar equipment. Figure 2.1 shows the SSR antenna mounted on an AN/CPS-1 primary antenna as used during its trials at London Heathrow Airport in 1952 and later in front of the US Civil Aeronautics Administration and Air Navigation Development Board. The antenna had a modicum of vertical aperture, although the "production" version (the Cossor type 1 SSR) employed a linear array, as can be seen in Fig. 2.2. The latter type of antenna had a small vertical aperture which resulted in a broad vertical pattern and formed the basis of SSR antennas produced throughout the world. It took nearly 30 years to appreciate the advantages of being able to shape the vertical radiation pattern with a larger vertical aperture.

The type 1 experimental equipment [2.4] was designed by Derek Levell, working under Ken Harris and with Duncan Terrington [2.5] of the UK National Air Traffic Services. The equipment used 1215 MHz for the ground-to-air interrogation of aircraft and 1375 MHz for the air-to-ground reply.

At this time other national authorities were beginning to recognize the value of a secondary radar system to assist civil ATC. The problem was achieving compatibility between the transponders, carried by aircraft, and the interrogators in all the countries that they were likely to visit. There was a need for international standardization.

Some early proposals favored the "dependent" SSR system, i.e. triggering of the airborne transponder by the primary radar pulse along the lines of the Mks I and II IFF. However, the problems of suppressing replies to

(a)

(b)

**Fig. 2.1** Original SSR trials equipment: (a) SSR antenna mounted on primary radar antennas; (b) airborne transponder system (from ref. 2.4).

sidelobe signals, polarization and the variety of radar signals and frequencies to be accommodated could not be solved satisfactorily. Dependent SSR was abandoned in favor of the "independent" SSR with similarities to Mk V IFF.

In the meantime IFF had developed to Mk X (SIF) with 1030 MHz being used for interrogation from the ground and 1090 MHz for aircraft replies, each with a frequency tolerance of $\pm 8.5$ MHz. Significantly, IFF Mk X was released from military security in late 1953 and made available for international civil use. The Mk X (SIF) reply signal format was very similar to the present SSR format in that two framing pulses, spaced 20.3 μs apart, were always transmitted with code pulse positions every 2.9 μs between the framing pulses. Only 10 reply codes, as shown in Fig. 2.3, were available, and they were indicated by various combinations of two code pulses which were readable on a raw video plan position indicator (PPI) display.

**Fig. 2.2** First production ground antenna.

At the ICAO Communication Division meeting in 1954 the Mk X frequencies were adopted for civil SSR as compatibility with military aircraft was considered to be important. Compatibility included specifying the civil Mode A interrogation mode to have the same characteristics as the military Mode 3. However, for some time, at least in the UK and France, Mode B was used as the principal operation mode. This was because of the limited performance of Mode A with its wide reply pulse tolerances and the limited number of reply codes, both necessary to accommodate the military equipment.

ICAO refined and developed the SSR standards and recommended practices in 1957 and 1962. In step with these developments Aeronautical Radio INCorporated (ARINC) produced characteristic 532 (in 1955) and 532B (in 1957) to describe the form and fit of the airborne transponder to meet the ICAO requirements. Richard Trim, from Cossor, took equipment to the USA to demonstrate interrogator sidelobe suppression to the Department of Defense and the Air Navigation Board and, following successful trials, the concept was adopted by ICAO.

Sidelobe suppression could be realized in two ways [2.6], one using two transmitted pulses and the other using three transmitted pulses. The UK favored the former and the USA the latter. For a while ICAO supported both, and transponders were designed that could operate with either. A disadvantage of the two-pulse method was the need for a much higher transmitter

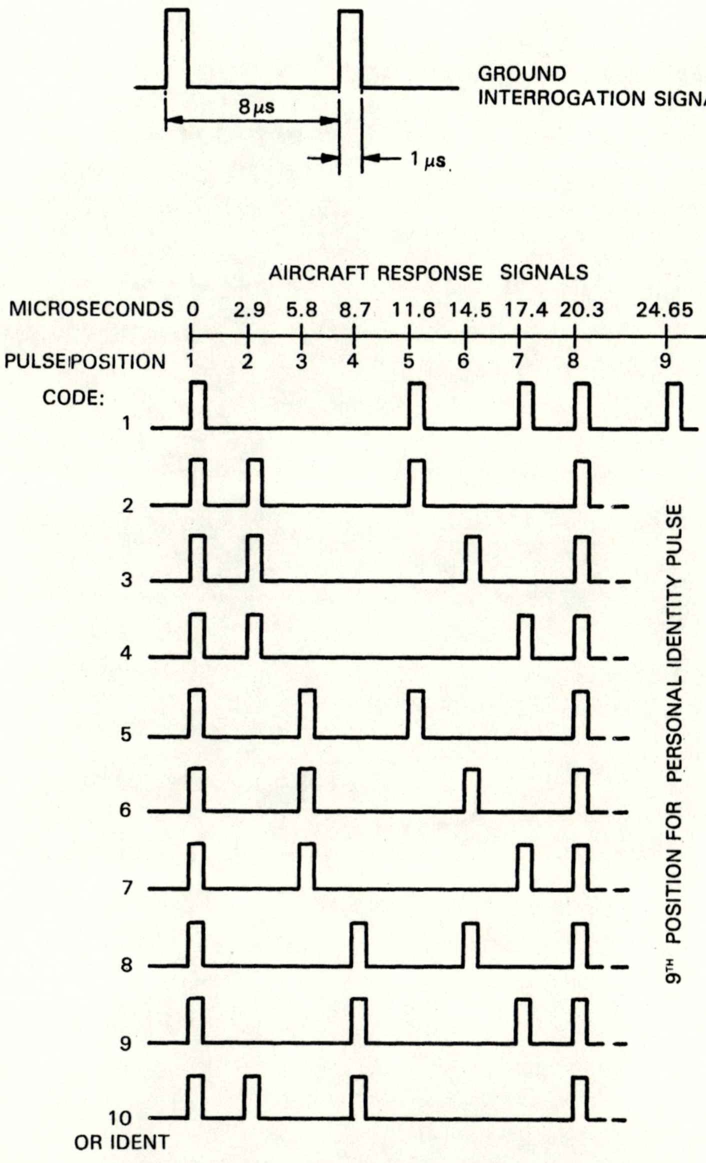


**Fig. 2.3** Original 10 IFF Mk X (SIF) reply codes.

power for one of the pulses. Eventually ICAO abandoned its support for the two-pulse method.

An experimental ground interrogator equipment, the SSR4G [2.7], was built to the ICAO specification and installed at both London Heathrow and Paris Orly. The SSR1251 transponder served as the airborne component. This

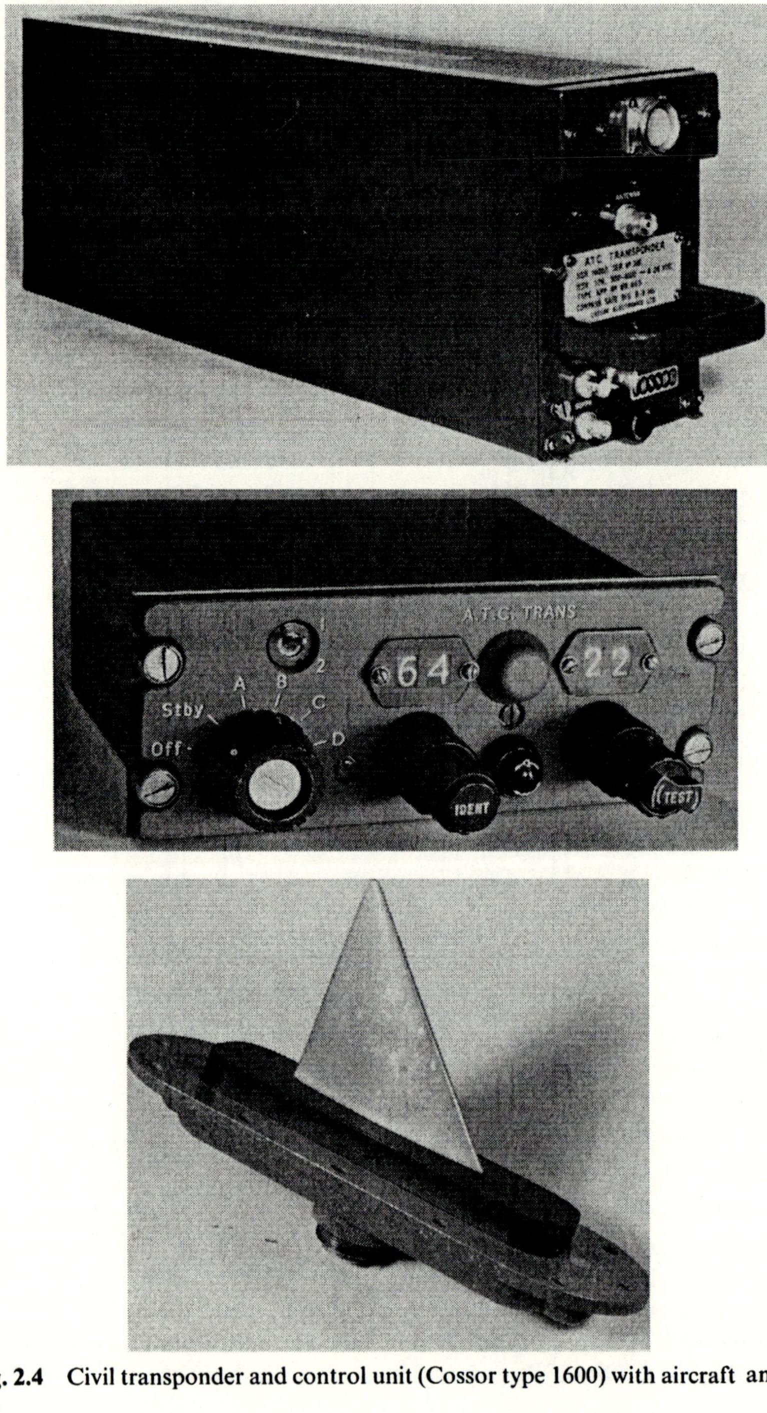


**Fig. 2.4** Civil transponder and control unit (Cossor type 1600) with aircraft antenna.

transponder used transistors for video processing but vacuum tubes for r.f. and i.f. amplification. Following further revised ICAO recommendations in 1962 the UK Ministry of Aviation purchased the SSR5G interrogator for operational use. Transponders were further developed and now incorporated transistors throughout except for the 1030 MHz transmitter. Figure 2.4 shows typical equipment of this era.

By the mid-1960s SSR had become operational and was obtaining increasing acceptance by national authorities. However, there was also some concern that SSR might become a victim of its own success, and in 1955–1956 both Vickers and Crippen in the USA [2.8] and Harris in the UK [2.6] published papers describing some of the problems. The problems foreseen at that time included sidelobe responses, loss of replies owing to capture, fruit, overinterrogation and lobing of antenna vertical patterns. Some mention of an improved system is given in a paper by Ullyatt of RRE in 1967 [2.9] with details in a second paper in 1969 [2.10].

In the USA the Air Traffic Control Advisory Committee, chaired by Ben Alexander, also proposed system improvements [2.11]. The following is an extract from an Airlines Electronic Engineering Committee (AEEC) newsletter [2.12] written by Bill Carnes in October 1969.

> "In view of the feared 'beacon overloading' it may come as a surprise that the Alexander Committee urges extensive R & D efforts to make it do many more tasks. This so called 'Super Beacon' envisaged by the Alexander Committee has been stated to be 'compatible' with the present SSR system and present airborne hardware but would simply require an additional feature in the aircraft to make it do those additional tasks. Insisting upon strict evolutionary compatibility but lacking the R & D background of proof the Alexander Committee may have retreated from their own ultimatum inasmuch as they seem to have suggested additional frequency channels to accommodate the Super Beacon!
>
> But the Alexander Committee apparently did not know of an R & D program in the UK already underway to make a 'Super Beacon' play in a completely compatible manner with existing beacons. Although publicized as early as 1963 in the United Kingdom's Royal Radar Establishment, it was not widely known that this program was still progressing with extremely encouraging results. To our knowledge no one in the RRE had been aware of the interest by the Alexander Committee in expanding the SSR system. But the US and UK people suddenly discovered this mutual commonality of interest and effort on Thursday September 18 at 11:22 a.m. in a technical session chaired by Tom McWiggan, Chief of Telecommunications of the UK Board of Trade. This session was part of the five day United Kingdom Symposium on Electronics for Civil Aviation at London's Royal Festival Hall. The occasion was the presentation of a paper by Mr C. Ullyatt of the Royal Radar Establishment entitled 'Sensors for the ATC Environment with Special Reference to SSR'.
>
> But only a small handful of the audience knew how that paper might relate to the presentation in the 1963 symposium held at RRE, and its

possible relationship to 'the Alexander Committee Super Beacon'. Suggestions for such a change were certainly not timely in 1963 and the UK government had slowed the development down to a mere crawl in the full knowledge that the SSR system needed to get full implementation in accordance with ICAO-agreed standards before thoughts of system change should be introduced in international circles. Perhaps, in retrospect, the suppression of the R & D (as well as the sidelobes!) served a very important purpose—it focused maximum attention by the RRE people on the absolute necessity of making any future developments completely compatible with the present system; they soon discovered that additions to SSR are nearly unsaleable even if compatible! Thus the UK RRE have produced what we believe to be some very ingenious techniques for maintaining compatibility—which the Alexander Committee had hoped for but not actually proven. The most intriguing example is that of transmitting a two-pulse interrogation that will trigger the airborne sidelobe suppression circuitry. This effectively shuts off all the old unmodified transponders and allows the interrogator to then send a discrete-addressed message to a particular transponder (using the same address as that aircraft had set up on his reply code selector). The addressed transponder can then reply, thereby allowing position fixing more accurately with less garble and less interference, through the use of mono-pulse radar techniques. This little trick of using the existing sidelobe suppression circuitry in everybody's transponders to shut them off would seem to be the key to continued usage of the same SSR frequencies in a compatible manner in the future without disturbing unmodified airborne sets. A number of other tricks have been discovered and are being tested with the very limited funds available at RRE for such work.

The paper by Chris Ullyatt does not mention the term 'data link' although it does refer to the possibility of additional telemetered data to facilitate the ATC data processor on the ground. The specific parameters eventually to be used are, of course, somewhat controversial at this time but the paper sees them as examples, suggesting that speed, rate of turn and rate of climb are easier to derive in the air than by computer processing on the ground. Deriving them in the air could reduce the cost and complexity of the ground ATC computer for a given SSR update rate. Although the data handling (*not* called data link, however!) capacity of the modified system is quite high, there is no suggestion being given within the United Kingdom that we are aware of to using the SSR system as a data link *instead of* the more conventional VHF data link. Considerable UK government interest is being shown in the expanded development on the VHF data link—notwithstanding the Super Beacon potential described in the Alexander Committee report."

The essence of Ullyatt's proposal was as follows.

(1) The SSR frequencies 1030 MHz and 1090 MHz should be used for the improved system in order to maintain compatibility with the existing system during a possibly protracted implementation period.

(2) The new system should include the SSR suppression pulse pair so as to inhibit the existing system from attempting to decode the interrogations and thereby generate unwanted interference.

(3) The interrogation should be contained within 30 μs to conform with the existing system's suppression period.

(4) The interrogation and the reply should include parity to avoid the problems of transmission errors due to interfering signals encountered in the existing system.

(5) Both aircraft identity and height data should be included in the one aircraft reply to avoid the problems of associating separate SSR replies.

(6) Necessary information to update an aircraft's track should be obtained from a single aircraft reply, and this would necessitate the use of monopulse direction-finding techniques.

In 1969 Cossor, using its own funds, built an experimental SSR monopulse receiver in order to test the major feature that was important to the success of the concept. Experiments were conducted and the results were published in 1971 [2.13]. This represented the first application of monopulse direction finding to SSR, although an earlier investigation had used an antenna difference beam to locate the reply closest to the beam center [2.14]. The experiment was successful and showed that better results could be obtained than Ullyatt had expected. It is probably true to say that if this experimental work had shown contrary results then that would have been the end of the UK program. Cossor received contracts from the Civil Aviation Authority (CAA) for equipment definition, development of trials equipment and trials for the new system which was called Adsel (ADdress SELective).

In the USA the new system, first called Super Beacon, was renamed DABS (Discrete Address Beacon System), and the system definition was placed in the hands of the Lincoln Laboratory of MIT. DABS included all Ullyatt's proposals with, in addition, a considerably expanded data capacity that the UK proposals had lacked.

The US Air Traffic Control Advisory Committee had thought that a target position update rate of 4 s, which was within the capability of a mechanically rotating antenna, would not be sufficient for the future requirement. The Hazeltine Corporation designed and built a prototype electronically scanned antenna which could achieve a scan rate of at least one rotation per second. The antenna (Fig. 2.5) proved to be very expensive and before its time and did not go into production.

The UK and US programs were different but remarkably complementary [2.15]. The UK was mainly interested in improving surveillance of aircraft and, while the system responsibility rested with RSRE, the bulk of the work was placed with industry for the design and production of the trials equipment. These trials contributed new knowledge of the problems of multipath propagation and multistation working.

In the US the interest included the development of the data-link potential of the system including its application to intermittent positive control (IPC) which was a ground-based collision-avoidance system. The system responsi-

**Fig. 2.5** E-scan electronically scanned antenna (bottom) and open array (top). (Reprinted with permission of Hazeltine Corporation, Commack, NY 11725.)

bility rested with the Lincoln Laboratory, under contract to the Federal Aviation Administration (FAA), who built their own experimental equipment first before work was placed with industry to build to the Laboratory's design.

Figure 2.6 shows the Lincoln Laboratory DABS experimental antenna. It measured 8 ft high by 20 ft wide and represented an early attempt to shape the vertical radiation pattern of the antenna beam, a development which was itself a significant step to improved performance. A further development of the large vertical aperture antenna was the 4 ft open array shown in Fig. 2.5 with a separate top-mounted omnidirectional antenna for the control function. A requirement of the new antenna was that it should present a windage similar to that of the standard SSR linear-array antenna so that it could be mounted on existing turning gears. Several prototypes of the 4 ft array were constructed before it was replaced by a 5 ft high version which had better discrimination against ground-reflected signals and also an integral control pattern. Antennas of the latter type are made by a number of manufacturers, and it has become the standard SSR antenna worldwide.

The development in the two countries was highlighted by the close collaboration and the free exchange of information, successes and discussions of problems between the engineers of the CAA, the FAA, the RSRE, the Lincoln Laboratory and Cossor. The two countries built their equipment to

**Fig. 2.6** DABS experimental antenna. (Reprinted with permission of Lincoln Laboratory, Massachusetts Institute of Technology, Lexington, MA.)

operate with both the new transponders with selective address and data-link facilities and the existing transponders responding to the current SSR Modes A and C. Experiments quickly showed that accurate bearing measurement could be achieved for SSR by using monopulse techniques, thereby overcoming one of the original concerns about the performance of the existing SSR system.

Work by a team led by Paul Drouilhet and Vincent Orlando of the Lincoln Laboratory [2.16, 2.17] showed that, by using monopulse direction measurements made on each SSR reply pulse, multiple overlapping SSR replies could be successfully unscrambled. This development largely overcame a second significant problem of the existing SSR system, that of garble.

The research into Adsel and DABS had led to solutions to the more significant problems of the existing SSR system without the need for separately addressing each aircraft. Some of the main reasons for Adsel and DABS had been achieved using techniques that could be exploited with the existing SSR system and, most significantly, without the need for new transponders in the aircraft [2.18].

While development was confined only to the UK and the USA, the use of the local names Adsel and DABS presented no problem. However, to obtain the international recognition that would be necessary before the system could be adopted, it was necessary to find an internationally acceptable name. For a variety of reasons neither Adsel nor DABS was suitable. The name proposed was Mode S.

The development of the new system had been regularly reported in published papers including the compatibility between the new and existing systems. At first the differences were of main interest, but as development progressed it became important to stress the evolutionary, as distinct from the revolutionary, nature of the program and particularly the absence of the need to carry two types of transponder on aircraft. At one of the regular meetings between the two countries the author remarked that there was a need to explain it as a new mode of existing SSR. John Langley, of the CAA, proposed that we should call the system "SSR Mode S", with the S standing for "selective", which would closely relate it to the existing SSR Modes A, B, C and D as well as indicating that it was a new enhancement.

The application of monopulse processing to SSR to relieve its current problems left only the data-link requirement unsatisfied. This undoubtedly meant that the implementation of the new system would be delayed until the operational requirements for the data link itself could be defined. Several countries decided that they could not wait for Mode S to pass through the lengthy stages of international definition and acceptance before they could benefit from the improvements that monopulse processing gave to the current SSR system. The UK CAA led the world by placing contracts for operational monopulse SSR equipment as part of its radar replacement program. Major contributors were John Shaw of RSRE, Tony Evans of the CAA and Alex McDevitt of Cossor, with the author acting as the Cossor system design authority. The target detection, code extraction and false target rejection of the equipment has proved to be very good [2.19], and it has since been implemented by many other countries. Similar equipment is now available from several other manufacturers.

With civil SSR evolving into Mode S, the military IFF is also evolving along somewhat different lines. IFF Mk XII has existed for some time to provide a secure system for use in times of conflict. A new system, known as IFF Mk XV, is under development by several NATO countries as a replacement for Mk XII.

## References


2.1 Swords, S. S. *Technical History of the Beginnings of Radar*, Peter Peregrinus, Stevenage, 1986.

2.2 Lord Bowden of Chesterfield. The story of IFF, *Proc. Inst. Electr. Eng., Part A (Special Issue on Historical Radar)*, **132** (6), 435–437, 1986.

2.3 1973 Pioneer Award, *IEEE Trans. Aerosp. Electron. Syst.*, **9**, 789–797, 1973.

2.4 Levell, D. A. Secondary surveillance radar, *Wireless World*, **60** (5), 227–230, May 1954.

2.5 Terrington, D. G. Development of secondary surveillance radar for air traffic control, *Proc. Inst. Electr. Eng.*, **122** (5), 861–871, 1965.

2.6 Harris, K. E. Some problems of secondary surveillance radar systems, *Brit. IRE*, **16** (7), 355–382, 1956.
2.7 Archer, D. H. R. Secondary surveillance radar, Parts 1 and 2, *Electron. Eng.*, **33** (401), 414–420, July 1961; **33** (402), 490–494, August 1961.
2.8 Vickers, T. K. and Crippen, D. S. Problems and applications of the ATC radar beacon system, *IRE Trans. Aerosp. Navig. Electron.*, **2** (2), 32–39, June 1955.
2.9 Ullyatt, C. Secondary radar in the era of auto-tracking, *IEE Conf. Publ.*, **28**, 140–143, 1967.
2.10 Ullyatt, C. Sensors for the ATC environment with special reference to SSR, *Electron. Civil Aviat.*, **3**, C1–C9, September 1969.
2.11 *Report of the Department of Transportation Air Traffic Control Advisory Committee*, US Government Printing Office, Washington, DC, December 1969.
2.12 Carnes, W. T. The perennial ATC transponder—the Super Beacon in the U.S. and the U.K., *AEEC Lett. 69-2-65*, 2 October 1969.
2.13 Stevens, M. C. Precision secondary radar, *Proc. Inst. Electr. Eng.*, **118** (12), 1729–1735, 1971.
2.14 Doty, J. C. ATCB1-2 linear system and ACTB1-3 logarithmic system, *FAA Rep. SRDS RD-65-28*, Syracuse University, Syracuse, NY, March 1965.
2.15 Bowes, R. C., Drouilhet, P. R., Weiss, H. G. and Stevens, M. C. ADSEL/DABS—A selective address secondary surveillance radar, *AGARD Conf. Prog. 188*, pp. 9/1–9/13, NATO, Brussels, May 1975.
2.16 Gertz, J. L. The ATCRBS Mode of DABS, *Lincoln Laboratory ATC-65, Rep. FAA-RD-76-39*, Lincoln Laboratory, MIT, Lexington, MA, 31 January 1977.
2.17 Nelson, R. G. and Nuckols, J. H. A hardware implementation of the ATCRBS reply processor used in DABS, *Lincoln Laboratory ATC-78, Rep. FAA-RD-77-92*, Lincoln Laboratory, MIT, Lexington, MA, 19 September 1977.
2.18 Stevens, M. C. New developments in secondary surveillance radar, *IEE Conf. Publ.*, **105**, 26–31, October 1973.
2.19 Plant, R. J. and Stevens, M. C. A practical application of monopulse to the solution of SSR problems, *Conf. RADAR 84*, pp. 55–60, Institute Français de Navigation, Paris, May 1984.

# Chapter 3

# Principles of Operation

## 3.1 The secondary surveillance radar system

Figure 3.1 shows a pictorial representation of an SSR system which includes both ground-based and airborne elements. The ground station emits pulses of r.f. energy via the directional beam of a rotating antenna using a frequency of 1030 MHz. When the antenna beam is pointing in the direction of an aircraft, airborne equipment, known as a transponder, detects the emitted interrogation signal and modulates its own transmitter in response. This reply, on a frequency of 1090 MHz, is detected by the ground station and processed by a plot extractor. The latter equipment measures the aircraft's range and bearing, decodes the aircraft's replies to determine its identity and flight level, and passes the data to radar displays at an ATC center.

The use of an airborne transponder permits the transponder reply frequency to be different from the ground transmitter frequency, thereby avoiding the problems of clutter returns experienced by primary radar. Much lower transmitter powers are required compared with primary radar, since only one-way path losses are involved. The presence of the transponder also enables the reply signal to be modulated so that the additional data of identity and flight level can be communicated by the aircraft. Of course SSR is completely dependent on aircraft carrying working transponders.

## 3.2 Antenna characteristics

The ground antenna used has two principal beams: the "interrogate" beam, which has a high gain and narrow main lobe with low sidelobes, and the "control" beam, which is broader but has a lower peak gain (Fig. 3.2). The essential feature is that the gain of the control beam is greater than that of the interrogate beam in all directions except that of the narrow main lobe.

Short-range aircraft can be sufficiently close to the antenna site for the signal from the sidelobes of the interrogate beam to be detected. Without special measures, excessive replies would be obtained from short-range aircraft, the number of aircraft present could be exaggerated and the direction of the real aircraft falsely measured. The control beam is used to prevent the aircraft from replying to signals from the interrogate beam sidelobes. This feature is described further in Section 3.3.

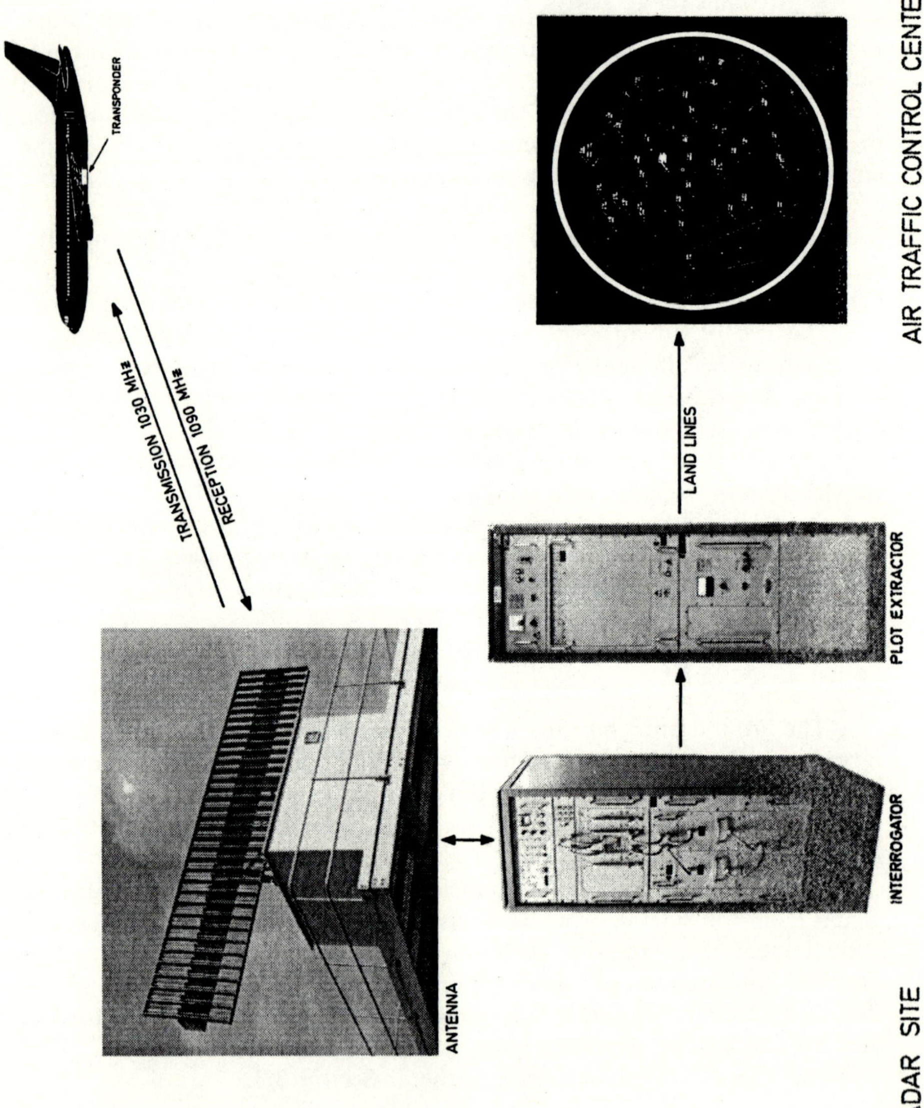


**Fig. 3.1** Components of an SSR system.

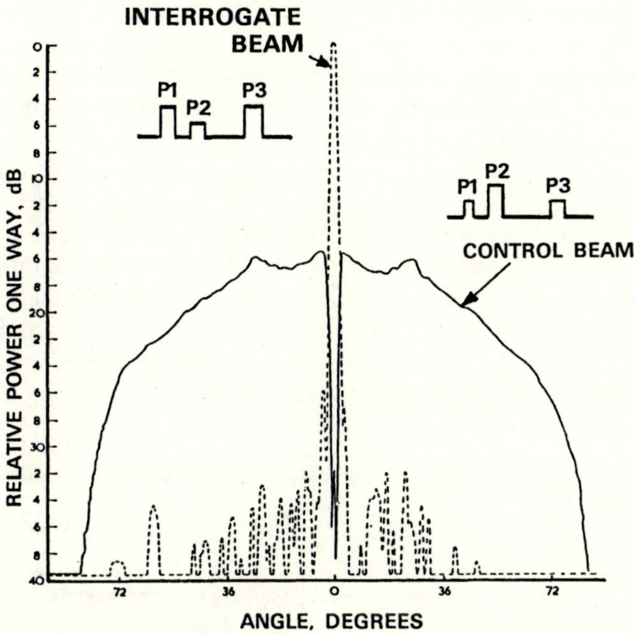


**Fig. 3.2** Antenna interrogate beam and control beam patterns.

## 3.3 Interrogation signals

The signal transmitted by the ground station is conventionally called an interrogation. Figure 3.3 shows the characteristic of the interrogation signal. Two pulses P1 and P3 are transmitted via the interrogate beam of the antenna, and the spacing of these two pulses will determine the data content of the transponder reply. A further pulse P2 radiates from the control beam.

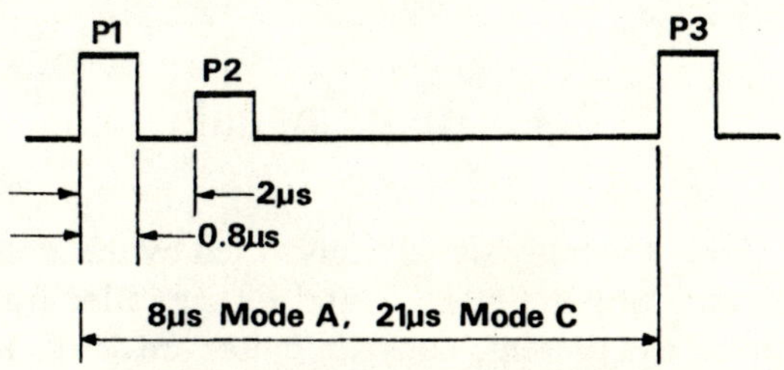


**Fig. 3.3** Interrogation-signal format.

Figure 3.2 shows how the relative amplitude of the pulses changes with antenna direction such that pulse P2 is greater in strength than P1 in all directions except that of the main lobe of the interrogate beam. By comparing the relative strengths of pulses P1 and P2, the airborne transponder can determine whether the signal originates from the main lobe or a sidelobe. It can therefore determine whether a reply is required or not. By this process, responses to sidelobes are suppressed. This facility is known as interrogator sidelobe suppression (ISLS or SLS).

The pulses P1 and P3, both of which are transmitted via the interrogate channel, can adopt a number of different spacings. Table 3.1 lists the P1–P3 spacings that are used. All the pulses have a duration of 0.8 μs. The detailed specification of the interrogation signals is given in Chapter 5 (Table 5.1). Originally, Mode 1 did not allow for the use of the SLS feature as inadequate spacing was provided between the 0.8 μs pulse P3 at 3 μs and the 0.8 μs pulse P2 at 2 μs. The close spacing requires careful design in both the ground interrogator and the airborne transponder.

**Table 3.1** Spacing of interrogator mode pulses

| *Mode* | *P1–P3 pulse spacing* (μs) | *Use* | *User* |
|---|---|---|---|
| 1 | 3 | Identification | Military |
| 2 | 5 | Identification | Military |
| 3/A | 8 | Identity | Civil/Military |
| B | 17 | Not used | Civil |
| C | 21 | Height | Civil |
| D | 25 | Not used | Civil |

In civil use the SSR is a radar system with the added ability to obtain positive aircraft identification and height indication. The military application is for identification of friend from foe, hence the military name IFF. There is a fourth military mode, called Mode 4, which is a secure mode designed to avoid the possibility of being copied by an unfriendly aircraft. The operation of Mode 4 is not included in this book.

## 3.4 Reply signals

Figure 3.4 shows the reply signal transmitted by the aircraft in response to an interrogation. The two pulses F1 and F2 are called framing pulses, or brackets, and are always present. The data pulses are designated A, B, C and D with a suffix 1, 2 or 4 to give a total number of 12. The pulse in the middle, the X pulse, is not currently used. The final pulse—the special position indicator (SPI)—is used only occasionally and is discussed below. The 12 pulses that are used give 4096 permutations and communicate the reply data. The data contained in the reply relate to the question being asked through the interrogation mode. Not all of the 4096 reply codes are used on all modes as indicated below:

| | | |
|---|---|---|
| Mode 1 | 32 codes | (B4, all C and D pulses not used)<br>(note: some national authorities use 4096 codes on Mode 1) |
| Mode 2 | 4096 codes | |
| Mode 3/A | 4096 codes | |
| Mode C | 2048 codes | (D1 pulse not used) |

The principal interrogation mode is Mode 3/A, the common civil/military mode. This mode is used for general identification with the identity number of the aircraft formed from the octal value of the reply pulses in the order ABCD; an example is shown in Fig. 3.5. By international agreement, specific sets of identity numbers can indicate the type of flight, or the destination or origin of the aircraft. Three particular codes are universally used to indicate emergency conditions:

7700 emergency
7600 radio failure
7500 hijack

These special codes are of great value for indicating particular difficulties to the ground authority in circumstances when the pilot may be unable to communicate using the normal voice channels. Special reply sequences are used to indicate military emergencies, and these are shown in Chapter 6 (Figs 6.3 and 6.4).

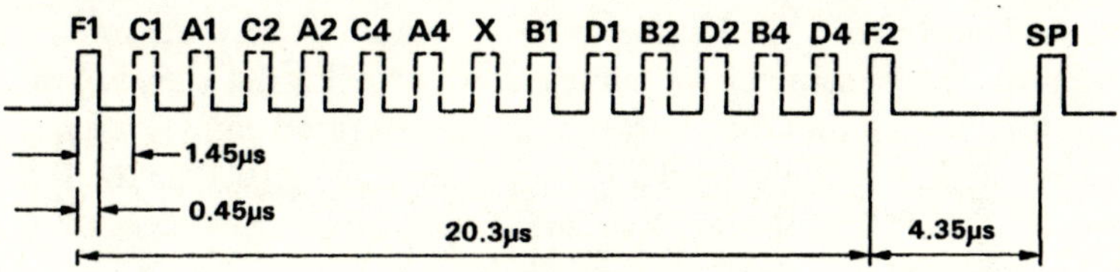


**Fig. 3.4** Reply-signal format.

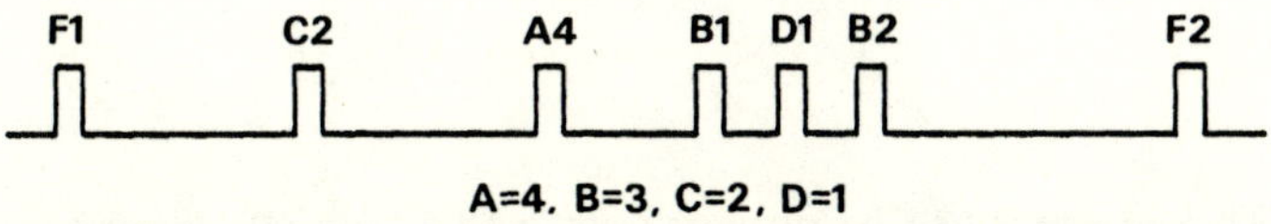


**Fig. 3.5** Mode 3/A reply indicating identity number 4321.

In the case of a Mode 3/A reply, a further pulse can be added to the reply pulse train. This is the SPI pulse which is located 4.35 μs after the F2 pulse. This pulse is controlled by the pilot by means of a switch on the transponder control unit. Pressing this switch activates the SPI pulse for approximately

20 s, and for this period all replies to Mode 3/A interrogations have this pulse added. The SPI pulse is usually transmitted only on the request of the ground air traffic controller who may use it for further identification.

The next most commonly used mode is Mode C which is used to communicate to the ground the aircraft's height as indicated by its aneroid barometer. Only 11 pulses are used in a Mode C reply (pulse D1 is omitted), but the 2048 permutations resulting are sufficient to indicate height in 100 ft increments from −1000 ft to +121 000 ft. Figure 3.6 gives an example of a Mode C reply. Since the height reported is obtained from the measurement of local air pressure, it is necessary to compensate for the normal changes in barometric pressure by adjusting the airborne instrument to the local ground barometric pressure as reported by ground ATC.

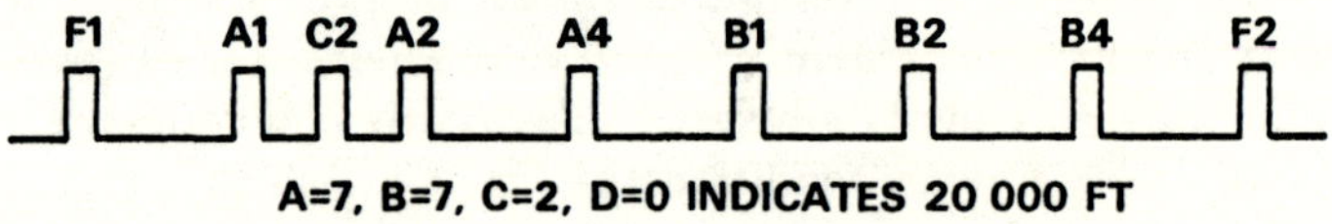


**Fig. 3.6** Example of a Mode C reply.

Above a transition altitude, normally around 6000 ft, height measurement is based on a standard barometric pressure of 1013.2 mbar regardless of surface atmospheric pressure. This process removes the necessity for long-distance aircraft to change their ground barometric reference pressure frequently as their journey progresses; such aircraft usually fly well above the transition level.

Above the transition level, the Mode C reply data do not relate to height above sea level but to flight level in units of 100 ft. The accuracy and reliability of the airborne barometers is sufficient to provide the safe separation of aircraft on the basis that their barometrically measured flight levels differ by at least 1000 ft. Above flight level 290 (approximately 29 000 ft) the separation widens to 2000 ft, but this is likely to be reduced after extensive worldwide trials have been concluded.

## 3.5 System problems

SSR has proved to be of very great value to ATC. However, it has its imperfections, and the more important that SSR becomes as an ATC tool, the less tolerant are users of its shortcomings. The SSR system problems can be broadly classed as aircraft detection errors and reply data decoding errors. The causes of these problems can also be placed in two classes: those originating from mutual interference effects and those resulting from multipath phenomena. Mutual interference occurs because SSR, of necessity, uses common frequencies—1030 MHz for the uplink and 1090 MHz for the downlink.

Obviously a transponder can only reply to one ground interrogation at a time. The reply itself takes 20.8 μs to transmit after which, for typically 60 μs and a maximum of 125 μs, the transponder transmitter and power supply are recovering and the transponder cannot produce a second reply. This effect is known as capture and results in a loss of replies to some interrogations. Typical transponder reply probabilities are in the range 90%–95%, although probabilities of 80% or lower are sometimes assumed for design purposes. A similar effect to capture occurs when a transponder receives an interrogation from an antenna sidelobe with pulse P2 being stronger than pulse P1: the transponder reply is suppressed for between 25 and 35 μs.

Transponders have only a limited ability to transmit replies given their size and the practicability of removing heat from airborne equipment. They contain a desensitizing circuit which comes into effect when the average reply rate exceeds a preset value, usually between 1200 and 2000 replies per second. When this rate is exceeded, the transponder receiver sensitivity is reduced to limit the reply rate and the transmitter dissipation. By desensitizing the receiver, the weaker, and therefore usually more distant, ground interrogators are denied replies while service to the closer stations is maintained.

National authorities usually regulate the use of interrogators so that overinterrogation and excessive fruit rates do not degrade the system for all users. Station operators are licensed and limited to a transmitter power which is no more than that necessary for the range required and limited to the minimum practical pulse repetition frequency. Figure 3.7 [3.1] shows a typical interrogation rate distribution for interrogations on Modes 3/A and C. Figure 3.8 shows the distribution for sidelobe suppression pulse pairs.

In a sidelobe direction the range at which pulse P2 can be detected is greater than that for pulse P1. The result is that the airspace experiences a large number of P2 pulses for which the associated P1 pulse is too weak to be detected. The distribution of these is shown in Fig. 3.9. These large numbers of pulses are not very damaging to the SSR system since, unless they form a chance valid pulse pair with another pulse, a transponder will ignore them.

Reasons why an interrogator may not obtain the expected number of replies are identified in the above discussion. There are also effects whereby extra replies are received. These extra replies are known by the acronym "fruit" (false replies unsynchronized in time). Fruit replies result from aircraft responses to the interrogations from other neighboring ground stations.

Nonmonopulse SSR equipment determines the bearing of replying aircraft in the following way. Aircraft replies are not received until the leading edge of the rotating antenna beam lies in the direction of the aircraft. The direction of the antenna is noted when the first reply is received (Fig. 3.10). As the antenna continues to rotate more replies are received. When the last reply

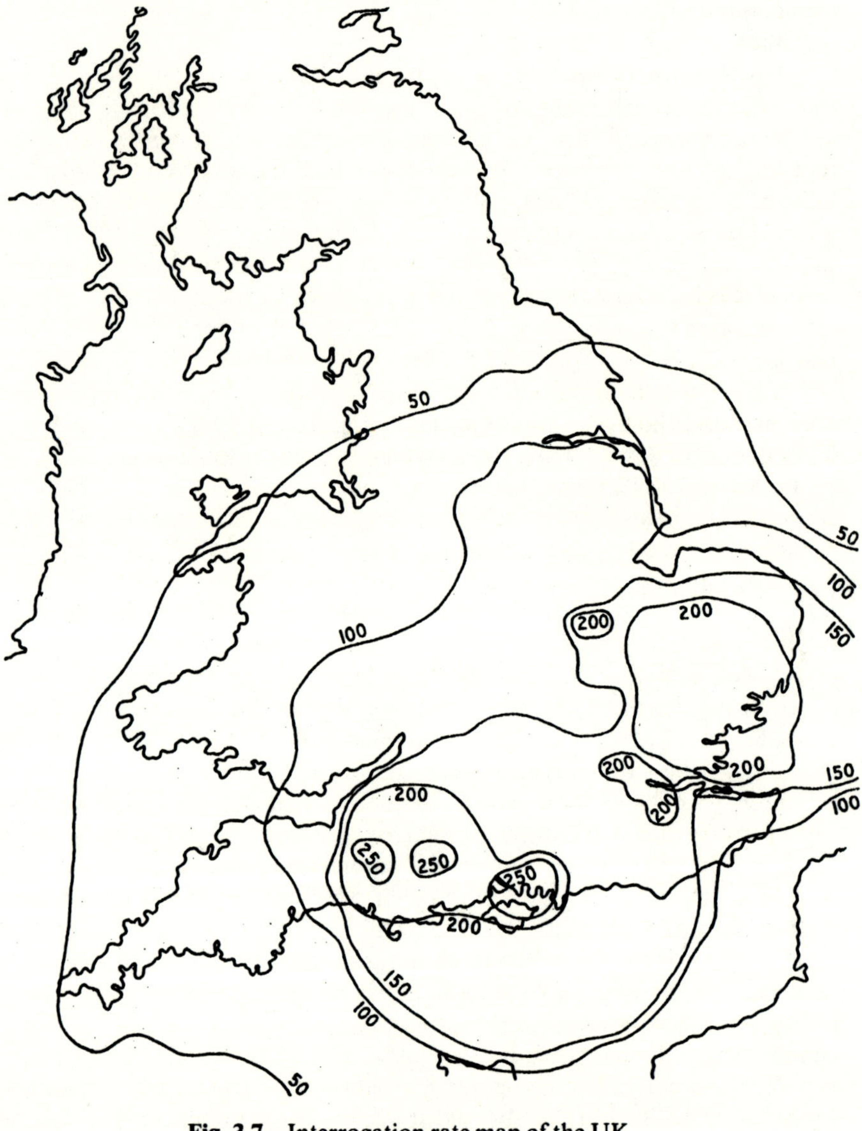


**Fig. 3.7** Interrogation rate map of the UK.

is received the antenna direction is again noted. The average of the two noted directions is taken as the true direction of the aircraft. The process works well as long as aircraft reply reliably to interrogations and there are not too many fruit replies.

When two or more replies from two or more aircraft overlap at the receiving station, the receiving equipment can be confused. Figure 3.11 shows

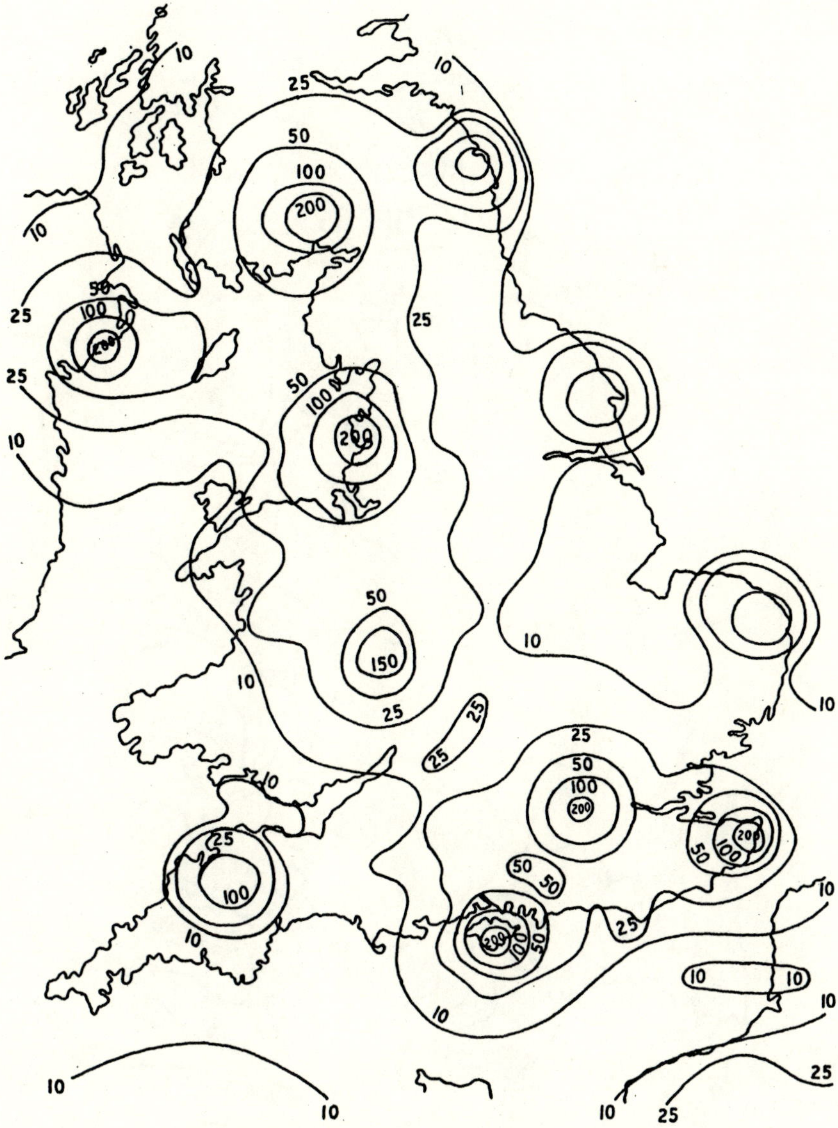


**Fig. 3.8** Sidelobe suppression rate map of the UK.

one such example where the number of replies is difficult to determine since the framing pulse pair separation F1–F2 can be found in four different places. The two spacings in the middle result from the chance timing between the data pulses of the true first and the last reply. These middle spacings can be confused as representing aircraft which, in fact, are not present. The false replies are known as phantoms.

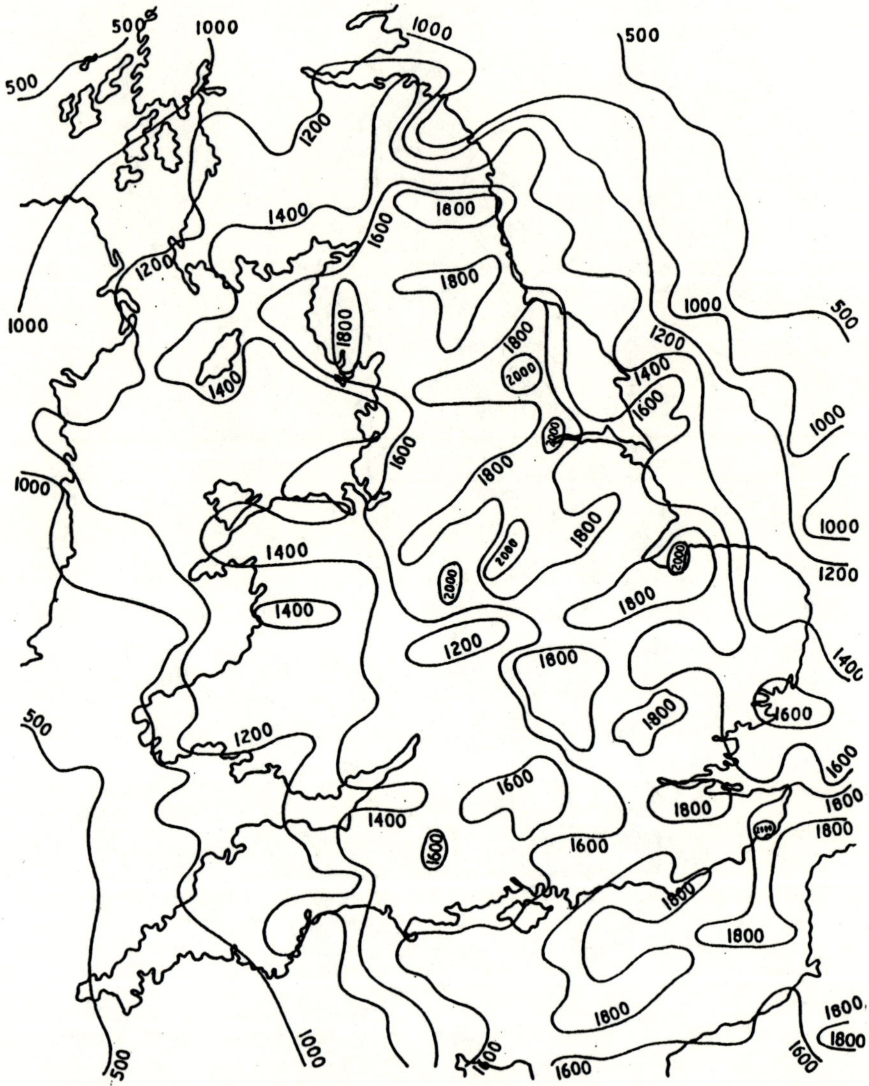


**Fig. 3.9** Single P2 pulse rate map of the UK.

Other forms of mutual interference do not involve separate aircraft replies but result from self-corruption caused by the presence of multiple reply paths. "Multipath" is a term used to describe multiple signal paths between the transmitter and receiver. Usually one of these paths includes the direct line-of-sight path from the aircraft to the ground or vice versa. Additional paths can occur owing to reflections from adjacent ground, large buildings, chain-link fencing or antenna masts.

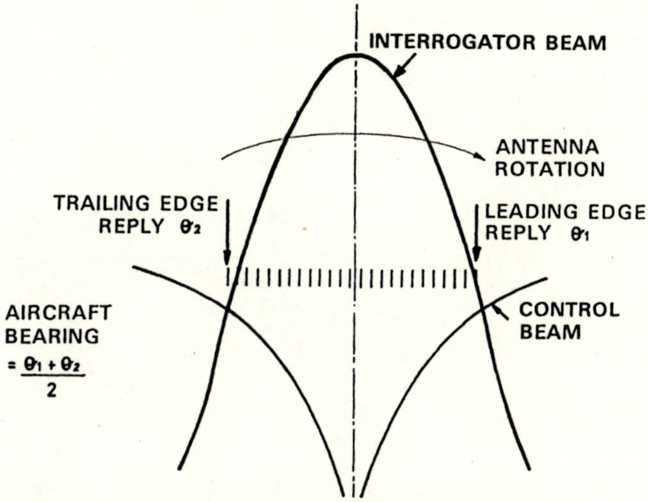


**Fig. 3.10** Sliding-window method of bearing measurement.

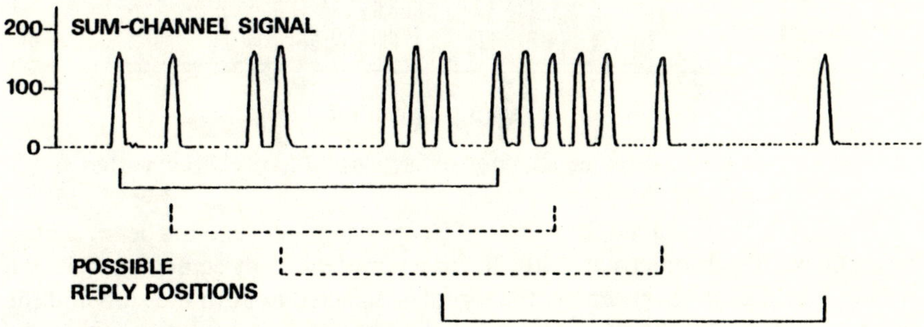


**Fig. 3.11** Example of a multiple reply (courtesy of the CAA).

## 3.6 Monopulse

Monopulse is a technique for the accurate measurement of the angle of arrival of a signal. The name arises from the ability to determine the angle of arrival from a single reply pulse. The technique was originally developed during the Second World War; monopulse has been applied to SSR only relatively recently [3.2].

Figure 3.12 shows the radiation patterns of an SSR antenna used for monopulse. It will be seen that a third beam has been added in addition to the high gain interrogate beam and the much broader control beam shown in Fig. 3.2. The third beam is known as the difference beam, in contrast with the interrogate beam which is also known as the sum beam. The difference channel output from the antenna is connected to a separate receiver channel.

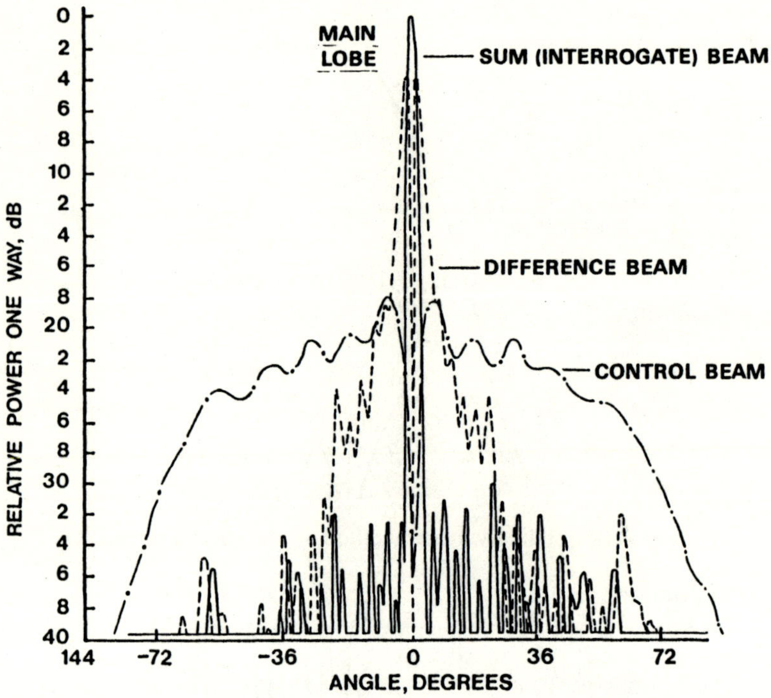


**Fig. 3.12** Antenna sum (interrogate), difference and control beam patterns.

Figure 3.13 shows the sum and difference beams close to the beam center. Note the central deep narrow null. If, for example, a reply signal is received a little to one side of beam center, the signal is detected in both the sum and the difference channels of the antenna. Measurement of the signal strengths in the sum and difference receiver channels enables the position of the reply in the beam to be determined. The antenna beams are symmetrical about the center and the same sum-to-difference signal ratio can be found in two places, one on each side of the center. The side of beam center where the signal is received is determined by measuring the relative phase between the sum and difference signals since this differs by 180° between the two sides. Alternative methods for making the bearing measurement are described in Chapter 5, Section 5.4.

The application of monopulse direction-finding techniques to SSR provides an accurate method of measuring aircraft bearing. Monopulse also assists in the decoding of overlapping replies by separately identifying the direction of arrival of each reply pulse. Figure 3.14 shows the same reply signals as Fig. 3.11 but with the difference-channel signal added. The additional data available from the antenna difference channel are of value in the determination of the number of replies present and of which pulse belongs to which reply. The processing of replies using monopulse data is described in Chapter 7.

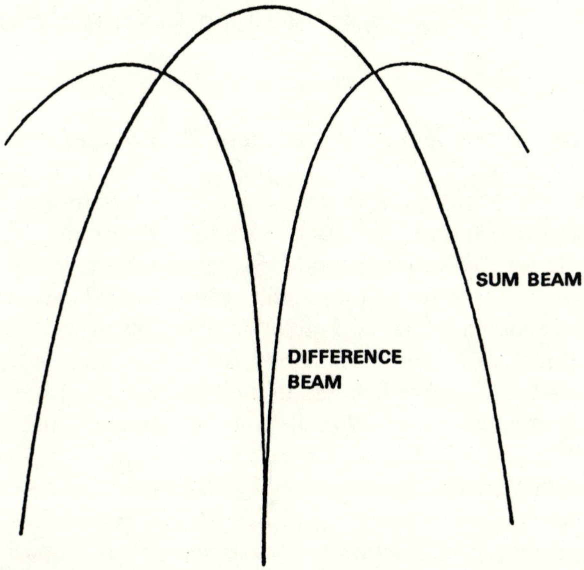


**Fig. 3.13** Detail of the sum and difference beam patterns.

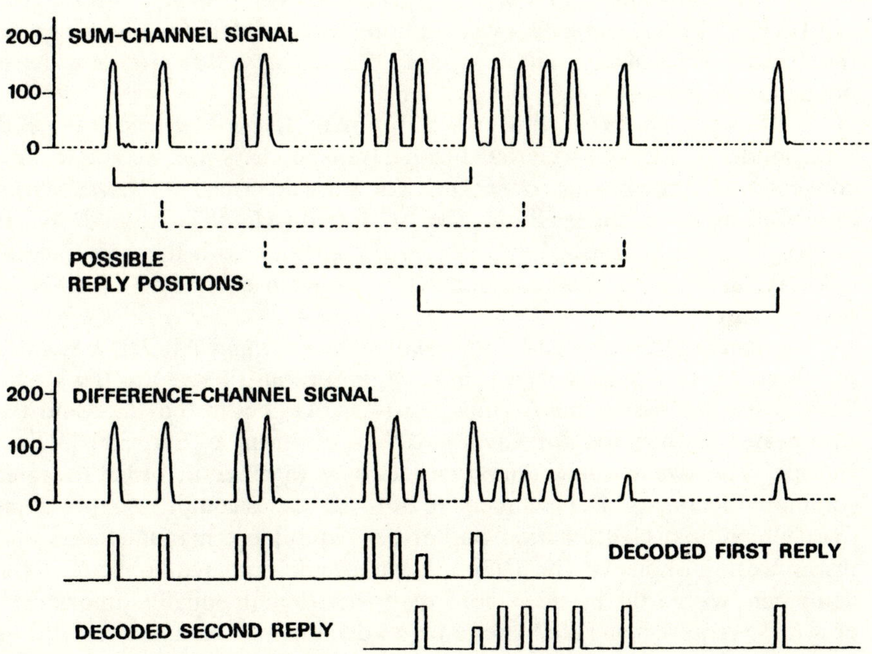


**Fig. 3.14** Use of the difference-channel signal in decoding complex replies (courtesy of the CAA).

## 3.7 Specifications and standards

SSR depends upon the fact that aircraft carry transponders both to detect the interrogations from the ground and to respond with replies carrying the requested data. Many aircraft are involved in international flights which can take them across many national boundaries. These aircraft will be interrogated by many different ground SSR stations made by many different manufacturers. It is obvious that such a system as SSR can only function if the equipment involved is built and operated to common agreed standards.

The principal international body for controlling and publishing these standards is the ICAO which has its headquarters in Montreal. The ICAO publication *Annex 10 to the International Convention on Civil Aviation* [3.3] describes all the navigation aids which have an international standing and is the most important document specifying SSR. Section 3.8 of *Annex 10* defines the SSR interrogation and reply pulses, their rise and fall times, and their duration, spacings and tolerances. Recommendations concerning transmitter powers are included so as to minimize the interference to other users of SSR, possibly in other countries.

A similar document to *Annex 10* is *STANAG 4193* which is published by NATO [3.4]. This document covers additional specifications concerned with the characteristics of the military Modes 1, 2 and 3, as well as Mode 4 which is not described in this book.

Although *Annex 10* includes a description of the performance of the SSR transponder, it does not seek to define the transponder's size, weight, fixings, connections, controls, plugs or sockets. These are described in *Characteristics* published by Aeronautical Radio INCorporated (ARINC) [3.5]. ARINC is an organization established and run by the airlines, with the assistance of industry, in order to ensure commonality between equipment of different manufacture.

*Minimum Operating and Performance Specifications (MOPS)* are also produced by two quasi-governmental organizations. These are the Radio Technical Commission for Aeronautics (RTCA) [3.6] in the USA and the European Organization for Civil Aviation Electronics (Eurocae) [3.7] in Europe. The two organizations work closely together in order to agree common documents. The membership of these bodies comprises representatives of national civil aviation authorities, equipment manufacturers and research establishments. The *MOPS* define the minimum requirements of the equipment within the broader scope of *Annex 10* and, equally importantly, how these requirements are to be tested to demonstrate compliance with the requirements.

The ARINC *Characteristics* and the RTCA and Eurocae *MOPS* are widely accepted as internationally applicable specifications.

## 3.8 Typical ground equipment performance specification

The performance specification given in Table 3.2 relates to a monopulse SSR ground equipment of the type obtainable from a number of manufacturers.

**Table 3.2** Typical secondary surveillance radar equipment performance specification

| | |
|---|---|
| *System* | |
| Maximum range | 250 n miles |
| Target capacity | |
| Total | 400 aircraft |
| Peak | 200 aircraft in 90° |
| Range | |
| Accuracy | 95 ft rms |
| Resolution | 360 ft (two aircraft with identical azimuth) |
| Discrimination | 0.01 n miles |
| Azimuth | |
| Accuracy | 0.05° rms (0.03° rms typically) |
| Resolution | 0.6° (two aircraft with identical range) |
| Discrimination | 0.022° |
| Detection probability | 98.7% typically (98.0% typically in garble) |
| Code validity | |
| Mode A | 99.7% typically (99.5% typically in garble) |
| Mode C | 98.8% typically (97.0% typically in garble) |
| Interrogation rate | 120 Hz typically (12 rpm antenna scan rate) |
| Fruit rate | |
| Mean | 4000 per second |
| Peak | 8000 per second |
| *Antenna* | |
| Antenna beams | Sum, difference, control |
| Antenna gain | |
| Sum peak at +0° | >27 dBi |
| Sum peak at +50° | >7 dBi |
| Difference peak | 2 dB below sum peak |
| Control | >6 dB above sum sidelobes |
| Difference-beam null depth | −38 dB typically below sum peak |
| Sum 3 dB beam width | |
| Azimuth | 2.45° |
| Elevation | 12.5° |
| Sidelobe level | |
| Sum | −27 dB max |
| Difference | −27 dB max |
| Vertical cut-off rate | >1.7 dB per degree at 0° elevation |
| Negative elevation sidelobes | −20 dB max |

*Interrogator*

| | |
|---|---|
| Transmitter | |
| Power | To meet range requirement (1000 W typically) |
| Frequency | 1030±0.01 MHz |
| Sidebands | |
| ±60 MHz | −50 dB max |
| ±100 MHz | −70 dB max |
| 1250–1350 MHz | −115 dB max |
| Sidelobe suppression | ISLS (IISLS optional) |
| Mode generator | |
| Interrogation rate | 50–450 Hz |
| Modes | 1, 2, 3/A, B, C, D, S |
| Mode programs | One, two or three mode interlace (third mode switchable) |
| Sector blanking | 11.25° sectors, programmable |
| Receiver | Monopulse and RSLS |
| Frequency band | 1090±5 MHz |
| Out-of-band rejection | |
| 1090±12 MHz | > −40 dB |
| 1090±25 MHz | > −80 dB |
| 1250–1350 MHz | Dependent on local primary radar |
| Dynamic range | −16 dBm to −80 dBm |
| Dynamic threshold | −6 dB to −30 dB for 25 μs |
| Gain–time control | |
| Fixed | Logarithmic |
| Programmable | 32 angle cells, 64 range cells |

*Plot extractor*

| | |
|---|---|
| Fruit rejection | Sidelobe and main beam |
| Reply decoding | Four simultaneous replies |
| Overlapping bracket discrimination | 360 ns |
| False response detection and rejection | Wide pulses, ground bounce echoes, phantoms, reflections, ring-around, Mode S preamble, code swaps |
| Undetected code errors | < 1 per scan |
| Minimum replies for detection | Two (any mode) |
| Special replies | 7700, 7600, 7500 codes; military emergency and special ident replies; code in transition |
| False target parameters | Range, signal strength, track duration, reflection zone, reflection and ring-around indicators |
| Stored reflecting surfaces | 50 permanent, 50 transient |
| Time in store | |
| < 10 n miles range | 625 ms |
| > 10 n miles range | 400 ms |
| Track coasting | Up to six misses, continuous through zenith cone of silence |

*Mode S enhancement*

| | |
|---|---|
| Transmitter | |
| Pulse capability | Short and long P6 |
| Duty cycle | One Comm-A interrogation per aircraft per scan |
| Phase modulation | Included or by simple addition |
| All-call P4 | Included |
| Other functions | By module addition only |

| *Control, monitoring and built-in test* | |
|---|---|
| Channel changeover | Automatic and by remote action |
| Equipment status | Local and remote reported |
| Faults | |
| Indication | Local and remote |
| Detection | >90% |
| Location | |
| One module | >80% |
| Multiple modules | >10% |
| Antenna patterns | Remote measurement |
| *Environment* | |
| Antenna | |
| Temperature | −55 to +65 °C |
| Wind, operational | 100 mph |
| Wind, survival | 150 mph |
| Interrogator and plot extractor | |
| Temperature | −10 to +50 °C |
| Humidity | 95% max |

# References


3.1 Stevens, M. C. and Jenkins, V. Model of the secondary surveillance radar environment of the United Kingdom, *Proc. Inst. Electr. Eng.*, **121** (5), 340–344, 1974.

3.2 Stevens, M. C. Precision secondary radar, *Proc. Inst. Electr. Eng.*, **118** (12), 1729–1735, 1971.

3.3 Aeronautical telecommunications, *Annex 10 to the Convention on International Civil Aviation*, Vol. 1, International Civil Aviation Organization, Montreal, April 1985.

3.4 Technical characteristics of IFF Mk XA and Mk XII interrogators and transponders, *Standardization Agreement, STANAG 4193, Edition 3*, NATO, Brussels, 1986.

3.5 Mark 3 air traffic control transponder (ATCRBS/Mode S), *ARINC Characteristic 718-3*, ARINC, Annapolis, MD, 10 September 1981.

3.6 *Minimum Operating Performance Standards for Air Traffic Control Radar Beacon System/Mode Select (ATCRBS/Mode S) Airborne Equipment, DO-181*, RTCA, Washington, DC, March 1983.

3.7 *Minimum Operating Performance Requirements for SSR Transponders and Altitude Measurement Encoding Systems, ED-43*, Eurocae, Paris, October 1983.

Chapter 4

# Ground Antenna Systems

The antenna, the rotating joint, the r.f. feed cable, the mounting of the SSR antenna on a colocated primary radar antenna and the properties of the radome, if used, are discussed in this chapter.

The SSR ground antenna is a most important element of the SSR system. The effects of low gain, high sidelobes or reflected signals received off the ground or via nearby objects are difficult to overcome by subsequent processing. Poor antenna performance means poor SSR performance.

The antenna has two principal properties—its characteristics in the horizontal plane and those in the vertical plane. The horizontal characteristics mainly dictate the accuracy of the aircraft bearing measurement, the ability to resolve different aircraft at similar ranges, and the minimization of interference caused by aircraft replies from outside the main beam. The vertical characteristics are important in reducing the effects of signals reflecting off the ground which can both reduce system performance and generate false aircraft returns.

## 4.1 Horizontal characteristics

The horizontal dimension most commonly used for SSR antennas is in the range 8–10 m. This aperture permits peak gains in excess of 23 dB to be realized which is adequate for ranges up to 250 n miles. Beam widths, as measured at the −3 dB points, are generally between 2.0° and 2.5°. Figure 4.1 shows an example of such an antenna.

SSR antennas are frequently mounted on top of primary radar antennas, but in some instances this latter structure is insufficiently robust to support a full-size SSR antenna. In this case a smaller antenna, with a horizontal dimension of about 4 m, can be used. Naturally such an antenna has a wider beam width and a lower gain, approximately 5° and 19 dB respectively, compared with the larger version.

Other still smaller antennas are used for IFF, where target identification is the main application and good horizontal resolution and long-range operation are less important. Such antennas are frequently constrained to conform to limited available space, and the problems thereby imposed are not a subject for this book.

**Fig. 4.1** Modern SSR antenna (Cossor CRS512).

### 4.1.1 Interrogate (sum) beam

The radiation pattern of the principal horizontal antenna is the interrogate beam which is also known as the sum beam. An example of an interrogate beam pattern is shown in Fig. 3.2 (Chapter 3) and in Fig. 4.2 together with the control beam which is discussed separately. It will be seen that the interrogate beam consists of a main lobe and a number of lower amplitude sidelobes. The art of the antenna designer is to maximize the gain of the main lobe and to minimize the gain of the sidelobes. It is recommended in ICAO *Annex 10* that the sidelobes should be at least 24 dB below the gain of the main lobe.

Most antenna designs use a horizontal array of radiating elements such as single or stacked dipoles. The principles of the design of the horizontal aperture distribution system are well established and usually employ a Dolph–Chebyshev or a Taylor function. With these distributions, most of the power is radiated near the center of the array and is successively reduced towards the ends of the array. The degree of power reduction controls the sidelobe level of the radiation pattern at the expense of the gain and width of the main lobe. Examples of the distribution are given later in Fig. 4.4.

### 4.1.2 Control beam

Figure 4.2 includes a second antenna beam known as the control beam. It can be seen that the gain of the control beam exceeds that of the interrogate

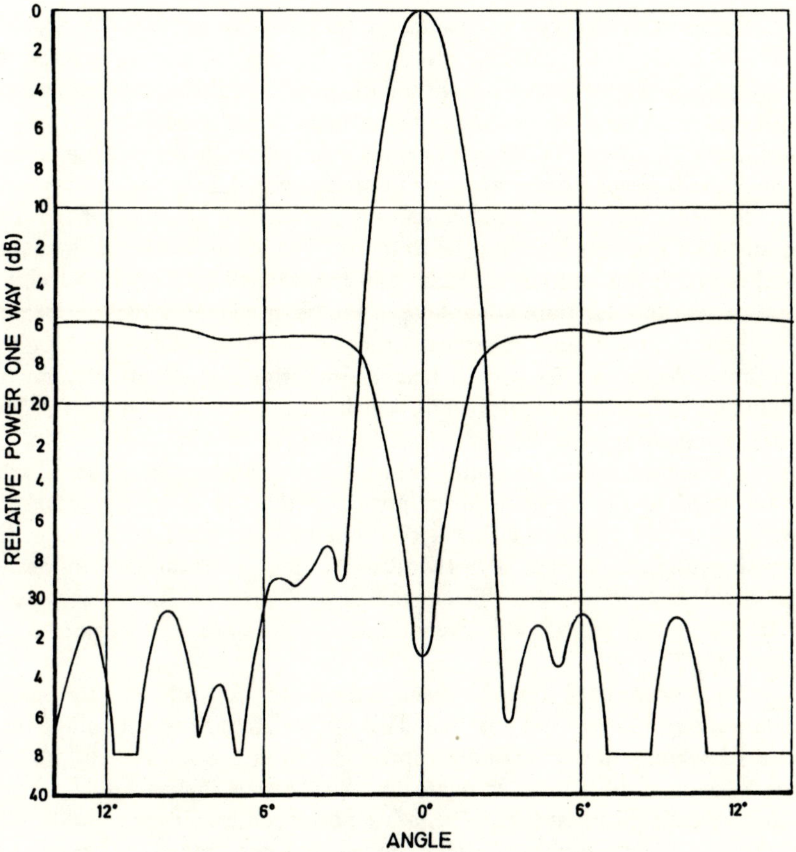


**Fig. 4.2** Interrogate and control beam patterns.

beam in all directions except that of the main lobe itself. Such a control beam can be formed by using the interrogate beam distribution but with the center element in the array driven in antiphase at an appropriate level.

The use of the control beam, in conjunction with the interrogate beam, is to inhibit transponder replies to sidelobes. The interrogation pulses P1 and P3 are transmitted by the interrogate beam and the sidelobe suppression pulse P2 is transmitted by the control beam. A transponder receiving these pulses compares the relative amplitudes of P1 and P2 and determines whether the signals are being received from the antenna sidelobes (P2 > P1) or from the main beam (P1 > P2). The transponder design rules state that it must reply to an interrogation if P1 is greater in strength than P2 by more than 9 dB but must not reply if P1 is weaker than P2. In the undefined region where P1 is between 0 dB and 9 dB stronger than P2 the transponder may or may not reply.

For aircraft at very close range the transponder receiver dynamic range may be inadequate, causing amplitude limiting of strong pulses. As a consequence the measured amplitude of pulse P1 can be reduced compared with that of P2 and the reply can be wrongly suppressed. To overcome this problem the gain of the control beam is reduced in the direction of the main lobe of the interrogate beam, as can be seen in Fig. 4.2.

In some countries it has been a long-established practice for the control beam to be generated within the antenna. The same radiating elements are used for both the interrogate beam and the control beam but with different feed networks. The main advantage of the practice is that the two beams are produced with the same phase center and any vertical pattern lobing due to ground reflections affects both beams similarly. The effects of main beam killing and sidelobe punchthrough, which are discussed in Chapter 9, Section 9.1.1, are thereby avoided.

A disadvantage of the combination of the two beams in the same antenna is the need to supply transmitter power to the two separate antenna feed systems, which requires either a dual-channel r.f. rotating joint or an r.f. switch within the antenna. Another disadvantage is the difficulty of designing a control beam which has sufficient gain at $\pm 90°$ to cover the interrogate beam sidelobes in this region. However, these sidelobes are very weak and have a short range.

An alternative practice has been to derive the control beam by means of a separate omnidirectional antenna. This approach has the advantage that the control beam gain can be made approximately equal in all directions to give good cover of all the interrogate beam sidelobes including the backlobe. However, the disadvantage of separate phase centers (if the omnidirectional control antenna is mounted above the interrogate antenna) or obstruction (if they are mounted side by side) frequently outweighs the benefits.

The use of an antenna with an integral control beam is becoming the common practice.

### 4.1.3 Difference beam

The difference beam is necessary when monopulse direction finding is employed as discussed in Chapter 5, Section 5.4. A difference beam can be used as a control beam in smaller antennas, although its cover of the interrogate beam sidelobes is less effective than a beam specifically designed for that purpose. The control application of a difference beam is not discussed further.

Figure 4.3 shows a simplified diagrammatic representation of the horizontal feed system in which the antenna has been divided into two halves. The halves are recombined in a hybrid ring which has a sum and a difference output with the sum output producing the original interrogate beam. (In a

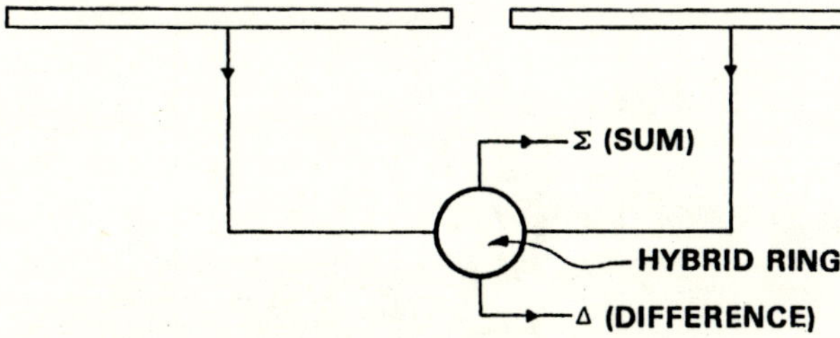


**Fig. 4.3** Production of sum and difference beams.

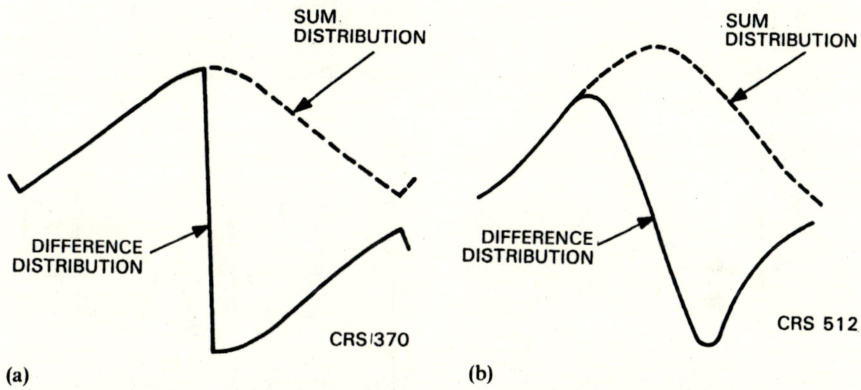


**Fig. 4.4** Antenna voltage distributions for interrogate (sum) and difference beams: (a) simple implementation; (b) improved implementation.

monopulse system the interrogate beam is usually called the sum beam.) Figure 4.4(a) shows the distributions and Fig. 4.5 shows the antenna patterns formed in this way.

The above describes a rather simple way of producing a difference beam in which the horizontal feed system of the antenna is designed principally to generate a sum beam with low sidelobes. The difference beam is not optimized and as a result it has rather high shoulders and a peak gain about 4 dB below the sum peak gain. The antenna shown in Fig. 4.6 is an example of this type which also conforms to the earlier practice of a small vertical dimension giving a broad vertical beam.

An alternative form of difference pattern is shown in Fig. 4.7 which requires a partially independent and more complex distribution (Fig. 4.4(b)). An advantage of this form of difference pattern is that the peak gain is greater, being only 2 dB below the sum peak gain, which contributes to a small increase in accuracy. Another advantage is that the patterns lack the high shoulders of the earlier form. These shoulders add nothing to the monopulse performance but could receive extraneous signals which marginally degrade performance. The antenna shown in Fig. 4.1 is an example with this type of difference

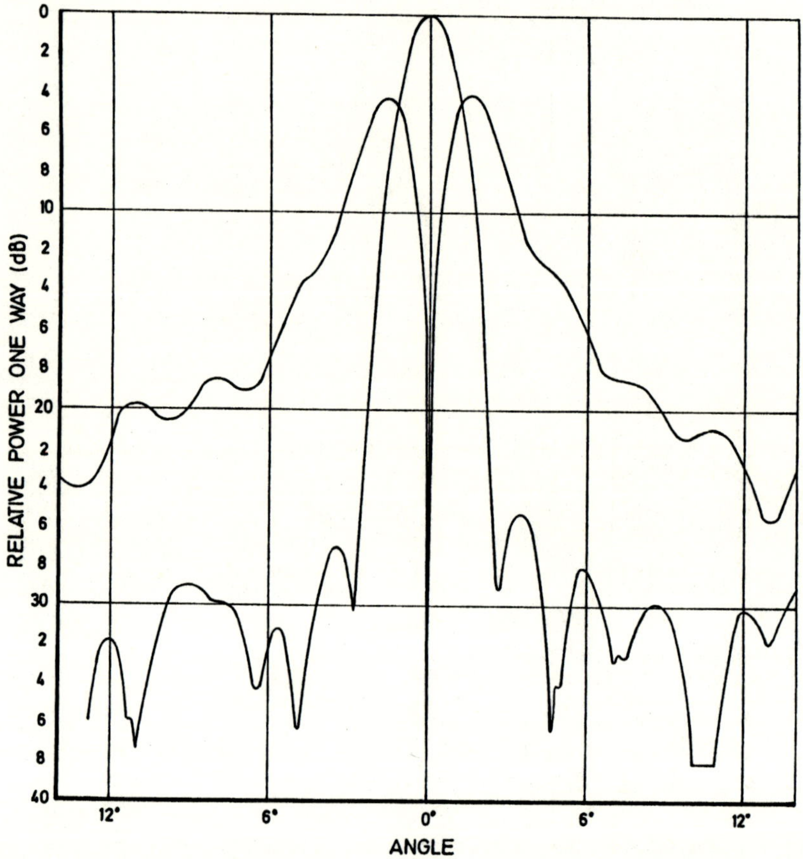


**Fig. 4.5** Beams produced by the distribution shown in Fig. 4.4(a).

pattern. This antenna also has a larger vertical aperture which has advantages for the vertical characteristics.

## 4.2 Vertical characteristics

The linear array antenna shown in Fig. 4.6 has a vertical dimension of about 0.5 m which provides little opportunity to focus the radiation in the vertical plane. The consequence is that the vertical beam shape is broad as shown in Fig. 4.8. To some extent this beam shape is satisfactory in that the maximum gain lies close to the horizon which is the elevation at which the most distant aircraft occur. The result is that aircraft which are subject to the most propagation loss are in the most sensitive part of the beam. Further-

**Fig. 4.6** Earlier form of SSR antenna (linear array or "hogtrough") (Cossor CRS370).

more, good high-angle cover is achieved for the detection of overflying aircraft. This type of antenna has given good service worldwide for the last 30 years since the original version used with the type 1 SSR in 1954. However, two problems have long been noted—one of signal fades and the other of false targets due to reflections.

Much of the radiated energy is directed towards the ground where it can be reflected back toward the sky. An aircraft in the radar beam is likely to receive two signals, one directly from the antenna and the other indirectly off the ground. The way in which these two signals combine at the aircraft antenna depends upon the relative phase of the two signals. If the two signals are in phase, they add and the result is a stronger signal. However, if they are in antiphase, they tend to cancel so that the resultant signal is much weaker than the direct signal alone.

Some regions of space will experience signal enhancement and others signal fading, depending on the height of the antenna above the reflecting surface. As aircraft fly through the fade regions signals become weak and contact can be lost. This effect is discussed further in Chapter 9, Section 9.1.1.

Antenna tower heights are often chosen so that the antenna is above nearby buildings and therefore not obscured by them. The broad vertical beam shape of the type of antenna shown in Fig. 4.6 can result from the reflection of signals from the surfaces of these buildings to interrogate unintended aircraft. The replies from these aircraft will appear to the radar to be originating at the bearing of the reflector and give rise to targets at incorrect locations.

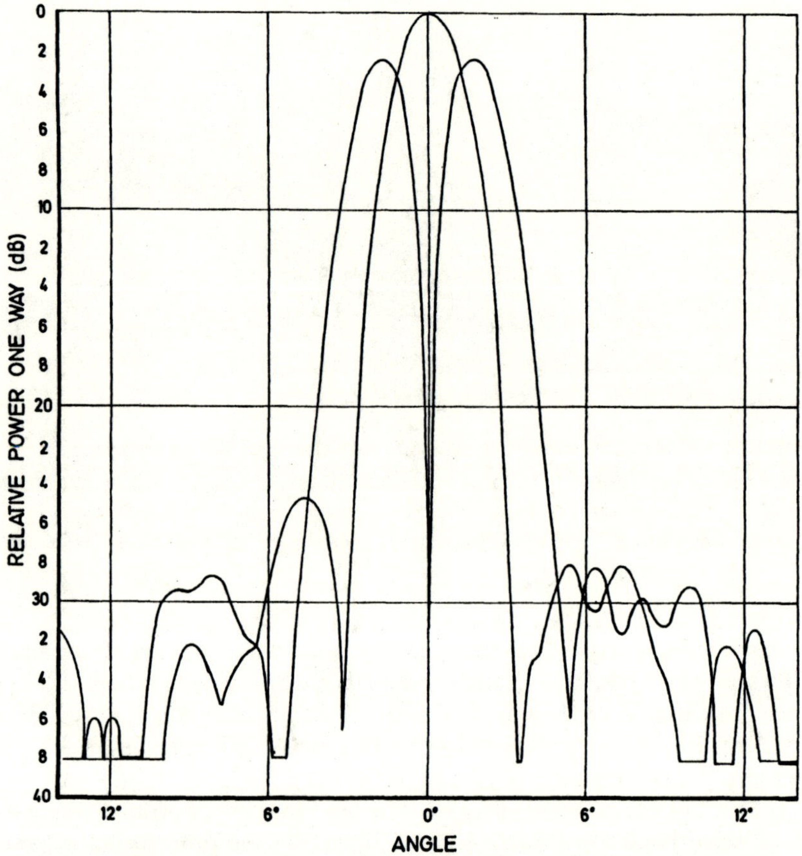


**Fig. 4.7** Improved form of difference pattern.

SSR is subject to increasing use as a principal ATC tool with consequent greater demands being placed upon its performance. The problems that arise from the ground reflections are no longer acceptable. This is particularly true of monopulse SSR systems because of the otherwise high precision that these systems demonstrate.

The performance now demanded from SSR has encouraged the development of better antennas. The use of a larger vertical aperture enables the vertical beam pattern to be shaped to reduce the energy directed at the ground. Figure 4.1 is an example of such an improved antenna which has a vertical dimension of 1.5 m stacked with 10 dipole radiators. The vertical beam pattern is also shown in Fig. 4.8 where it can be compared with the pattern of the earlier type of antenna. The essence of the new vertical pattern is the rapid rate of cut-off close to zero elevation angle. This aspect has a significant influence in reducing the extent of the fade regions and the incidence of false targets due to reflections.

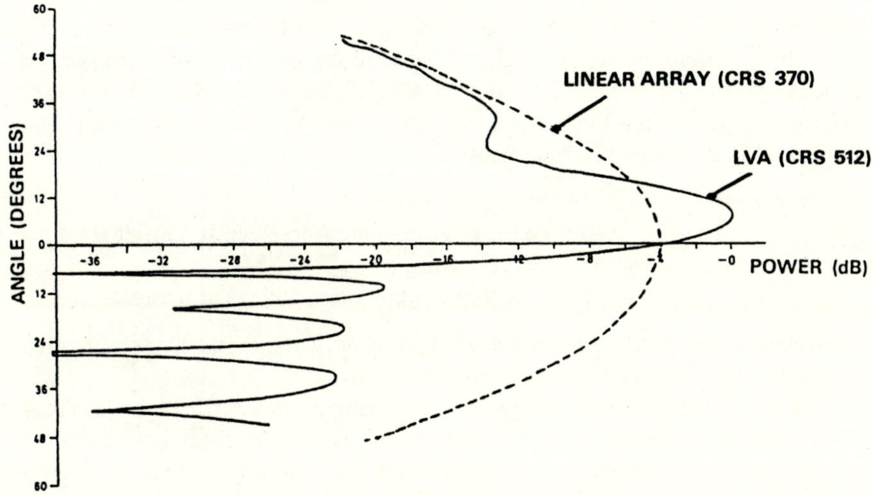


**Fig. 4.8** Vertical beam shapes of small vertical aperture and large vertical aperture (LVA) antennas.

## 4.2.1 Vertical tilt angle

The generation of nulls in the antenna vertical radiation pattern due to reflection from the ground is discussed in Chapter 9, Section 9.1.1. The selection of the optimum tilt angle to minimize the depth of such nulls is included in the discussion.

Two conflicting requirements affect the choice of the amount of radiated power to be emitted by the antenna at low elevation angles. One is that the power must be sufficient to reach long-range aircraft which, by their nature, will present a low angle to the radar. The other requirement is the need to minimize the power radiated into the ground. This can only be realized if the gain of the antenna reduces rapidly close to the horizon. Inevitably there will be some loss of gain, compared with the peak value, at small positive elevation angles where long-range targets are present.

## 4.2.2 Gain at high angles

SSR interrogators are frequently sited near airports or close to airways so that aircraft fly overhead. Constant coverage of overflying aircraft is usually desirable, although inevitably there is a cone of silence immediately above the antenna.

Aircraft fly relatively close to the ground—even an altitude of 50 000 ft is only a distance of 8 miles above the ground. Consequently an aircraft presenting a high elevation angle to the antenna is also of short range. Only modest gains are required to detect aircraft at high angles. The classical cosecant law, which relates signal strength to elevation angle for an aircraft at a given altitude, applies equally to SSR:

received signal voltage at a given angle
$= \text{voltage at zero angle} \times \text{cosec(elevation angle)}$
received signal power at a given angle
$= \text{power at zero angle} \times \text{cosec}^2\text{(elevation angle)}$

However, greater antenna high-angle gain is required than suggested above to overcome receiver short-range gain reduction.

Most aircraft at short range are at low angles and in the high-gain region of the antenna and so present a strong signal at the receiver. There is an advantage in temporarily reducing the gain of the receiver so as to reduce reception of extraneous replies due to fruit and reflections. This receiver gain reduction is known by the abbreviation GTC or STC (gain–time control or sensitivity–time control) and is discussed in Chapter 5, Section 5.3.4. Typically, the receiver gain will be reduced by 18 dB for targets at 8 miles range.

### 4.2.3 Changes in the horizontal characteristics with elevation

For a perfectly constructed antenna in which each of the three horizontal beam patterns originates from a common phase center, the three beams maintain their relative shape with elevation. Even if this is realized in practice the three beam shapes will appear to broaden with elevation angle when used in a radar. The reason for this broadening is that a radar assumes that bearing measurements are made in the horizontal plane even when the incident signal is arriving from a significant elevation angle. To illustrate the point, consider an antenna pointing north with an aircraft overhead but slightly to the east. The antenna may detect the aircraft 1° from its boresight direction, and the radar would determine its azimuth as 1° from north whereas its true direction is 90° from north when projected to the horizontal plane.

The broadening effect of the antenna pattern is only of real importance for monopulse direction finding and for elevation angles above about 20°. Figure 4.9 shows the error that occurs when the bearing of an aircraft is measured at 1° and 2° from boresight. The error can be corrected by dividing the measured angle by the cosine of the elevation angle as determined from the slant range and the Mode C height. Correction is rarely worthwhile as only short-range aircraft are affected and little error in plan position results. When monopulse direction measurements made either side of boresight are averaged, the error is much reduced.

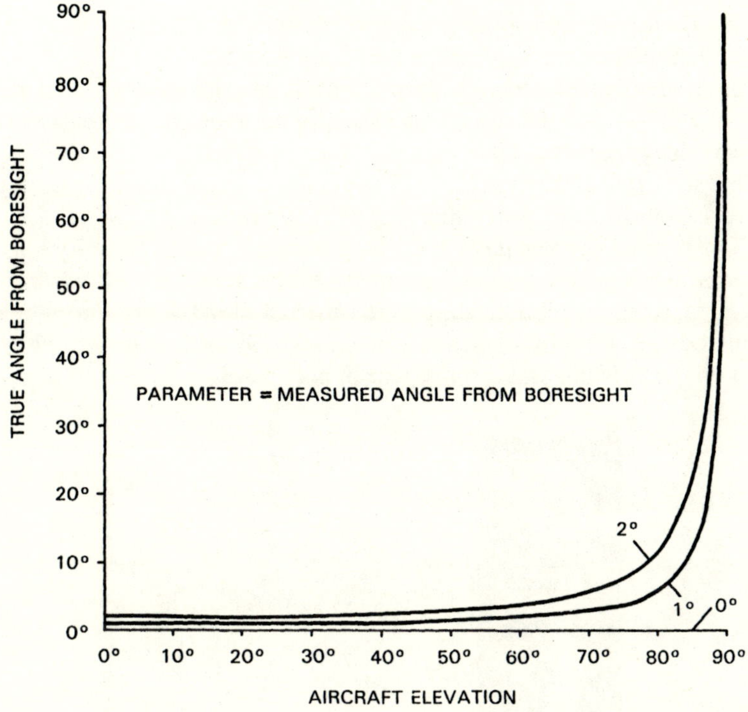


**Fig. 4.9** Monopulse bearing error due to beam broadening at high elevation angles.

The beam patterns produced are critically dependent on the distribution of power across the antenna aperture. This same relative power distribution must also be maintained at all elevation angles, which is difficult to achieve in practice. Adequate cover of interrogate beam sidelobes by the control beam may not be realized in some cases and erroneous replies due to "ring-around" result.

## 4.3 Backlobe

Most SSR antennas are constructed as either linear or planar arrays with the phase of the feed system designed to focus the main beam perpendicular to the face of the array. While most of the radiated r.f. energy is directed in the forward direction it is inevitable that some will spill over to the rear direction, and the phasing of the distribution system will also focus the energy in this rear direction to form a backlobe.

A probe-fed waveguide feeding a horn is a convenient radiating element for use with a linear array antenna. The horn assists in shaping the beam of the individual element and gives good directivity and a front-to-back ratio of about 30 dB, which is adequate for the requirement.

Such an element design is inappropriate for the large vertical aperture array of Fig. 4.1 because the vertical spacing between the elements must be a half-wavelength or less. It is convenient to print a vertical stack of dipoles on a suitable substrate and to use that substrate for a stripline distribution system. Figure 4.10 shows such a stripline feed with 10 printed dipoles.

The printed dipoles have a front-to-back ratio near to unity and other measures are required to minimize the backlobe. One possibility is to place a solid metal screen behind the dipoles to obstruct the back radiation; however, this introduces substantial windage which would impose severe mechanical forces on the antenna and turning gear in high winds.

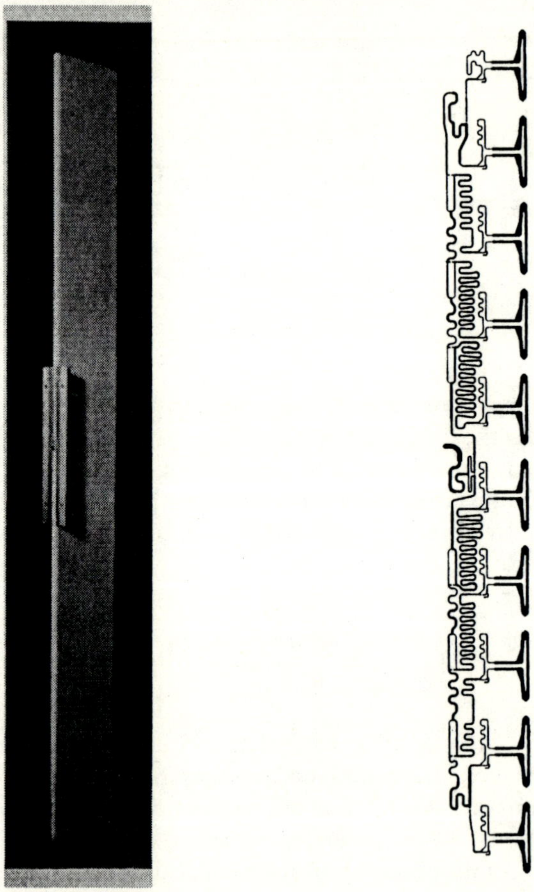

(a) (b)

**Fig. 4.10** Vertical column of a large vertical aperture antenna.

A more practical possibility is to use an open metal mesh instead of the solid screen which would reduce the windage without substantially increasing the back radiation. The solution is not perfect, since the mesh still exhibits some windage and would readily collect snow and ice. A better solution is the use of an open structure which provides maximum backward attenuation at the frequencies used. These structures take the form of vertical rods or plates spaced by approximately a quarter-wavelength. The rods can be constructed of half-wavelength vertical resonators or continuous metal elements. An example of the isolation properties of the latter is shown in Fig. 4.11.

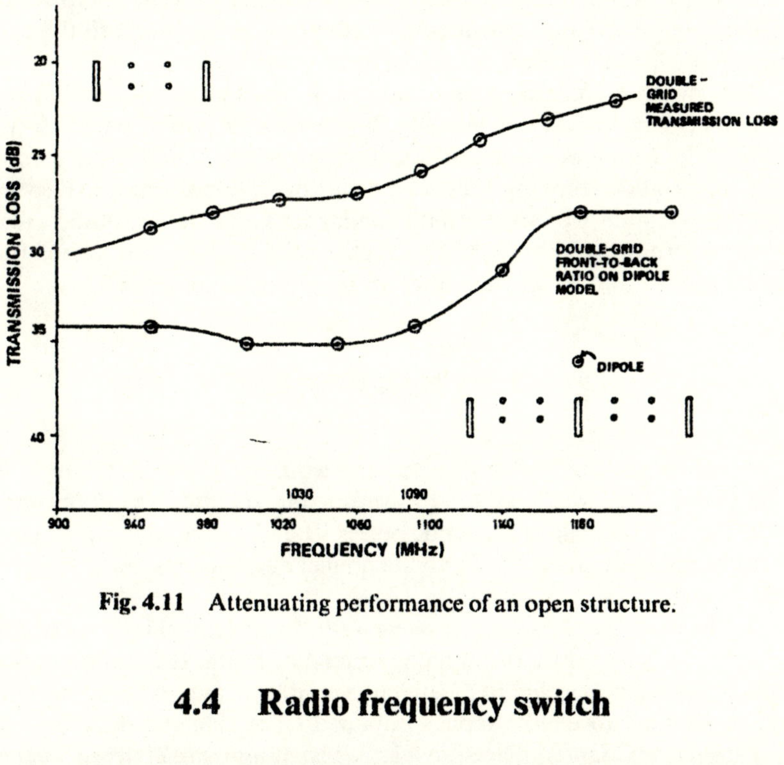


**Fig. 4.11** Attenuating performance of an open structure.

## 4.4 Radio frequency switch

The transmitter and receiver in the interrogator are connected to the antenna at the top of the tower by means of r.f. coaxial cables. At the turning gear the run of the cables is broken so that a rotating joint can be included in order that communication with the antenna can be maintained whilst the antenna rotates. The number of channels needed in the rotating joint depends upon the type of SSR system being employed.

In a minimum system only one channel is required. This is sufficient to carry the interrogation pulses and the reply signals. Of the three interrogation pulses P1, P2 and P3, the two mode pulses P1 and P3 are radiated via the interrogate beam (see Fig. 3.2). However, the suppression pulse P2 must be switched to the control beam at the appropriate time. This can be achieved by means of an r.f. switch mounted with the antenna above the rotating joint.

Clearly the switch must operate by coupling the appropriate r.f. cable to the interrogate beam at the time of P1, to the control beam at the time of P2 and back to the interrogate beam in time for P3, where it remains for the duration of the replies from the aircraft. The switch operation can be controlled by means of signals sent to the switch through slip rings in the turning gear. However, this is not usually preferred as slip rings are not ideal for the transmission of high speed signals and can cause false operation due to contact noise. A preferable technique is for the switch to control its own action by detecting pulse P1 and switching briefly to direct pulse P2 to the control beam after which it relaxes back to the interrogate beam.

R.f. switches are quite simple and can be made to be highly reliable. The reliability can be further improved by duplicating the critical parts. However, failure of the switch would necessitate its repair which would require stopping the antenna and removing the radar from operational service. Depending upon the availability of an alternative radar service, a better solution may be the use of a rotating joint with two r.f. channels, of which one is used for the mode pulses P1 and P3 and the other for the suppression pulse P2.

## 4.5 Rotating joint

A nonmonopulse SSR system requires a dual r.f. channel rotating joint to avoid the need for an r.f. switch in the antenna, as discussed in Section 4.4. One channel is needed for the mode pulses P1 and P3 and the second channel for the suppression pulse P2. The first channel also conveys the SSR replies to the receiver.

In the case of a monopulse system a third r.f. channel is required for the reply signals detected by the antenna difference beam. It is common for the second channel, connected to the antenna control beam, to be connected to a receiver channel to enable replies obtained from sidelobe directions to be recognized (see Chapter 5, Section 5.3.5). The monopulse system places more stringent requirements on two of the channels of the rotating joint: those used for the sum and the difference reply signals. Both the amplitude and the phase match between the sum and difference signals must be accurately maintained. Table 4.1 shows a typical specification for the rotating joint for a monopulse system.

The design technique used for realizing a channel in the rotating joint is worth some discussion. Sliding contacts can be used for some applications, but these have a limited life owing to the constant wear that they experience. Since SSR is normally required to provide a continuous service, the preferred technique is to use a section of coaxial line where the mechanical joint takes the form of a contactless choke. Figure 4.12 shows a cross-section of such a design; its electrical equivalent circuit is given in Fig. 4.13. Multiple channels can be realized by stacking single channels and passing the output cables of the lower channels through the center of the upper channels. When an SSR antenna is comounted with a primary radar antenna the rotating joint will also include channels for the latter.

**Table 4.1** Specification of a typical rotating joint

| | |
|---|---|
| Frequency | 1020–1100 MHz |
| Number of channels | Three (sum, difference, control) |
| Peak power | 10 kW |
| Mean power | 50 W |
| VSWR[a] | 1.25 max |
| Isolation | 60 dB min between channels |
| Insertion loss | 0.65 dB |
| Wow | 1.08 max |
| Phase tracking | 2° max over 360° rotation |
| (1090 MHz sum and difference channels) | 5° max over temperature range |
| Operating temperature | −20 to +70 °C |

[a]VSWR, voltage standing-wave ratio.

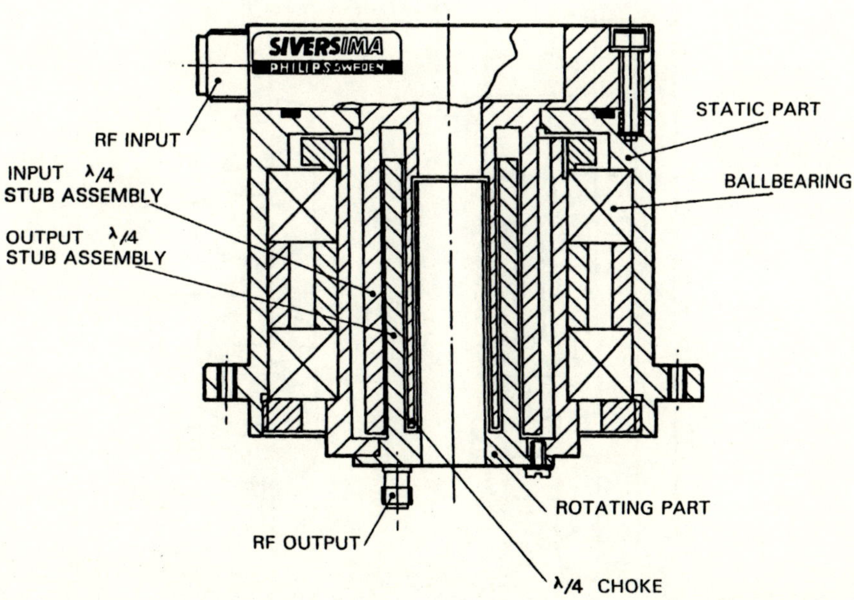


**Fig. 4.12** Construction of a commercial rotating joint (Sivers IMA type SL17100).

## 4.6 Radio frequency cables

Although antennas are normally mounted on the top of towers, for reasons of accessibility the associated transmitters and receivers are usually housed on the ground. The two sets of equipment are interconnected by r.f. cables. It is of major importance that the electrical loss of the r.f. cables is kept to a minimum. Coaxial cables produce a typical loss of about 3 dB including the loss due to the rotating joint. Losses of less than 1 dB could be realized if waveguides were employed, as in the case of primary radar, but the expense of this is rarely warranted for SSR with its inherently better signal-to-noise ratio.

In the case of monopulse SSR the phase match between the sum and difference channels is also important, and typically this must be held to within ±5° at 1090 MHz. Typical cable lengths involved can exceed 50 m, or 60 000 electrical degrees, so that the (differential) stability required is better than 1 part in $10^4$. Coaxial cable can provide this performance provided that it is cut from the same length and phase stabilized by temperature cycling. Care must be taken with the cable installation to ensure that neither of the phase-matched cables is subjected to mechanical stress or to differential heating or cooling due to sun and wind.

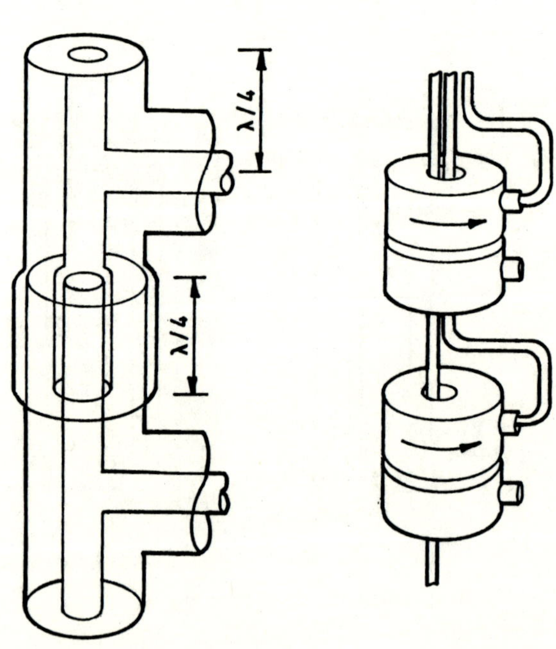


**Fig. 4.13** Electrical circuit of a rotating joint channel (Sivers IMA).

Both low attenuation and good phase stability require the employment of dielectric material principally composed of air with the use of either spacers or foam. Table 4.2 shows the performance of typical commercially available coaxial cable.

**Table 4.2** Characteristics of typical commercially available coaxial radio frequency cable used for secondary surveillance radar (Andrew Heliax®)

| Type | LDF4-50A | LDF5-50A | LDF7-50A |
|---|---|---|---|
| Diameter (in) | 0.64 | 1.10 | 2.0 |
| Impedance (Ω) | 50 | 50 | 50 |
| Dielectric | Foam | Foam | Foam |
| Peak power (kW) | 19 | 44 | 145 |
| Mean power (kW) | 0.55 | 1.5 | 3.69 |
| Loss at 1000 MHz (dB per 100 ft) | 2.34 | 1.31 | 0.82 |
| Minimum bending radius (in) | 5 | 10 | 20 |
| Weight (lb per ft) | 0.15 | 0.33 | 0.92 |

## 4.7 Colocating with primary radar

Primary and secondary radar can be constrained by different requirements of range, optimum rotation rate, clutter areas, optimum height above the ground etc. These different needs can best be met with separate installations. Frequently, however, issues of the availability of suitable sites and the costs involved dictate that the two systems must be colocated and share a common tower and turning gear.

### 4.7.1 Top mounting

The most common solution is to mount the SSR antenna on top of the primary radar antenna which may be operating in either the S band or the L band. Figures 1.1 and 4.14 show examples. Clearly, the turning gear and the antenna support structure must be sufficiently strong to support both antennas under all anticipated environmental conditions. The use of a radome (Section 4.8) may be advantageous.

In mounting the SSR antenna care must be taken that the primary radar reflector does not intrude into the near field of the antenna and distort its radiation patterns. The SSR antenna will be illuminated by the edge of the main lobe or the sidelobes of the primary radar feed horn. R.f. energy from the primary radar will be received by the SSR antenna, but the latter's performance at the primary radar frequencies will almost certainly not have been measured and the gain at these frequencies will be estimated on a theoretical basis. Leakage between the primary radar and SSR channels in the rotating joint will also be a source of cross-coupling, and experience suggests that this will be of a similar magnitude to the coupling between the antennas. Extra filters in the SSR receivers may be required to protect them from blocking by the primary radar transmitter pulse, particularly where multi-megawatt transmitters are involved.

## 4.7.2 Back mounting

Mounting the SSR antenna on top of the primary radar antenna may present severe mechanical problems including that of the overturning moment on the total structure. The SSR antenna can be mounted closer to the turning gear by mounting it on the back of the primary radar reflector. The two antennas will then be pointing in opposite directions. The backlobe of the SSR antenna is obstructed by the primary radar reflector and is reflected in directions which are difficult to predict. However, the resulting reduction in backlobe directivity appears to reduce the significance of the backlobe to a negligible level such that rear radiation of the control pattern is not required.

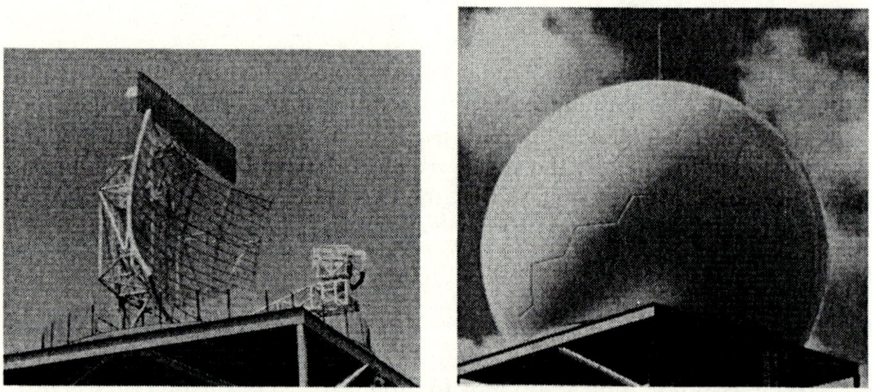

**Fig. 4.14** Radome 55 ft in diameter housing an SSR antenna mounted on top of an L-band primary radar reflector (Raytheon).

This technique is not widely used and some operational aspects have to be resolved. Since SSR and primary radar detection are performed at different times, plots from the two systems will appear to leapfrog one another. This performance may be irritating and distracting to the controller. When both systems feed plot extractors, the two plots for the one aircraft can be combined before presentation to the controller. Should a target be detected by means of primary radar only, then tracking can predict the aircraft's position half an antenna revolution into the future so that it can be presented in synchronism with the detection of other aircraft by SSR.

### 4.7.3 Back-to-back mounting

Some long-range primary radars use low antenna turning rates so as to increase the number of hits on the target and maximize detection. SSR is not constrained in the same way and could provide more frequent position updates on long-range targets. The effective doubling of the SSR scan rate can be accomplished by operating two SSR antennas in a back-to-back configuration. The technique has not been widely used and little system experience exists. Apart from attention to the possible backlobes of the antennas, no particular problems are foreseen.

### 4.7.4 Integrated feed systems

The availability of a large reflector used for a colocated primary radar offers the prospect of a low-cost high-gain SSR antenna system. The reflector is designed to provide very high discrimination against signals from the ground. Numerous attempts have been made, over a period of at least 25 years, to design an integrated feed system that would permit SSR system operation at the same time as primary radar operation. The modification normally takes the form of a quartet or octet of SSR dipoles equally disposed on each side of the primary radar antenna feed horn, although some designs make common use of the same horn. A number of these attempts have gone into operational service. Few have survived the test of time, at least in a civil ATC role.

Most primary radar reflectors are doubly curved to shape the antenna beam in both the horizontal and vertical dimensions. This aspect presents difficulties for the generation of the approximately omnidirectional SSR horizontal control pattern. To provide a well-controlled antenna pattern, the sidelobes of the feed system must not illuminate the reflector. The spillover that results has little impact on the performance of the primary radar antenna. In the case of SSR, the frequencies used are lower than those for primary radar and the feed antenna tends to have a broader beam so that a higher proportion of the radiated power misses the reflector. Aircraft behind the reflector can be interrogated either directly or via multipath propagation, and give rise to erroneous target reports.

The double curvature of the reflector causes polarization rotation at high angles off boresight resulting in a variation of copolar beam shape with elevation. Accurate monopulse characteristics, suitable for SSR, are difficult to achieve [4.1]. The problems are less severe in the case of a single curved reflector where the vertical pattern is shaped and the horizontal pattern is determined by the linear feed antenna at the focus. However, this type of primary radar antenna is less common because of its extra cost.

## 4.8 Radomes

The climatic conditions at some sites are such that it is advantageous to protect the antenna from the effects of wind, rain, snow or ice. The protection is provided by an enclosure known as a radome. The radome usually takes the form of a hollow sphere, fixed to the top of the tower, which totally encloses the antenna and in which the antenna rotates. An example is shown in Fig. 4.14.

In principle, an antenna and turning gear can be designed to withstand the effects of wind and rain so that a radome is not needed. Under extreme conditions snow and ice can dielectrically load the antenna, thereby altering its electrical properties. Satisfactory operation is then not possible without a radome. Even if the protection required is only for high winds, the cost of the radome can sometimes be justified if it avoids the need for a mechanically strong and heavy antenna structure and turning gear with powerful motors, and reduces the total life-cycle costs.

A most important requirement of the radome is that it must be substantially transparent to the radar signals and not perturb the antenna beams unduly. At the same time it must be strong enough to resist the forces of nature. The skill of the radome designer is in meeting both these requirements at an acceptable cost. The electrical performance of the radome is of prime importance, the more so with monopulse SSR because of the higher accuracy requirements that usually accompany that system. A radome must be self-supporting either by means of internal air pressure or through the strength of its own structure.

One method of construction places all the strength in a space frame in the form of a geodesic spheroid with the panels filling the frames made of thin dielectric material. A second type of radome uses thick panels formed as a sandwich and joined to their neighbors at their edges. The panels themselves contribute much of the strength of the structure. The third type of radome is made from a synthetic rubber, such as neoprene or hypalon, reinforced with man-made plastic or glass fibers. Support is provided by internal air pressure.

Undoubtedly the inflatable radome is the best for electrical performance. The shell is quite thin, typically 3–6 mm, and is made of a low-loss material. The lap joints between panels introduce little extra thickness. The material can be made sufficiently transparent to the r.f. signals not to affect the radar performance to a measurable extent. By their nature inflatable radomes must be filled with air at a higher pressure than outside, and this requires the use of air blowers. The blowers can be duplicated for reliability but there is a concern, particularly in the mind of the ultimate customer, that the electricity supply could fail and inflation be lost. Practical experience has shown that a deflated radome can be destroyed in high winds, even though the rotation of the antenna is stopped. The exposure to sunlight and the erosion and flexing caused by the wind will cause the surface of the material to deteriorate with time. The surface will need repainting at intervals, although with modern materials this maintenance may only be required every 5 or 10 years.

The use of rigid radomes avoids the above concerns, and the alternative materials available appear to give long life although some periodic surface treatment may be required. However, rigid radomes must provide their own strength, which requires the use of electrically thick material which compromises the transparency to r.f. energy. A common form of rigid radome is the space-frame radome. Figure 4.15 shows a photograph of a model of a space-frame radome. The thin panels filling the space between the frames have been omitted from the model. Figure 4.16 shows a cross-section of the structure and the membranes between the frames. The membrane material is usually of glass-reinforced plastic, but the frames can be either metal or plastic. Since the frames must supply the entire strength of the structure they are necessarily of a substantial size. The presence of the frames can have a serious effect on the antenna beam sidelobes and the direction of the difference-pattern null. The

**Fig. 4.15** Model of a space-frame radome (Transport Canada).

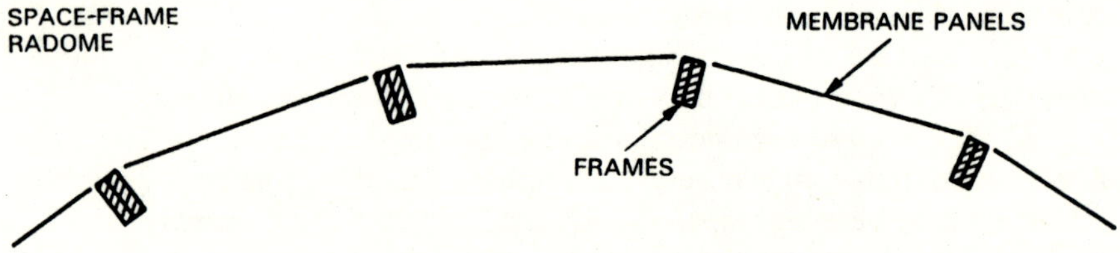


**Fig. 4.16** Detail of typical frame struts and panels.

sidelobe level can be increased to the extent that the gain of the sidelobes exceeds the gain of the control beam in that direction so that a transponder fails to be suppressed. On some sites [4.2] it has been found necessary to add an additional omnidirectional antenna on top of the radome to supplement the control beam and prevent false aircraft plots from being generated.

Antenna engineers will be well aware of the difficulties of making accurate measurements of antenna beam patterns even when using a calibrated test range. The problems and expense of making these measurements with the antenna installed in a radome are compounded. Even if the measurements with the radome are made, the results only apply to the line though the radome in the direction of the transmitter source. There is no guarantee that the impact of the presence of the radome will be the same in a different direction with an alternative disposition of the frame struts relative to the antenna boresight.

Attempts to make before and after measurements at a radar site during construction are often frustrated by pressures of project time-scales, and when they are made they can be ambiguous owing to site reflections and limitations of the positioning of the remote source. A more practical approach to the problem of determining the impact of a radome on the antenna performance is the use of scale models. The test site required is smaller and can be better controlled, and, more importantly, the radome can be rotated to investigate pattern perturbations in all directions. Figure 4.15 shows a model used for this purpose by Canadian Astronautics Ltd [4.3].

Measurements were made using a frequency of 9.375 GHz to give a scale factor of 8.84 to the mean SSR frequencies. Both metal and dielectric space frames were used, but the results for the dielectric material were only marginally better than those using metal frames. Figures 4.17–4.19 show the ripple caused to the omnidirectional control pattern and the impact of the sum-beam sidelobes caused by the radome. Figure 4.20 shows the change in depth and direction of the difference-beam null as the radome was rotated. These results indicate that space-frame radomes are unsuitable for use with monopulse SSR systems.

In the case of dielectric space-frame radomes, the capacitive nature of the material can be compensated by embedding a wire grid which is inductive at

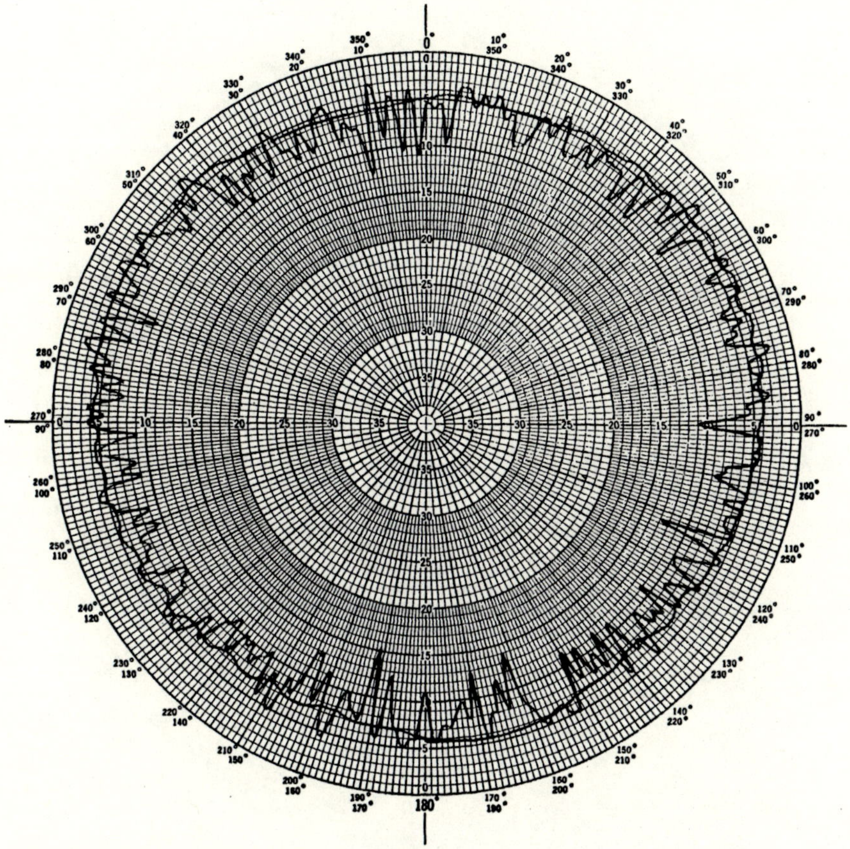

**Fig. 4.17** Omnidirectional control pattern with and without a radome (Transport Canada).

the frequency of interest [4.4, 4.5]. This technique is reported to reduce the scattering loss caused by the frame members significantly.

A better form of rigid radome uses panels with a sandwich construction (Fig. 4.21). The sandwich is formed by two thin high-dielectric sheets which are separated by approximately a quarter-wavelength thickness of foam or honeycomb material of lower dielectric constant. This type of sandwich is sometimes known as an "A sandwich". The sandwich panel itself is very strong, which reduces the need for the joints to provide all the strength of the structure. Lap joints which reduce the joint cross-section are used, and the material of the joints is chosen so that its electrical phase delay is similar to that of the panel. The result is a structure which minimizes the impact of the radome on the antenna beam patterns.

No model results are available for the sandwich construction to compare with those for the space-frame radome. However, some mathematical analysis

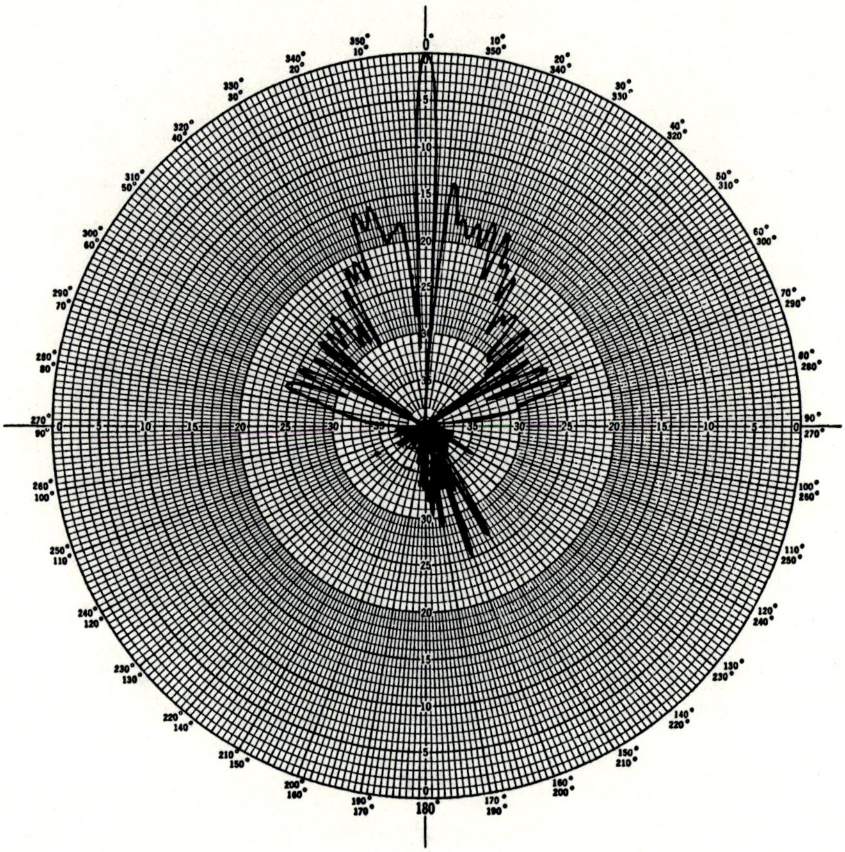

**Fig. 4.18** Sum pattern without a radome (Transport Canada).

is possible [4.6–4.9]. Simply explained, the joints and panels are treated separately and are each projected onto the antenna aperture. The two projected figures are called blocking areas. The two blocking areas are also regarded as equivalent radiation sources which produce fields in the forward direction. The ratio of the field due to the radome to that of the antenna without the radome is known as the induced field ratio (IFR), and this can be thought of as a figure of merit with a higher figure indicating worse scattering. The IFR for a sandwich construction radome is about 20 dB less than that for the space-frame radome.

Table 4.3 gives some calculated parameters for three types of radome. The inflatable radome has a negligible impact on performance, and the sandwich radome occupies a very good second place.

While the effects of ice and snow on antennas and radomes can be readily

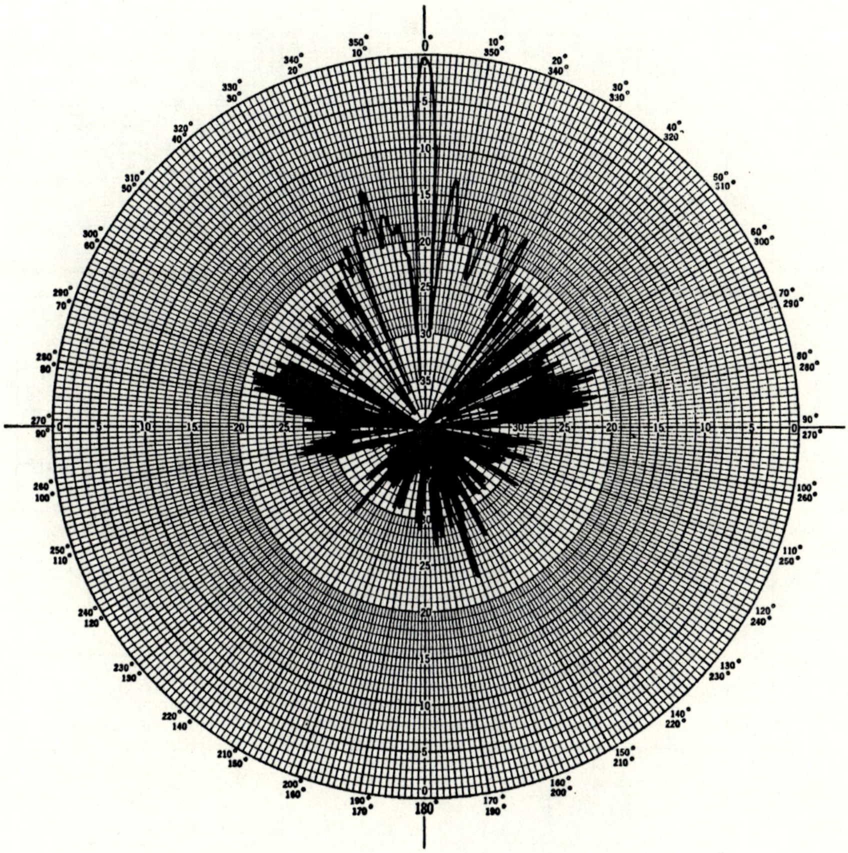


**Fig. 4.19** Sum pattern with a radome (Transport Canada).

calculated, very few data are available on how it accumulates on these objects. Snow or frost is of little concern since its high air content makes it a low-loss material. However, freezing rain can precipitate as almost pure ice and can produce a very significant dielectric load. The spherical shape of a radome encourages the air carrying the precipitation particles to flow around the radome instead of impinging upon it so that the buildup of snow and ice is much reduced. Intuitively, it would be expected that the flexible nature of the inflatable radome would break off any ice that might accumulate on the surface. However, there are unverified reports of heavy snow settling on the top of an inflatable radome and dimpling it—a subsequent rise in temperature leaves a pool of water that is only removed by evaporation.

The surface of the material used for the radome must not absorb water if attenuation of the r.f. signals is to be avoided. Moisture due to rain or

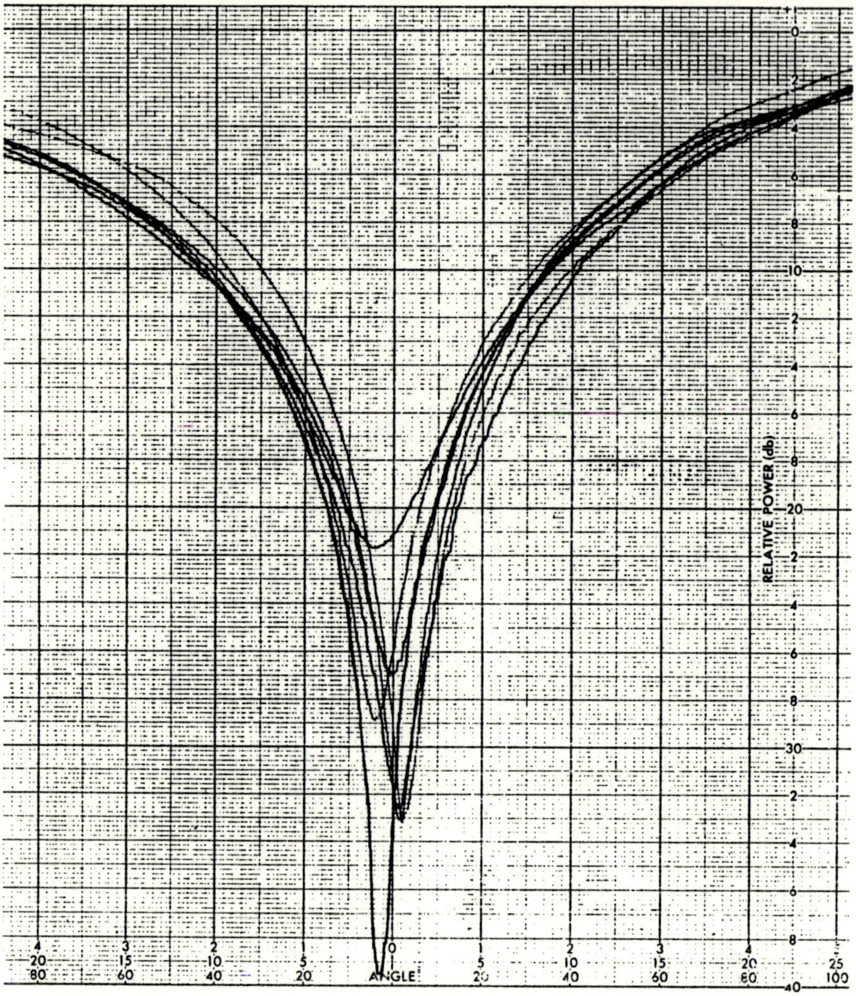


**Fig. 4.20** Change in depth and direction of the difference-beam null with radome rotation (Transport Canada).

condensation produces a sufficiently thin film of water for its effects to be insignificant at SSR frequencies. Internal condensation is a common occurrence with radomes, although its effect on performance is negligible. The large surface area of the radome makes it impractical to heat the inside with a view to melting internal frost or drying the surface.

While a radome provides a better environment for the maintenance engineer, should any repairs to antenna or turning gear be necessary, the internal noise level can be very high owing to the reflection of sound from the

**Table 4.3** Comparison of the performance of radome types

| | *Metal space frame* | *Fiberglass space frame* | *Inflatable* | *Sandwich* |
|---|---|---|---|---|
| Difference-null shift (deg) | 0.5 | 0.5 | 0 | <0.005 |
| Transmission loss (dB) | <0.5 | <0.5 | <0.02 | <0.2 |
| Near-in sidelobe increase (dB) | >2 | >2 | Small | Small |
| Scattered field (dB) | −35 | −35 | −50 | −42 |

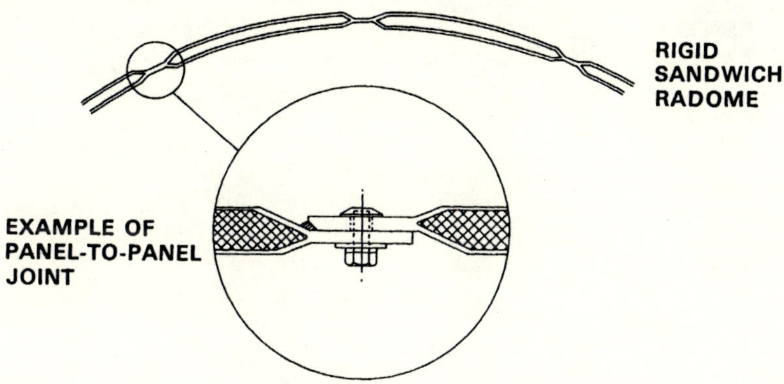


**Fig. 4.21** Detail of a radome with sandwich panels.

radome. The reverberation often makes it impossible to speak to a colleague more than 2 or 3 ft away.

Radomes can attract their own hazards. Apparently the hunting fraternity in the North American continent find radomes irresistible objects for target practice. Usually the bullet holes have a negligible effect on the mechanical strength or the air leakage of the radome. Fortunately antenna designers are not yet required to include resistance to shotgun damage in their design specifications.

# References

4.1 Guichard, P. Monopulse SSR antennae, *Eurocontrol Seminar, The ATC Radar of the Future*, Paper I-6, Eurocontrol Institute, Luxembourg, October 1985.

4.2 Cohen, A. SSR antennas: side-lobe suppression can be assured, *ICAO Bull.*, **33** (9), 34–35, September 1978.

4.3 Whelpton, J. *Final Report on ASR-803 Radome Study*, Canadian Astronautics Ltd, Ottawa, 1980.

4.4 Bennett, J. C., Chambers, B. and Munro, A. M. A comparative study of metal, G.R.P. and wire-loaded G.R.P. struts for use in space frame radome structures, *4th Int. Conf. on Antennas and Propagation, ICAP 85, IEE Conf. Publ.*, **248**, 153–157, April 1985.

4.5 Chang, K. C. and Smolski, A. P. The effect of impedance matched radomes, *IEE Conf. Proc. RADAR 87*, pp. 155–159, IEE, London, October 1987.

4.6 Bond, K. Radome effects, *Cossor Internal Memo. 853/KB/6960/DE*, November 1983.

4.7 Bergh, K. Sandwich shell radomes, *ELAB Memo 12-83*, Norges Tekniske Hogskole, Trondheim, July 1983.

4.8 Kay, A. F. Electrical design of metal space frame radomes, *IEEE Trans. Antennas Propag.*, **13**, 188–202, 1965.

4.9 ESSCO, A radome for air traffic control SSR radar systems, *Conf. RADAR 84*, Paper C-2, Institute Français de Navigation, Paris, May 1984.

# Chapter 5

# Interrogators

The word "interrogator" is now widely used to describe the ground transmitter–receiver equipment of the SSR system. Originally, and more correctly, it meant just the transmitter part, the function that generated the "interrogation", with the receiver part being called the "responsor". However, in deference to current practice the word will be used to refer to total transmitter–receiver function.

The following are described in this chapter:

(a) the mode generator which produces the transmitter modulation waveforms, the sequence of interrogation modes and the basic timing control of the interrogator;

(b) transmitter frequency control and the receiver local oscillator;

(c) the transmitter low power amplification and the modulator;

(d) the high power output stages of the transmitter;

(e) the receiver functions including gain–time control, reply pulse detection and receiver sidelobe suppression;

(f) the special requirements of a monopulse receiver.

## 5.1 Mode generator

The mode generator controls the timing of the transmitter and receiver functions of the interrogator and, through them, that of the plot extractor. Frequently the mode generator obtains its time reference from its own internal master clock but, alternatively, it can be synchronized to an external trigger source such as an associated primary radar. It is not unusual for a primary radar to operate with a higher interrogation rate than that applicable to the SSR, in which case the synchronizing external trigger must be frequency divided, or counted down, for SSR use.

### 5.1.1 Mode interlace programs

Aircraft provide different reply data in response to the various SSR interrogation modes, and all these data may be required. The interrogation modes normally used are as follows:

Mode 3/A (common civil and military identification)
Mode C (flight level)
Mode 1 or 2 (military identification)

Obviously only one mode can be transmitted at a time, but multiple modes can be accommodated by alternating the modes according to some preset interlace program.

A commonly used interlace program is the alternating transmission of Modes A and C. This program will obtain the two items of reply data for civil aircraft, i.e. aircraft identity and aircraft altitude, with equal probability. When military identification is also required, a three-mode interlace program can be used, i.e. A, C, 2, A, C, 2, A, . . . etc.

If reply data on both military identification modes 1 and 2 were required, a four-mode interlace program could be used. However, only a small proportion of aircraft would reply to all modes and insufficient replies might be obtained from most aircraft for their presence to be reliably detected. The normal solution is to employ three-mode interlace but to switch the third mode after every antenna rotation as follows:

| | |
|---|---|
| $n$th rotation | A, C, 2, A, C, 2, A, . . . |
| $(n+1)$th rotation | A, C, 1, A, C, 1, A, . . . |
| $(n+2)$th rotation | A, C, 2, A, C, 2, A, . . . |

For special applications it may be desired to interrogate only over a limited arc and not continuously over the full 360°. This may be the case where aircraft of interest are confined to certain sectors and interrogating elsewhere could degrade the SSR system for other users.

### 5.1.2 Transmitter pulse specifications

Table 5.1 gives the interrogator mode pulse specifications. The pulse specifications apply to the resulting transmitted r.f. pulse, and allowance should be made for delays in the modulator and r.f. stages. The pulse spacing

**Table 5.1** Interrogator mode pulse specifications

| | | |
|---|---|---|
| P1, P2, P3 pulses | | |
| Pulse duration[a] | | $0.8 \pm 0.1$ μs |
| Rise time | | $\geqslant 0.05$ μs, $\leqslant 0.1$ μs |
| Decay time | | $\geqslant 0.05$ μs, $\leqslant 0.2$ μs |
| Mode spacing | | |
| Mode | P1–P2 separation | P1–P3 separation |
| 1 | $2 \pm 0.1$ μs | $3^{+0.1}_{-0.05}$ μs |
| 2 | $2 \pm 0.1$ μs | $5 \pm 0.1$ μs |
| 3/A | $2 \pm 0.1$ μs | $8 \pm 0.1$ μs |
| B | $2 \pm 0.1$ μs | $17 \pm 0.1$ μs |
| C | $2 \pm 0.1$ μs | $21 \pm 0.1$ μs |
| D | $2 \pm 0.1$ μs | $25 \pm 0.1$ μs |

[a] *STANAG 4193* [5.2] specifies the P2 pulse duration as $0.8^{+0.05}_{-0.1}$ μs.

tolerances shown in Table 5.1 are more stringent than the signal spacing tolerances required by ICAO *Annex 10* [5.1] in order to allow some margin for degradation by noise and multipath interference.

The pulse rise and decay times can be verified by direct measurement using a calibrated detector and an oscilloscope. An alternative test, permitted by ICAO *Annex 10*, is to determine that the resulting sideband radiation is no greater than that produced by the minimum permitted rise and decay times.

Mode S transmissions impose other requirements on transmitter pulses and permitted sideband radiation. These are discussed in Chapter 12. The design of SSR equipment intended for possible future upgrading to Mode S should include consideration of the additional requirements of Chapter 12.

### 5.1.3 Interrogation repetition frequency

SSR interrogations can be transmitted with a steady pulse repetition frequency with equal spacing between each subsequent mode transmission. However, this simple arrangement can give rise to "second-time-around" and "group capture" problems.

Second time around occurs when an aircraft replies beyond the operating range of the SSR. Its replies will be considerably delayed from the interrogation triggering the reply and may be received after a subsequent interrogation has been transmitted. The replies will be in synchronism with the subsequent interrogation and the aircraft will appear at an artificially close range to the confusion of the air traffic controller.

When interrogated, a transponder is occupied by the reply and by the time necessary for its transmitter to recover. This capture period, during which the transponder is unable to reply to another interrogation, typically lasts for 70 μs but can persist for a maximum of 145 μs. If an aircraft is within range of a second SSR sensor with a similar interrogation frequency, then one SSR sensor can be denied replies for a substantial number of interrogations.

The problems of synchronism with a later interrogation with the second-time-around reply and of group capture by near synchronism with another interrogator can both be avoided if the interrogation period is subjected to a small randomizing or stagger. A simple interrogation stagger program is to vary the period between interrogations by 0%, −5%, +5%, 0%, −5% etc. A more complex program could switch between several different periods using a pseudo-random sequence.

### 5.1.4 Use of a radio frequency switch

Interrogation pulse P2 is transmitted via the antenna control beam, while pulses P1 and P3 are radiated by the main beam. All three pulses can be produced by a single transmitter amplifier chain and the various pulses switched to the appropriate antenna beam (see also Section 4.4).

If desired the r.f. switch can be placed above the antenna rotating joint, thereby requiring only a single SSR channel in the joint. Although r.f. switches can be made with very low failure rates, some users prefer to include two rotating joint channels so that the r.f. switch can be placed in the interrogator cabinet where it is protected from environmental extremes and access is easier. In the case of a monopulse SSR, three rotating joint channels are normally required for reply processing.

A single transmitter amplifier and an r.f. switch may not be practicable if "improved" interrogator sidelobe suppression (IISLS) (see Chapter 9, Section 9.1.3) is employed or if allowance is to be made for future inclusion of Mode S.

Control of the r.f. switch is another function of the mode generator although, where the switch is sited above the rotating joint, switch operation triggered from the detected r.f. transmitter pulses can be employed.

### 5.1.5 Triggers for external use

The mode generator also produces a number of triggers for use by other equipment. The most important of these is the system trigger to the plot extractor.

The plot extractor is mainly involved in detecting and processing aircraft replies, and in order to perform this task synchronism with the transmitter is necessary. Following the receipt of the system trigger from the mode generator, the plot extractor measures the spacing between the transmitter pulses P1 and P3 to determine the mode of the interrogation and to

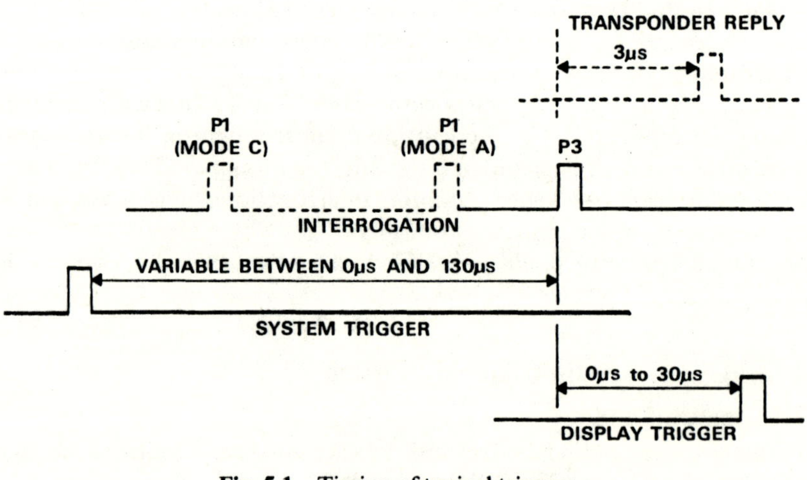


**Fig. 5.1** Timing of typical triggers.

synchronize its range clock to the leading edge of pulse P3. A typical time for the system trigger is 30 μs before the transmission of P3 so that it precedes the transmission of P1 even for a Mode D interrogation.

In simple SSR systems it may be sufficient to view the reply video directly on a PPI display. Correction for the 3 μs nominal transponder reply delay, and for the 21 μs decoding delay if F1 and F2 framing pulse detection is employed, may be desired. Such correction can be achieved by an appropriately timed display trigger from the mode generator.

Although ATC displays which present SSR reply data in video form are no longer in common use, a video display is often employed at the SSR site for checking the performance of the equipment. The mode generator normally provides a display trigger with a delay timed to the transmission of P3 which is variable between at least 0 μs and +30 μs.

The relationship of typical triggers produced by the mode generator is shown in Fig. 5.1.

## 5.2 Transmitter

### 5.2.1 Frequency control

Control of the transmitter frequency invariably uses a crystal-controlled exciter of some kind. The required frequency of 1030 MHz is too high for a quartz crystal oscillator to operate directly, although surface acoustic wave devices are available where this is possible.

One method of generating a stable 1030 MHz source is to employ a crystal oscillator at a lower frequency and to multiply the frequency to 1030 MHz. A typical approach would be to start at 114.444 MHz and to multiply by a step recovery diode. A bandpass filter then selects the ninth harmonic at 1030 MHz. Obviously care must be taken in the design of the filter to reject the other harmonics produced, particularly the eighth harmonic at 915.5 MHz and the tenth harmonic at 1144.4 MHz, since their presence could give rise to spurious receiver responses.

An alternative method of generating the required 1030 MHz frequency is by a voltage-controlled oscillator operating at 1030 MHz which is then frequency divided and phase locked to a lower crystal-controlled source. A block diagram is shown in Fig. 5.2. This method avoids the stringent filter requirements of the multiplier technique. Care must be taken to minimize the reference frequency sidebands.

Both methods, particularly that using the phase-locked loop, can be subjected to frequency pulling during the transmitter pulses if inadequate isolation is provided between the exciter and the modulator. The frequency tolerance of the 1030 MHz transmission required by ICAO *Annex 10* is

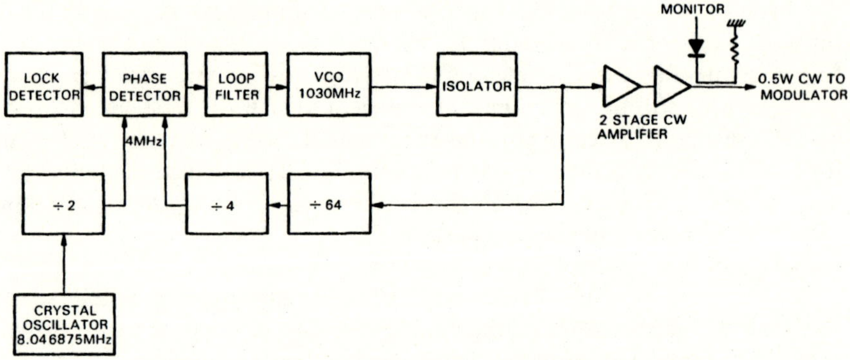


**Fig. 5.2** Block diagram of a phase-locked loop exciter.

$\pm 0.2$ MHz, although a tighter tolerance of $\pm 0.1$ MHz is well within the capability of a crystal-controlled source. Equipment intended to include a Mode S capability must meet a frequency accuracy of $\pm 0.01$ MHz, and this normally requires a temperature-controlled or temperature-compensated frequency reference.

The exciter output is at a low power level and is amplified before being modulated with the required pulse waveforms. Outputs are also required to act as local oscillator inputs to the receiver mixers: three are required for a typical monopulse receiver. The receivers are tuned to 1090 MHz, the transponder reply frequency, and typically use an i.f. amplifier frequency of 60 MHz, therefore needing a local oscillator at 1030 MHz which can be conveniently provided by the transmitter exciter. The local oscillator outputs must exhibit high isolation at 1090 MHz and 60 MHz so as to avoid any cross-coupling between the three channels of a monopulse receiver.

## 5.2.2 Modulation

Figure 5.3 shows a stripline implementation of a modulator which uses pin diodes in the arms of a hybrid ring to switch the r.f. power either to an internal load or to the output port. Other pin diodes on the same arms can be used to apply phase modulation to the output as required by a Mode S interrogation.

An essential requirement of the amplitude modulator is a good on–off ratio. ICAO *Annex 10*, paragraph 3.8.11, recommends that the radiated signal at 1030 MHz should not exceed $-76$ dBW when the modulator is off. Transmitter power output stages usually use limiting amplifiers for maximum power efficiency and hence may amplify low level signals by a greater amount than the high level pulses, thereby placing greater requirements on the performance of the modulator.

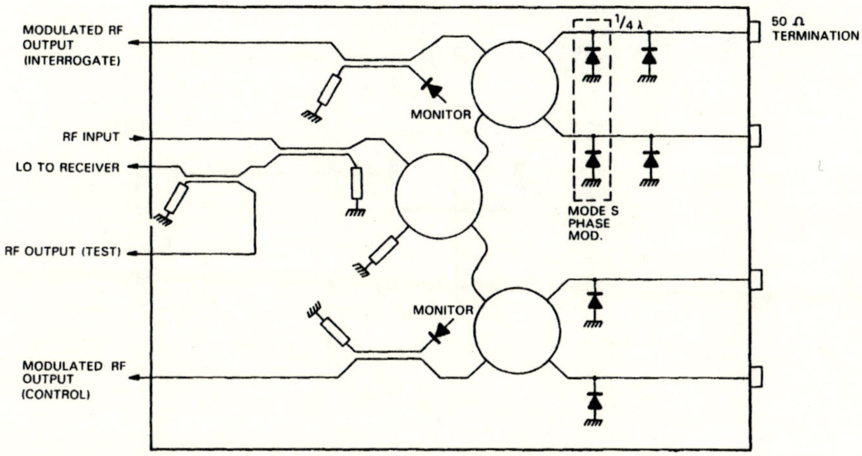


**Fig. 5.3** Typical amplifier modulator including a phase modulation capability for Mode S: LO, local oscillator.

## 5.2.3 Power output stage

As with most other modern electronic equipment, the SSR interrogator design makes extensive use of semiconductor technology. One exception to this is the high power stages of the transmitter where, until recently, vacuum triodes were still widely used owing to the difficulty of generating the required power using transistors. This situation is in the process of change, and both technologies are discussed but with the emphasis on solid state power generation.

An important parameter in the design of a transmitter is the required output power. This in turn is dependent upon other factors of the SSR system such as the maximum range, the antenna gain and r.f. feeder losses. These aspects of the system are discussed more fully in Chapter 10, but it can be assumed for present purposes that powers in the range 0–2 kW are typical requirements.

Other important parameters are the r.f. pulse length and duty cycle. For normal SSR the pulse length is 0.8 μs and typical duty cycles are less than 0.1%. For Mode S pulse lengths of up to 30 μs and short-term duty cycles of up to 67% may be required.

**Vacuum triode transmitters**

Until recently the only practical method of generating the r.f. power levels needed for SSR was by means of vacuum triodes. The normal method of using these triodes is in the grounded grid mode with a tuned cavity at both the anode and cathode. A circuit diagram is shown in Fig. 5.4 and the construction of a cavity to house a triode is illustrated in Fig. 5.5.

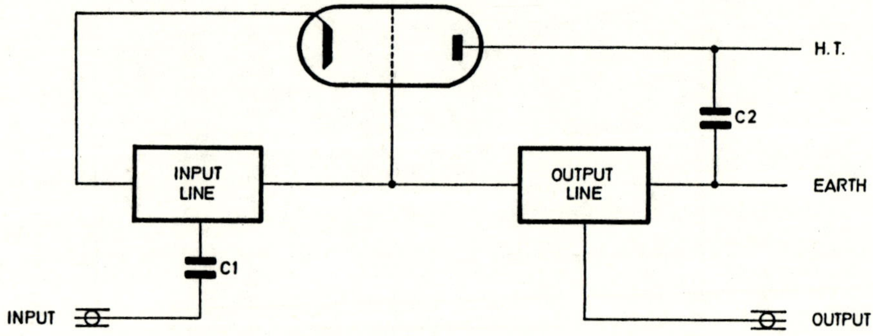


**Fig. 5.4** Circuit diagram of a triode transmitter.

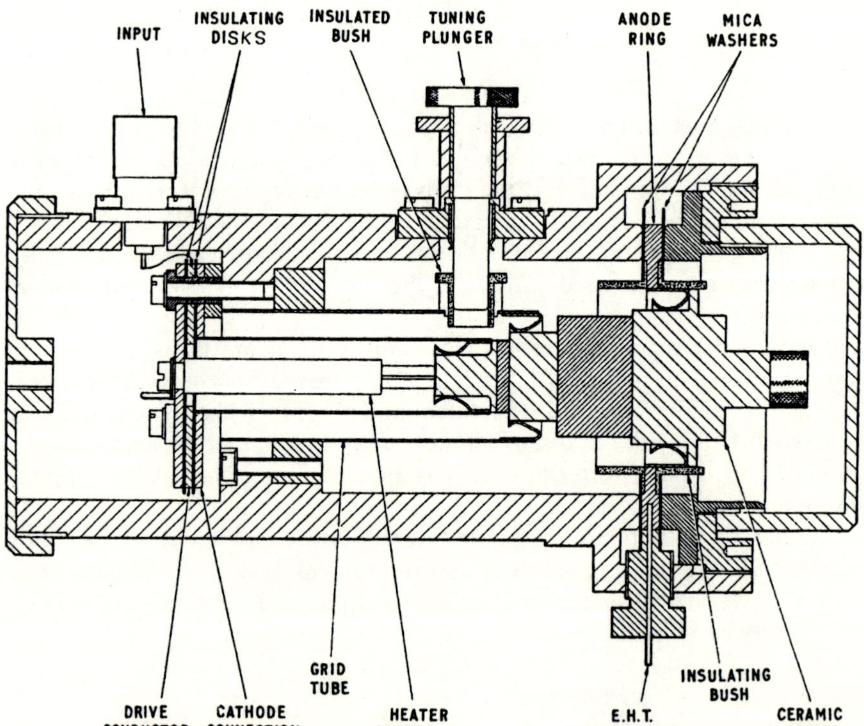


**Fig. 5.5** Triode cavity for a power transmitter.

Some early implementations used the triode as a self-oscillating power stage with the frequency determined by the cavities. Although this scheme is still used in some transponders, which are permitted a wide frequency tolerance of ±3 MHz, the method proved difficult to apply in ground interrogators which are required to operate within a frequency tolerance of ±0.2 MHz. More recent practice has used a crystal-controlled oscillator followed by a series of amplifying stages.

There are differing claims concerning the reliability obtainable from these vacuum thermionic devices. However, if they are conservatively driven and well cooled by the use of solidly constructed cavities, they are able to give service well in excess of the 2000 h normally claimed by the manufacturers. The usual failure mode is a fall in output power as a result of falling cathode emission. Routine measurement of output power has enabled the timely replacement of these triodes before catastrophic failure occurs.

Conventional wisdom holds that solid state devices are more reliable than vacuum thermionic devices, and there is considerable market pressure for the utilization of r.f. power transistors in place of the triodes. As yet there is little prolonged field experience to demonstrate that r.f. power transistors are indeed more reliable than triodes.

**Solid state transmitters**

While the output power requirements are within the capabilities of a single vacuum triode, the same is not yet true for r.f. power transistors. A typical transistor will have a manufacturer's rating of 600 W peak pulse power. It is normal to derate the transistor in order to reduce the stress on the device to obtain high reliability. Because of these factors a transistor rated at 600 W may be operated at an output power of 400 W. It is then necessary to operate a number of the devices in parallel to produce an output of 2 kW. Table 5.2 shows the typical characteristics of an r.f. power transistor available from a number of sources.

Nearly every aspect of r.f. power transistors makes them difficult to use. It is critically important to provide good cooling so as to remove heat generated at the collector junction and limit the junction temperature rise. Heat kills. A good design for high reliability limits the junction rise to 150 °C maximum for a silicon device. Figure 5.6 shows the junction temperature rise during a normal SSR pulse 0.8 μs wide and also during a Mode S pulse 30 μs wide. The transistor input and output impedances are highly reactive and the input impedance is usually less than 1 Ω, making impedance matching a problem. Unlike the planar triode, the construction of the transistor is not such that high-$Q$ cavities can be used for tuning and matching. Instead, the transistor package is more appropriate to the use of stripline techniques which, while being readily reproducible using printed circuit methods of manufacture, have higher losses.

**Table 5.2** Characteristics of a typical radio frequency power transistor

| | |
|---|---|
| Maximum rating | |
| Collector–emitter voltage $V_{ces}$ | 78 V |
| Peak collector current | 40 A |
| Peak power dissipation $P_D$ | 1350 W |
| Junction temperature $T_j$ | +250 °C |
| R.f. performance | |
| Input | |
| Frequency | 1030 MHz |
| Supply voltage | 43 V |
| R.f. input peak power | 75 W |
| R.f. pulse duration | 32 μs |
| Duty cycle | 2% |
| Output | |
| R.f. output peak power | 420 W min |
| R.f. pulse droop | 0.6 dB max |
| Insertion phase | ±15° |
| Derating | |
| Supply voltage | 55% $V_{ces}$ |
| Dissipation | 70% $P_D$ |
| Junction temperature | 70% $T_j$ |

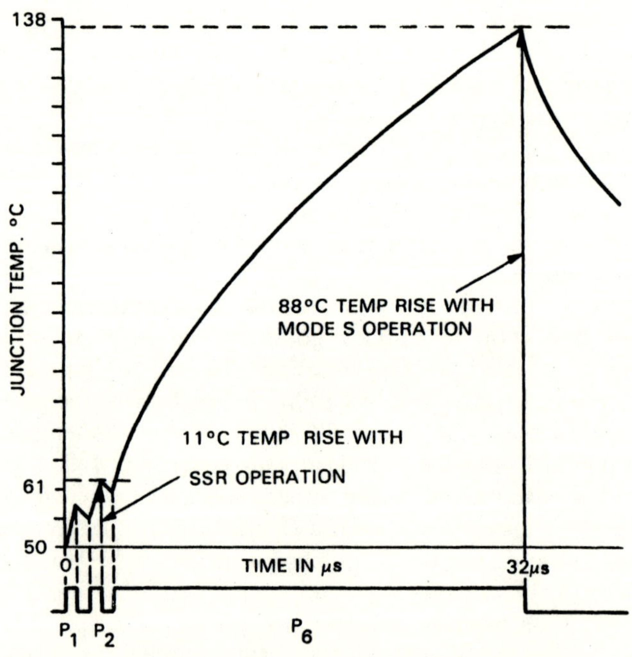


**Fig. 5.6** Junction temperature rise (SSR and Mode S).

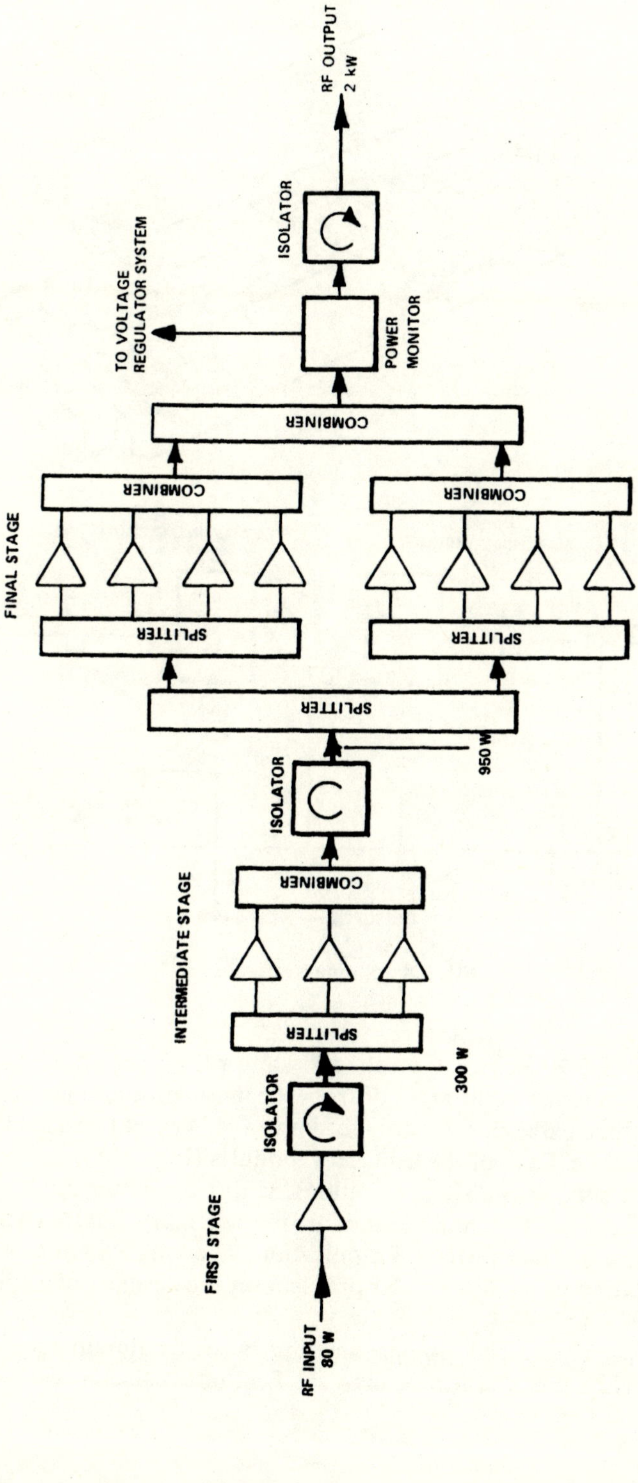


**Fig. 5.7** Block diagram of a solid state power amplifier.

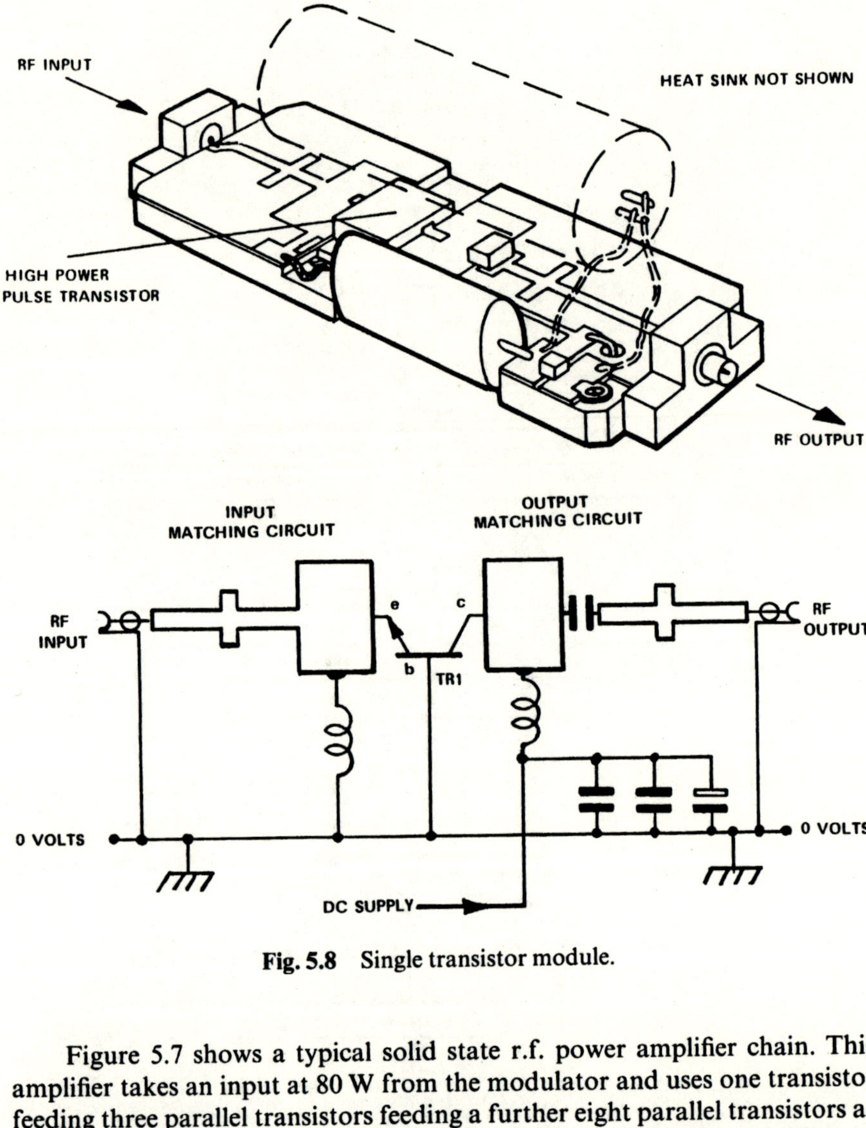


**Fig. 5.8** Single transistor module.

Figure 5.7 shows a typical solid state r.f. power amplifier chain. This amplifier takes an input at 80 W from the modulator and uses one transistor feeding three parallel transistors feeding a further eight parallel transistors as the final stage. Each of the transistor modules (Fig. 5.8) is similar and has its own heat sink. For SSR applications, with transmitter duty cycle requirements of about 0.1%, the heat sinks are not necessary, but this is not the case if the transmitter has a Mode S application. The long P6 pulse of the Mode S interrogation poses a particular problem for the designer of a solid state high power r.f. amplifier.

At present, silicon bipolar junction transistors dominate r.f. power solid state devices, but the future may see field effect silicon transistors, with a

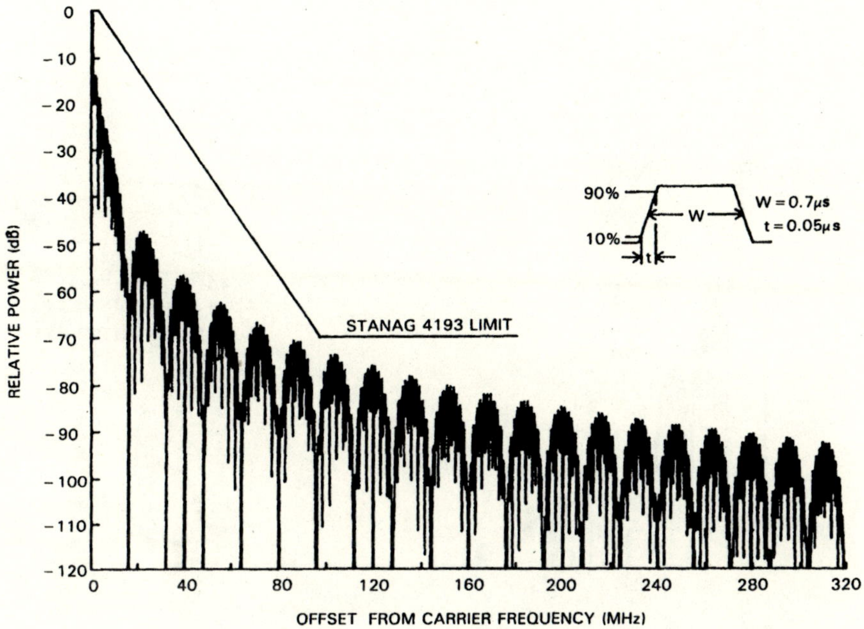


**Fig. 5.9** Permitted transmitter sideband spectrum.

higher and more convenient input impedance and resistance to secondary breakdown, and gallium arsenide devices offering a higher temperature capability.

### 5.2.4 Transmitter sideband spectrum and output filtering

ICAO *Annex 10* [5.1] permits the SSR sideband spectrum to be consistent with that produced by transmitter pulses with rise and fall times of 0.05 μs. *STANAG 4193* [5.2] places an upper limit on the sideband amplitude which is consistent with the ICAO requirement. Figure 5.9 shows these permitted spectra which are quite generous.

If the SSR is used with a colocated primary radar operating in the 1250–1350 MHz band, interference with the primary radar can occur. If an SSR effective radiated power of 52.5 dBW at 1030 MHz is assumed a typical requirement is for the SSR sideband spectrum to be at least −115 dB below the carrier in the primary radar operating band.

A second-order Butterworth filter can be employed at the output of the transmitter to contain the sideband spectrum without appreciably degrading the rise and fall times of the transmitted pulses. The resulting spectrum, after the addition of the filter, is shown in Fig. 5.10. The filter used will also restrict

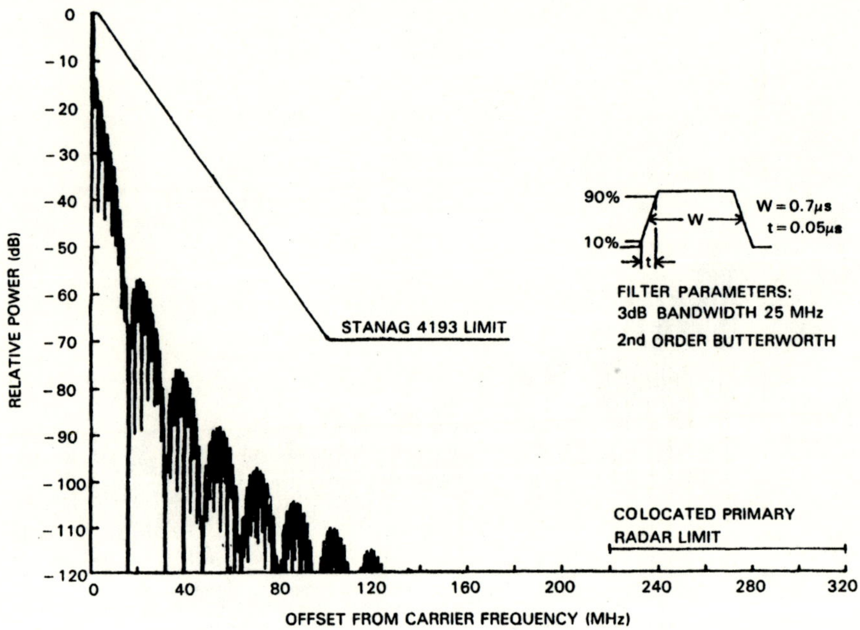


**Fig. 5.10** Transmitter sideband spectrum with filter.

the sidebands of a Mode S interrogation to meet the same requirement (see Chapter 12, Fig. 12.5).

# 5.3 Standard receiver

The standard receiver used in the SSR system is discussed in this section. The monopulse receiver, which provides an additional direction-finding facility, is described in Section 5.4.

## 5.3.1 Receiver sensitivity

The receiver is required to have a sensitivity sufficient for the weakest transponder reply signal reasonably likely to occur within the SSR system in which it is to be used. The system design is discussed in Chapter 10, Section 10.2.2, and a typical minimum reply signal of −80 dBm can be assumed for most cases. This minimum reply signal is consistent with the ICAO *Annex 10* recommendation of a receiver tangential sensitivity of −85 dBm or better. Tangential sensitivity can be defined as that signal which raises the observed noise level by its own height, as shown in Fig. 5.11.

(a)

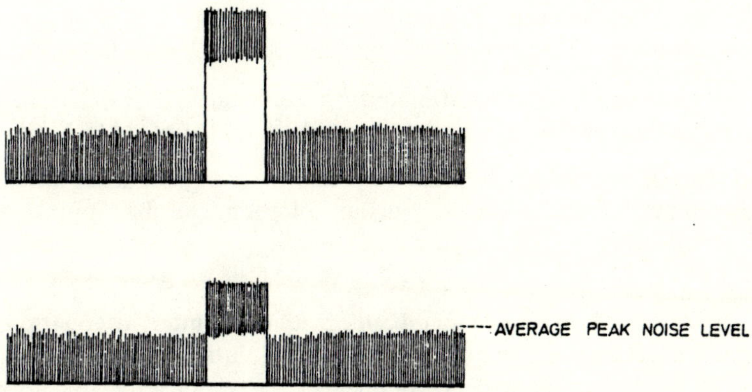


(b)

**Fig. 5.11** Measurement of the tangential sensitivity of a receiver: (a) normal signal; (b) tangential signal.

Tangential sensitivity is a convenient measurement to make as it only requires the use of a calibrated signal generator and an oscilloscope. The definition of tangential sensitivity assumes a receiver with a square-law detector. However, it is a very subjective measurement and is influenced by the brightness setting of the oscilloscope and the judgment of the level of typical noise peaks. Most modern receivers are designed with logarithmic characteristics which compress the vertical scale and further complicate the measurement. Different people tend to obtain very different results. Tangential sensitivity can be taken to correspond to a signal-to-noise ratio of about 8 dB with a measurement uncertainty of at least $\pm 2$ dB.

A more practical measurement of sensitivity is the pulse detection efficiency at a level corresponding to the minimum reply signal. A calibrated r.f. pulse generator and a counter can be used to perform the test in a way that avoids the subjective element.

The receiver must also be capable of correctly processing the strongest reply signals that would be detected from a nearby target. A realistic maximum signal is determined as $-18$ dBm (Chapter 10, Section 10.2.2).

## 5.3.2 Receiver bandwidth

The receiver bandwidth must be limited so as to reject adjacent channel signals and out-of-band noise but pass the transponder reply pulses with little degradation. It is normal to include filters in both the r.f. and i.f. sections of the receiver. The main purpose of the r.f. filter is to block signals at the image frequency and to prevent high power out-of-band signals, which could possibly come from a nearby primary radar, from saturating the front end of the receiver. The i.f. filter usually has greater selectivity than the r.f. filter in order to reject adjacent channel interference. The following out-of-band attenuation for the total receiver is specified in *STANAG 4193* [5.2]:

more than 40 dB attenuation outside the band 1078–1102 MHz
more than 60 dB attenuation outside the band 1065–1115 MHz

The type of i.f. bandpass filter used has a strong influence on the pulse performance of the receiver. Further filtering can be applied at video frequency following the detector.

The ideal bandpass filter would have the following characteristics:

(i) a sufficiently wide passband, with minimum attenuation, to accommodate the carrier frequency and the modulation sidebands;

(2) a constant group delay (or linear phase response) across the passband so that the pulse envelope is undistorted;

(3) maximum attenuation outside the passband to reduce the effects of out-of-band signals.

No single filter is able to provide all these properties and a compromise has to be made. The requirement for limited pulse shape distortion can be met by a linear phase or Gaussian class of filter amongst others. Sufficient sideband signals are passed so as not to compromise pulse shape seriously if the 3 dB bandwidth is at least $0.75/t$ where $t$ is the pulse duration. For an SSR reply pulse, with a minimum duration of 0.35 μs, the bandwidth is 2.1 MHz. This bandwidth, while effective for pulse detection, is not adequate to allow for the transmitter frequency tolerance of $\pm 3$ MHz, and when this further requirement is included the minimum bandwidth becomes 8.1 MHz.

The need to include transponder transmitter frequency tolerance increases the time–bandwidth product of the filter from 0.74 to 2.8, which produces a loss of about 4.2 dB in signal-to-noise ratio. This loss can be partly recovered by further band limiting after video detection where the transponder frequency tolerance is no longer an issue. An i.f. bandwidth of 8.1 MHz results in dominant noise with a frequency of about the same value, and a sharp cut-off video filter which rejects this noise without significantly degrading the pulse would provide some signal-to-noise improvement.

The Gaussian class of filter has relatively wide skirts to the passband which results in inadequate rejection of adjacent signals. Such signals can arise from distance-measuring equipment (DME) which shares the aeronautical L band with SSR. A hybrid class of filter, such as the transitional Gaussian filter [5.3], provides a practical solution to the problem. This filter type has a Gaussian-like characteristic in the passband and a Chebyshev-like characteristic in the stopband below about 6 dB of attenuation. The amplitude and group delay response of such a filter is shown in Fig. 5.12.

It is sometimes maintained that logarithmic amplifiers "stretch" the pulse width because of the different appearance of the output video pulse compared with the r.f. pulse input. Figure 5.13 shows examples of video pulses at

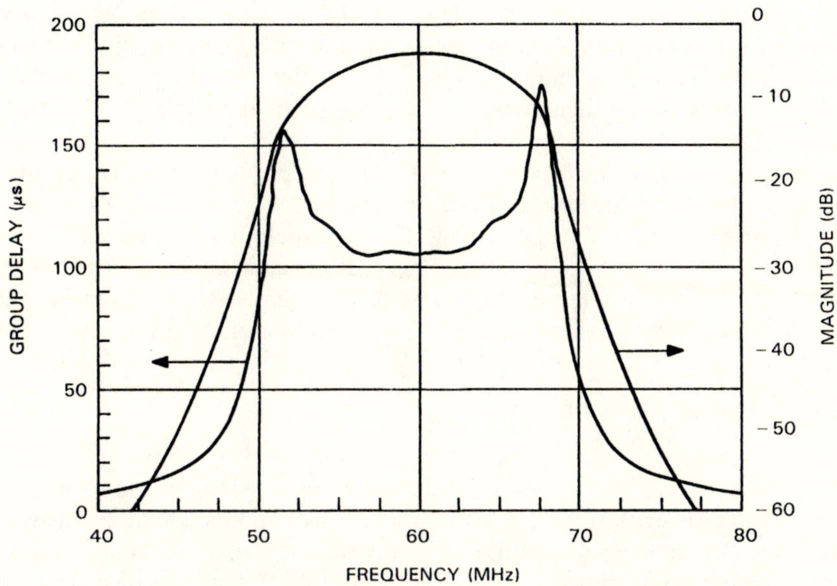


**Fig. 5.12** Response of a transitional Gaussian filter.

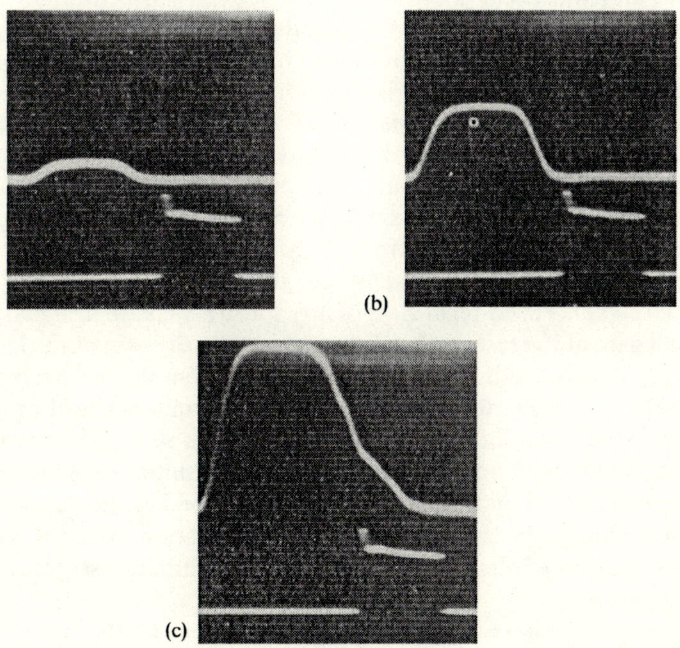
(a) (b) (c)

**Fig. 5.13** Pulse detection over a wide dynamic range: (a) −85 dBm; (b) −60 dBm; (c) −15 dBm.

different signal levels. Some small amount of stretching can occur owing to greater propagation delays at low level compared with high level signals which also tend to speed up the leading edge. However, the linear trailing edge of the pulse is mainly the mathematical consequence of an exponential pulse decay subjected to a logarithmic compression. Failure to realize a filter with a perfectly flat group delay can be seen in the trailing edge of the −15 dBm pulse in Fig. 5.13 where some low level extension is evident.

A possible future development in filter technology is the use of surface acoustic wave devices which offer the promise of independent shaping of the amplitude and phase responses. At present the high insertion loss (more than 20 dB) and triple pass reflections limit the application of these devices.

### 5.3.3 Pulse detection

Transponder reply pulses must be detected over the full dynamic range of signal strengths expected from the system. This dynamic range can extend from the largest expected signal, possibly at −15 dBm, down to at least −85 dBm, which is a range in voltage terms of 3000. The most convenient way of processing signals with this dynamic range is to use a receiver which incorporates a logarithmic characteristic so as to compress the output video signal to a more practical range.

A common design approach to realize a receiver with a logarithmic characteristic is to use an i.f. amplifier based on successive detection, an example of which is shown in Fig. 5.14. Each amplifier stage has a gain of approximately 12 dB, which drops to 0 dB when a preset level of input signal is exceeded. Each stage also has a video detector output so that all stages contribute to the total video signal produced by the receiver.

A very weak signal will not saturate even the final amplifier stage, and in this state the receiver is similar to a linear receiver. However, with a small increase in signal strength, the final stage saturates and its contribution to the total video signal is fixed regardless of any further increase in signal; in effect the overall gain of the receiver has been reduced by the gain of the last stage.

Larger signals result in more amplifier stages limiting with a consequent further reduction in overall amplifier gain. The dynamic range of the amplifier can be extended beyond the level at which the input stage limits by including parallel stages but with an attenuated input. The resulting effect is a piecewise approximation to a logarithmic characteristic. For best performance small delay lines can be inserted in the video summation path to match the propagation delay of the i.f. signal through the amplifier stages, but this is rarely warranted.

A logarithmic character is beneficial for another reason. The transponder reply pulse width is defined at the half-voltage points of the modulated r.f. output and is undefined at any other part of the pulse. Since one half-voltage

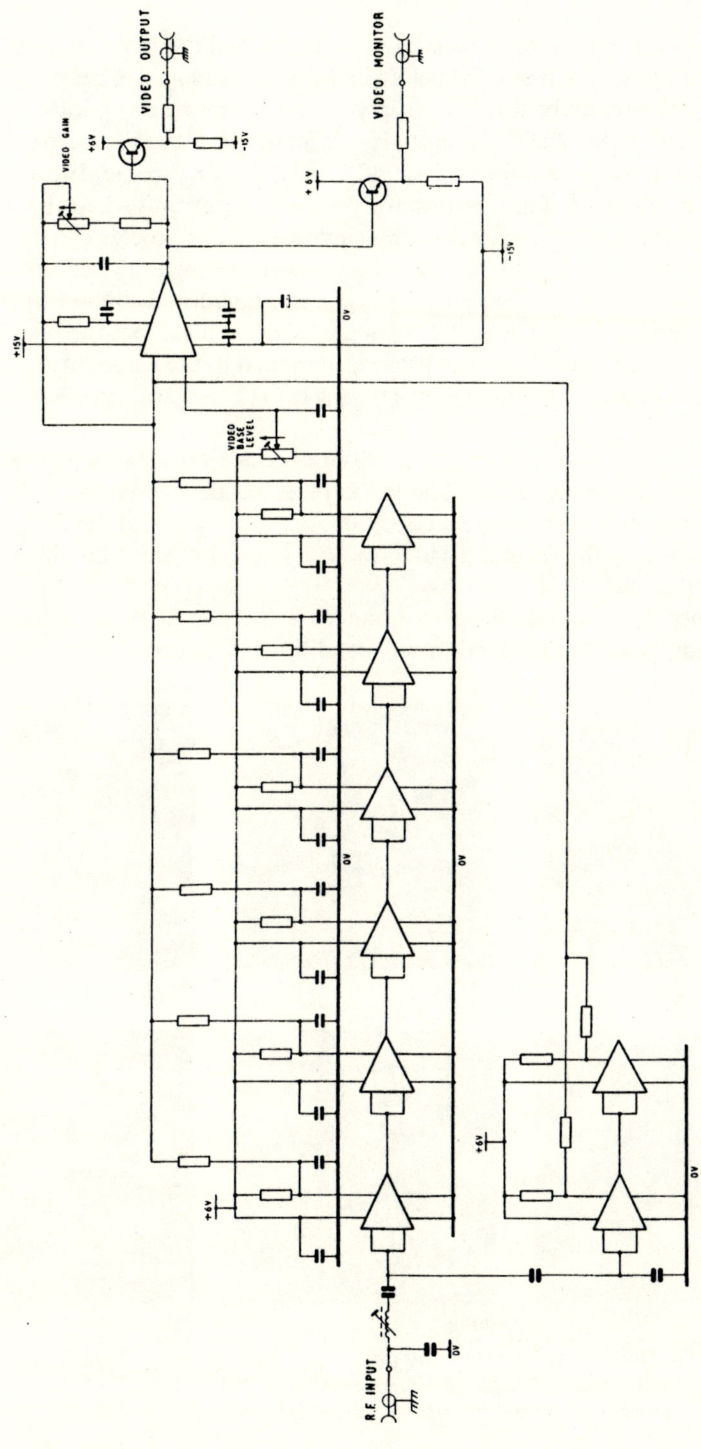


**Fig. 5.14** Successive detection logarithmic i.f. amplifier.

corresponds to a 6 dB reduction, it is possible to find the received pulse width by measuring its width at 6 dB below its peak value. The −6 dB point of the pulse rising edge can be found by applying the pulse to a comparator directly and via a delay line. The direct input is subject to a level shift corresponding to a −6 dB amplitude change and sets the comparator threshold to the pulse peak value less 6 dB. The comparator produces an output when the delayed front edge of the pulse exceeds the threshold. A similar process can be used to determine the −6 dB point at the pulse trailing edge, with the difference that the threshold is set via the delay line. The pulse width at the −6 dB level is found by measuring the delay between the two outputs. Since the receiver has a logarithmic characteristic, the level shift applied to the comparator input to reduce the threshold by the equivalent of 6 dB is constant regardless of input level.

Figure 5.13 shows photographs of pulses detected by such a process using widely different signal levels. The video pulse, as detected at its half-input-voltage point, is shown on the second line; the delay is due to the delay line. It can be seen that the resulting video pulse remains constant in width over a wide input signal range.

Figure 5.15 shows the performance of the circuit in the presence of overlapping pulses. Such overlapping conditions often occur when two replies

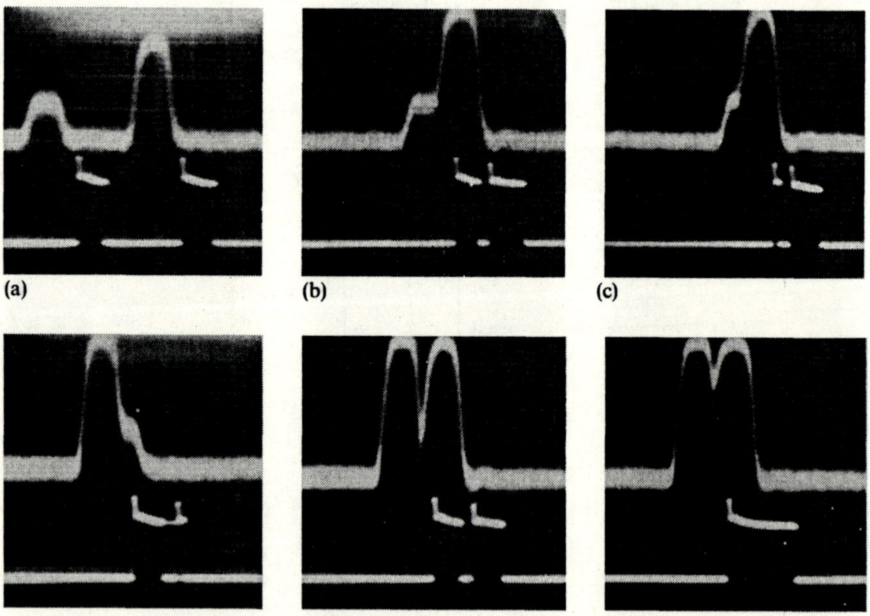

(a) (b) (c) (d) (e) (f)

**Fig. 5.15** Detection of overlapping pulses: (a)–(c), separate −85 and −70 dBm pulses ((a) well spaced, (b) spaced 470 ns, (c) spaced 250 ns); (d)–(f), two −70 dBm pulses ((d) spaced 250 ns, (e) spaced 560 ns, (f) spaced 500 ns).

are received from aircraft which are at a similar range. In such conditions it is important to be able to identify both aircraft since it is just when aircraft are close that identification is particularly required. The performance of the circuit in such circumstances can be seen for a number of degrees of overlap and for two relative signal strengths between the interfering pulses. When the two pulses overlap substantially, the video output of the weaker pulse is reduced in width since its peak reference has been disturbed by the stronger pulse. However, the ability to determine the location of the outer edge enables the pulse position to be known. The ability to deduce the leading and/or trailing edges of overlapping pulses is important for the detection of overlapping replies. This feature is exploited in the plot extractor and is described in Chapter 7.

An alternative technique for determining the leading and trailing edges of the received pulse is described in the FAA Engineering Requirement for Mode S [5.4]. This alternative technique measures the rate of change of the video signal, and if this value exceeds a preset amount a pulse edge is assumed.

### 5.3.4 Gain–time control

Under free-space conditions, i.e. ignoring the effects of antenna pattern lobing due to ground reflections, the received signal strength of a transponder reply is inversely proportional to the square of the range of the aircraft. Every doubling of range reduces the signal amplitude by 6 dB. Assuming the minimum transponder output power permitted by ICAO *Annex 10* and making due allowance for the practical variation in aircraft antenna gain and ground antenna lobing results in an estimation of the minimum transponder signal strength for any given range.

There is little point in detecting signals weaker than this signal strength as they are unlikely to be required replies from aircraft but fruit replies or echoes of replies arriving from some unwanted reflection path. The phenomenon of echoes or reflections is described in more detail in Chapter 9, and their presence is likely to confuse the detection of the true responses from aircraft. It is therefore advantageous to reject these weak replies where possible by reducing the receiver gain for ranges where strong replies are expected.

Receiver gain reduction can be effected by means of attenuators at either the r.f. or i.f. stages of the receiver. However, a particularly convenient place to reduce the receiver gain is at the video output of the receiver with a logarithmic response.

The varying of receiver gain is called gain–time control (GTC) or alternatively sensitivity–time control (STC). Immediately after the transmission of the interrogator pulse P3 the receiver sensitivity is depressed by a predetermined attenuation, after which the sensitivity is increased at a rate equal to 6 dB per doubling of range until full sensitivity is achieved. Table 5.3 shows a typical sensitivity variation.

**Table 5.3** Typical gain–time control law

| *Range* (n mile) | *Sensitivity depression* (dB) | *Minimum detectable signal* (dBm) |
|---|---|---|
| 1 | 36 | −44 |
| 2 | 30 | −50 |
| 4 | 24 | −56 |
| 8 | 18 | −62 |
| 16 | 12 | −68 |
| 32 | 6 | −74 |
| 64 | 0 | −80 |

At some sites, particularly airports, there can be many reflecting objects and a higher GTC sensitivity reduction is desirable for some antenna directions. At those same sites an aircraft approaching a runway may have its transponder antenna partly obscured by the airframe so that its reply signal is weaker than normal thereby requiring a lower GTC sensitivity reduction.

These conflicting requirements of GTC sensitivity reduction in some directions but not others can be met by using a programmable GTC facility. For example, such a facility can divide the antenna azimuth into 32 sectors each of 11.25° and, within each of those sectors, range steps of 1 n mile. A programmable read-only memory (PROM) is used to hold the instantaneous receiver sensitivity required in each sector–range cell. The receiver sensitivity can now be controlled to suit the detection requirements at any range or bearing.

Figure 5.16 shows a programmable GTC map that has been used for the SSR installation at Heathrow Airport [5.5]. Most of the map entries contain receiver sensitivity reductions according to Table 5.3. The shaded areas contain a further 6 dB sensitivity reduction in excess of these values in order to discriminate better against reflections for only those directions and ranges where false targets due to reflections are experienced.

### 5.3.5 Receiver sidelobe suppression

Fruit is the term given to transponder replies that have been triggered by other interrogators in the area. These replies are not synchronous with the interrogations emitted by the ground transmitter and cannot be associated with an aircraft in a useful way. The majority (about two-thirds) of detected fruit replies are received by the antenna sidelobes.

The reader will recall that the antenna has both an interrogate (or sum) beam and a control beam, with the characteristic of the control beam being that its gain is greater than that of the interrogate beam in all directions except

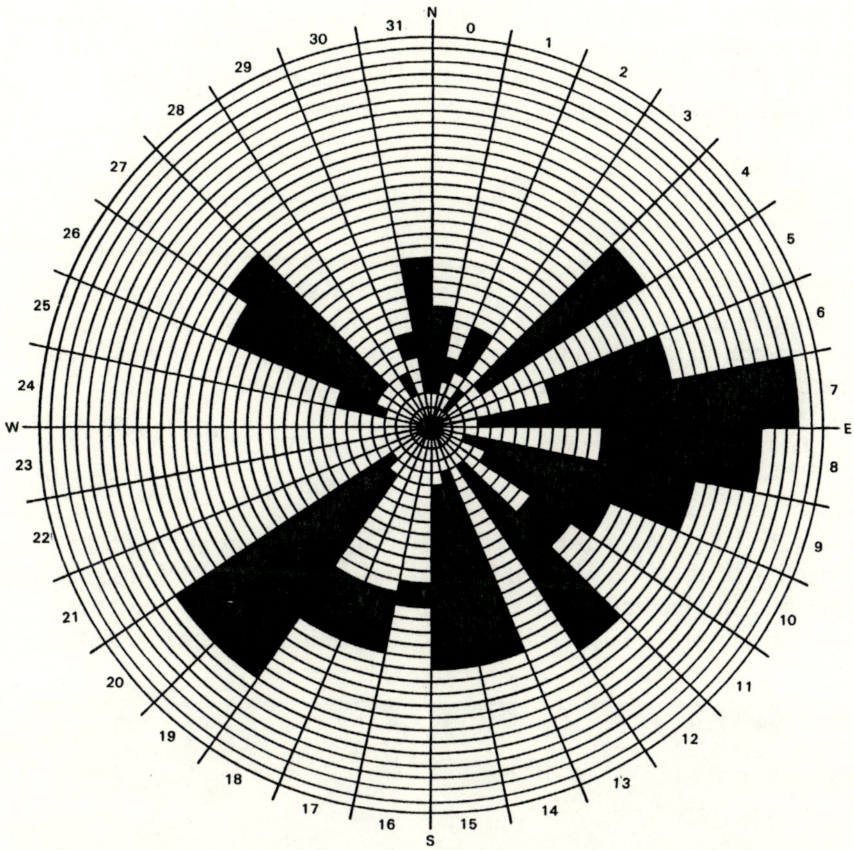


**Fig. 5.16** Programmable GTC map for Heathrow Airport: each range division equals two programmable range cells (1 n mile); each azimuth division equals 11.25°; unshaded area, values as Table 5.3; shaded area, additional 6 dB attenuation on the values in Table 5.3.

that of the interrogate beam main lobe. Specifically, the gain of the control beam is greater than that of the interrogate beam sidelobes and backlobe.

If a second receiver channel with a logarithmic characteristic is connected to the antenna control beam port then it is possible to compare the instantaneous signal strengths in the interrogate and control channels. When a reply pulse is found that has a greater signal strength in the interrogate channel than in the control channel, that pulse has arrived in the main lobe of the antenna and should be retained. However, if the control channel signal is stronger than the interrogate channel signal, the pulse has arrived from a sidelobe direction and can be discarded.

The sidelobe pulse could be gated out in the video circuit of the receiver but there is a danger that it may be masking a weaker main lobe pulse which should not be eliminated. It is better practice to mark the sidelobe pulse by an appropriate signal on a separate line, and to pass this additional information to the plot extractor where decisions can be made on a reply basis rather than an individual pulse basis.

## 5.4 Monopulse receivers

An important recent development in SSR is the addition of a direction-finding function based on monopulse techniques, and this is discussed here. Other receiver functions such as video pulse processing, GTC and bandwidth requirements, as described in Section 5.3, are also applicable.

### 5.4.1 Principles

The term monopulse is derived from the ability to measure the arrival direction of an incoming radar signal using a single pulse of that signal. Until recently, the method of determining target bearing has been to scan the antenna across the target and to note the direction at which incoming signals

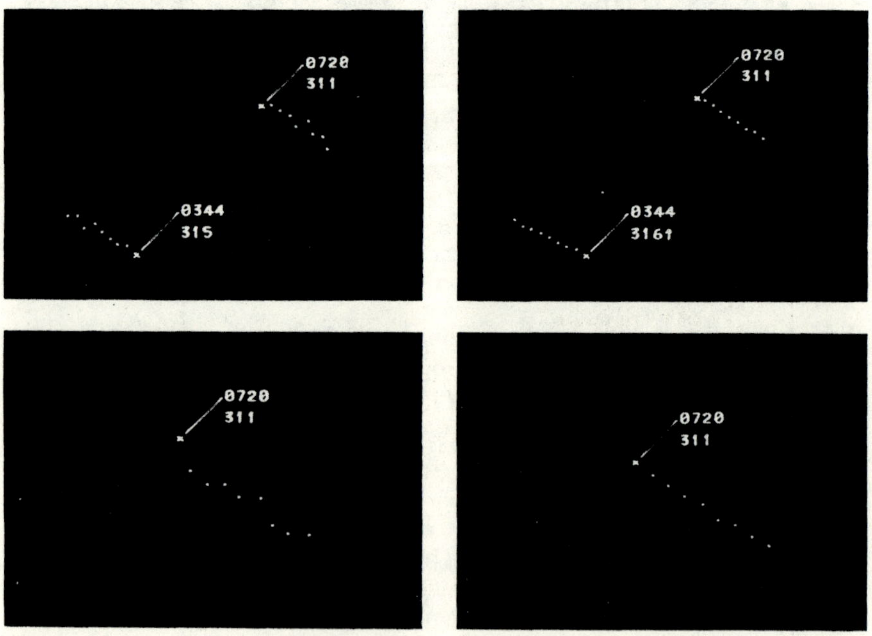


(a) (b)

**Fig. 5.17** (a) Sliding window and (b) monopulse tracks.

are detected and then, later, the direction when they cease. The target bearing is taken as the midpoint between these two directions. This "sliding window" technique is used in a nonmonopulse system and the principles are discussed further in Chapter 6. The benefits of monopulse direction finding, compared with the alternative method, can be appreciated in Fig. 5.17 which shows aircraft tracks detected by both methods. Further comparisons are made in Chapter 11, Section 11.2.

Monopulse techniques have been used for many years in association with primary radar and in particular homing missile systems. A common approach uses an antenna pattern consisting of two squinted beams which can be realized by two horns, one on each side of the focus of a doubly curved parabolic reflector. The result is two beams angled on either side of the antenna centerline (or boresight). A target lying on the centerline will present equal signals in the two beams. This type of monopulse is of little value to SSR.

The other type of monopulse system is similar to interferometry. In such a system the angle of arrival of a signal is determined by measuring the phase delay between two separated points; the two signals are in phase when the direction of arrival coincides with the antenna centerline. This second approach is better suited for SSR which is required to perform the dual function of a radar and a communication system.

An important function of SSR is the receipt of reply data from one aircraft in a way that minimizes interference due to replies from other aircraft. This requirement is assisted by an antenna with a single narrow high-gain beam with low sidelobes. Such an antenna can be constructed with the two phase centers required of an interferometer monopulse system. A simplified horizontal feed system for such an antenna is shown in Fig. 4.3 (Chapter 4). The amplitude distribution across the horizontal aperture is designed to produce a narrow high-gain beam with low sidelobes. Conceptually this horizontal feed system is divided electrically into two equal parts and recombined in a hybrid ring (or similar device). The sum output of the hybrid ring produces the original high-gain beam but the difference output produces a separate function.

The antenna shown in Fig. 4.6 (Chapter 4) is constructed along these lines, and its sum and difference patterns are shown in Fig. 4.5, with Fig. 3.13 (Chapter 3) giving a detail close to the beam center. The two half-antennas shown in Fig. 4.3 can be regarded as two antennas in their own right. In this case the two half-antennas form an interferometer, and signals arriving at them from any angle will have equal signal strengths. However, the two signals will have a relative phase difference depending upon the extra distance to be travelled to the more distant half-antenna. This is illustrated in Fig. 5.18.

The signals from the two half-antennas are processed by the hybrid ring to produce the vector sum and difference of the two signals as shown in Fig. 5.19. When the arrival direction of the signal is aligned with the antenna

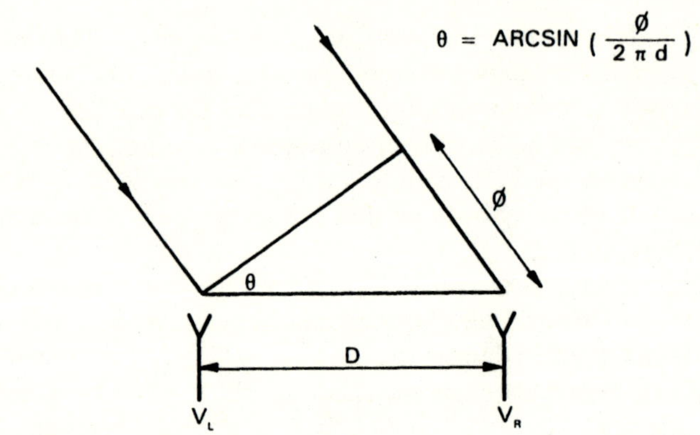


Fig. 5.18 Relative phase delay of the signal at the antenna.

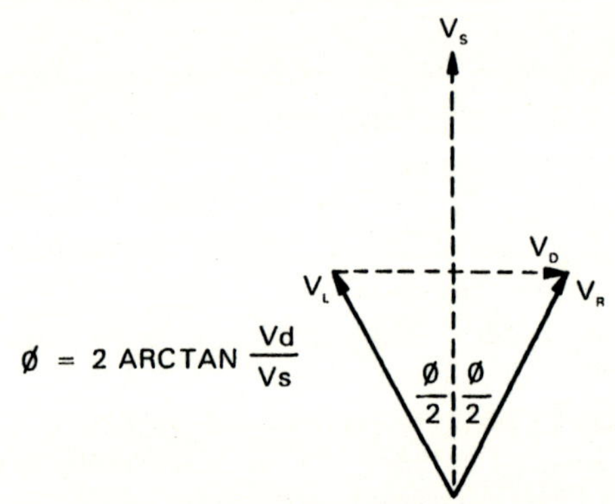


Fig. 5.19 Sum–difference phase diagram for an interferometer.

centerline the outputs from the two half-antennas are exactly in phase to give the maximum sum signal and zero difference signal. Away from the center there is a finite phase difference between the signals from the two half-antennas so that the sum output is reduced and the difference output is finite. Measurement of the ratio of the sum to difference outputs indicates the direction of arrival of the signal.

Given the presence of the sum and difference outputs from the antenna, the system designer can choose alternative processing methods to extract the angle data. He can either use them in their existing form, whereby the angle information is contained in the signal amplitudes, or he can convert them again to a phase form and determine the angle data by phase measurement. In either case the designer has two choices of amplitude processing and two choices of phase processing. Jacovitti [5.6] designates the four types of monopulse processing L1, L2 (amplitude) and P1, P2 (phase) as shown in Fig. 5.20. L1, P1 and P2 receivers use an antenna with sum and difference outputs. L2 receivers use an antenna with squinted beams as described earlier.

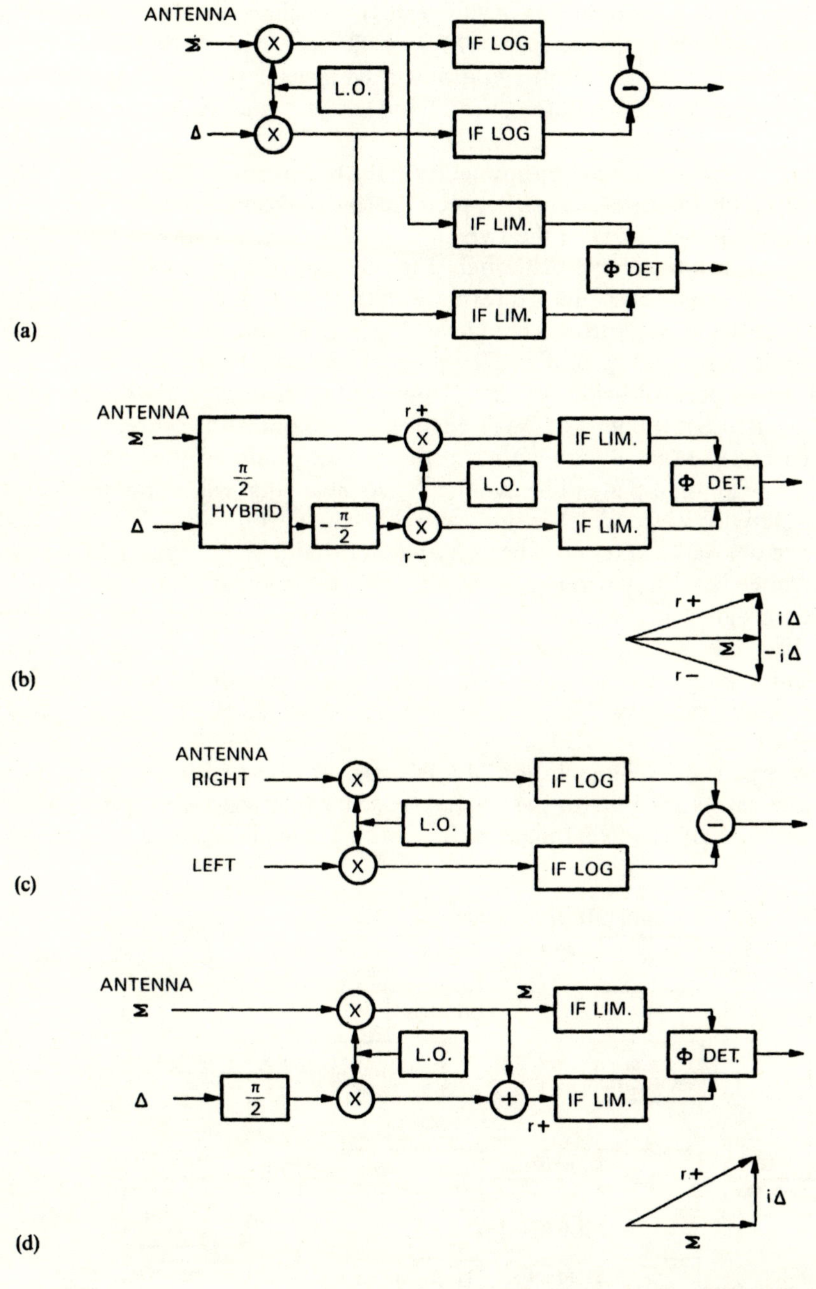




**Fig. 5.20** Four types of monopulse receiver: (a) L1 amplitude processor; (b) P1 phase processor; (c) L2 amplitude processor; (d) P2 phase processor (from ref. 5.6).

In practice each receiver type possesses an ambiguity point in its bearing characteristic which, in the case of L2 and P2 receivers, severely restricts the angle over which the aircraft azimuth can be measured. These two receiver types also perform less well than the L1 and P1 types and will not be discussed further.

L1 receivers [5.7] have an ambiguity at the beam center. This is resolved by measuring the phase difference between the sum and difference channels since the relative phase between the two signals undergoes an abrupt 180° phase change at beam center. Although Fig. 5.20(a) shows separate receiver channels to provide the phase information, this function is normally included as part of the i.f. logarithmic channels. A practical implementation of an L1 receiver is shown in Fig. 5.21 and includes an additional channel connected to the antenna control beam for the purposes of receiver sidelobe suppression (RSLS). RSLS is not part of the discussion of monopulse direction finding.

P1 receivers have ambiguities at the $\pm 3$ dB points of the antenna sum beam. The ambiguities can be avoided by constructing, within the receiver, a third reference signal which is similar to that which would have been detected at the center of the antenna. The right- and left-hand phase signals, formed in the arrangement shown in Fig. 4.3 (Chapter 4), are compared with the derived reference signal instead of directly with themselves. Thus the interferometer used has only half the baseline of the original receiver and therefore the ambiguity points are separated by twice the angular distance than would otherwise be the case. This arrangement is known as the half-angle phase processor.

A practical implementation of a P1 receiver is shown in Fig. 5.22. Two phase detectors are used for the two half-baseline interferometers synthesized. The outputs of the two phase detectors are added together to improve

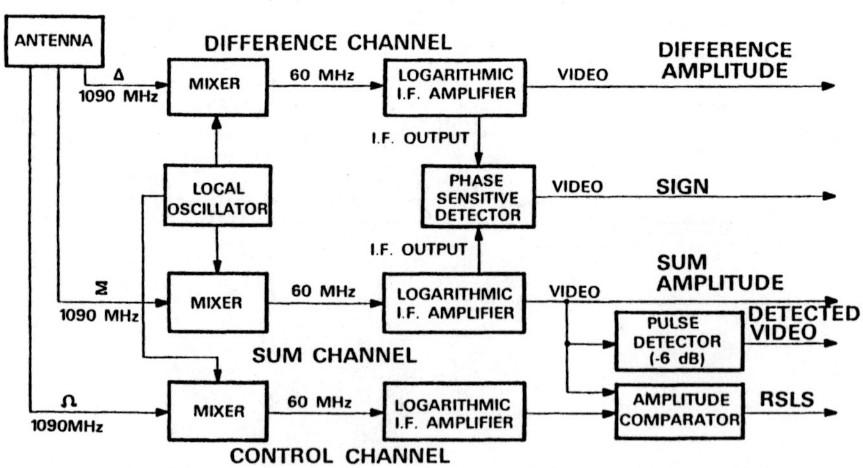


**Fig. 5.21** Practical implementation of receiver L1.

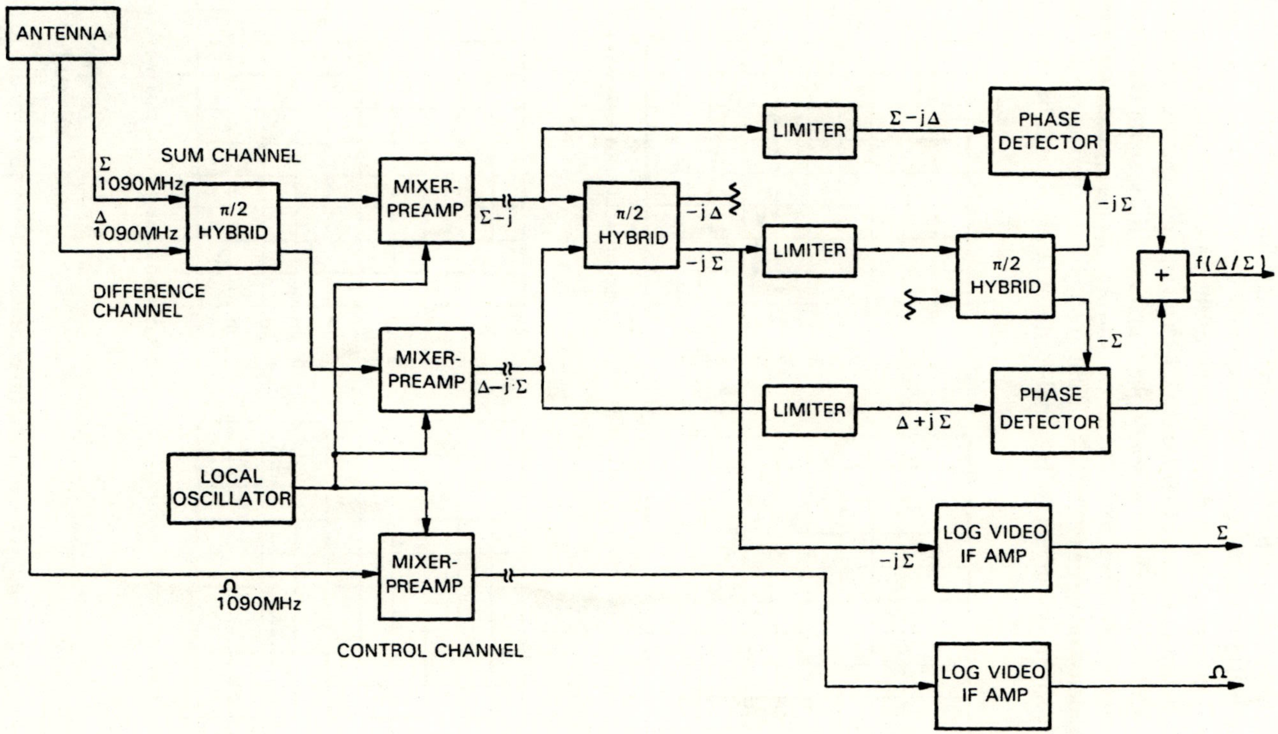


**Fig. 5.22** Practical implementation of receiver P1. (Reprinted with permission of Lincoln Laboratory, Massachusetts Institute of Technology, Lexington, M.A.)

accuracy, a process which largely eliminates any phase errors in the synthesized reference channel. Also shown in Fig. 5.22 is an additional receiver channel connected to the control channel for the purposes of RSLS.

Jacovitti [5.6] has calculated the susceptibility of the four receiver types to various sources of error at different parts of the antenna beam. These results are reproduced in Fig. 5.23. Since the L1 receiver only measures phase to the nearest $\pm 90°$, it is immune to small phase errors in the system. However, the P1 receiver is sensitive to both phase and amplitude system errors. The coherent detector of the P1 receiver, compared with the incoherent detector of the L1 receiver, can give a 3 dB signal-to-noise improvement in weak signal

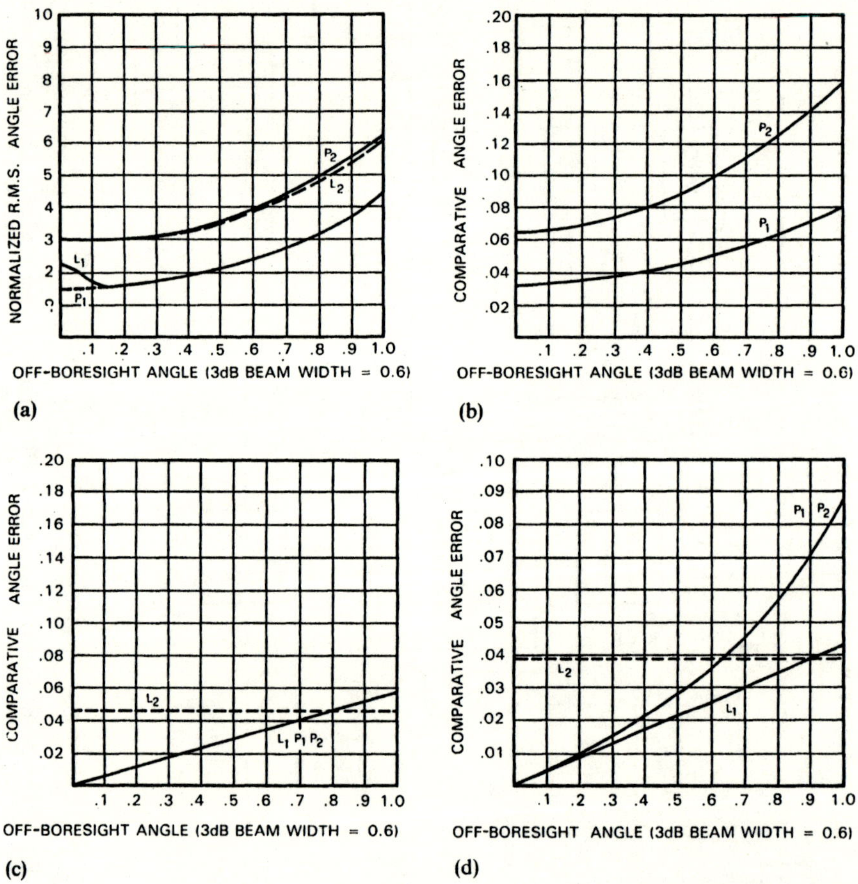




**Fig. 5.23** Monopulse receiver errors: (a) error due to noise; (b) error due to i.f. phase error; (c) error due to r.f. amplitude error; (d) error due to i.f. amplitude error (from ref. 5.6).

conditions as can be seen in Fig. 5.23(a) where the L1 receiver suffers a small degradation close to beam center.

Majerus [5.8] has suggested a modification to the L1 design which modifies the sum-to-difference ratio according to the sine of the relative phase angle of the two signals. The use of phase in the angle determination will introduce a sensitivity to receiver phase errors such as that possessed by the P1 receiver but not by the pure L1 receiver. This modification is used in one practical implementation [5.9].

A compromise to the Majerus modification, which is used in another practical implementation [5.10, 5.11], is to employ a multilevel decision phase detector to identify when the phase is approximately 0° or 180°. These phase relationships cannot relate to a real reply signal situation since such phases can only be ±90°. A measured phase of 0° or 180° can only arise if the dominant signal in the difference channel is receiver noise, in which case a smaller off-boresight angle can be assumed than the difference-channel signal level would indicate.

It is possible to determine the effect of interference on the accuracy of the monopulse receivers [5.12, 5.13]. The results for L1 and P1 receivers are shown in Fig. 5.24. The sum–difference crossover angle $\theta_c$ is approximately equal to the −3 dB position of the sum beam.

Interference can take the form of fruit replies or multipath. Fruit replies are most likely to affect only some of the reply pulses, and pulses showing atypical characteristics are ignored in the monopulse plot extractor for the purposes of bearing measurement (see Chapter 7). Multipath interference is potentially more significant since it can affect all pulses in a reply and its existence will not be recognized by the plot extractor. Fortunately the improved antennas of the type shown in Fig. 4.1 (Chapter 4) largely reject the damaging multipath signals by discriminating against ground reflections. Figure 5.24 shows that, for L1 and P1 receivers, the bias error resulting from an interfering signal is small. The standard deviation of the error is less for the P1 receiver than for the L1 receiver for strong interference-to-signal ratios but similar for weak ratios.

The L1 receiver has two important practical advantages. The antenna boresight direction is directly measured from the antenna shaft and the monopulse angle is added to that direction to obtain the bearing of the aircraft. Consequently the electrical stability of the antenna boresight compared with the mechanical antenna centerline is important. In the L1 receiver the difference signal has a null at boresight so that no receiver error can disturb its direction. In contrast the P1 receiver relies on the stability of the phase tracking in its receivers to maintain boresight direction.

Transponders reply with a transmitter frequency tolerance of ±3 MHz which can be exceeded in practice. A frequency offset from the nominal 1090 MHz can introduce a corresponding bearing measurement error in the receiver. In the L1 receiver, signal detection is incoherent and only signal

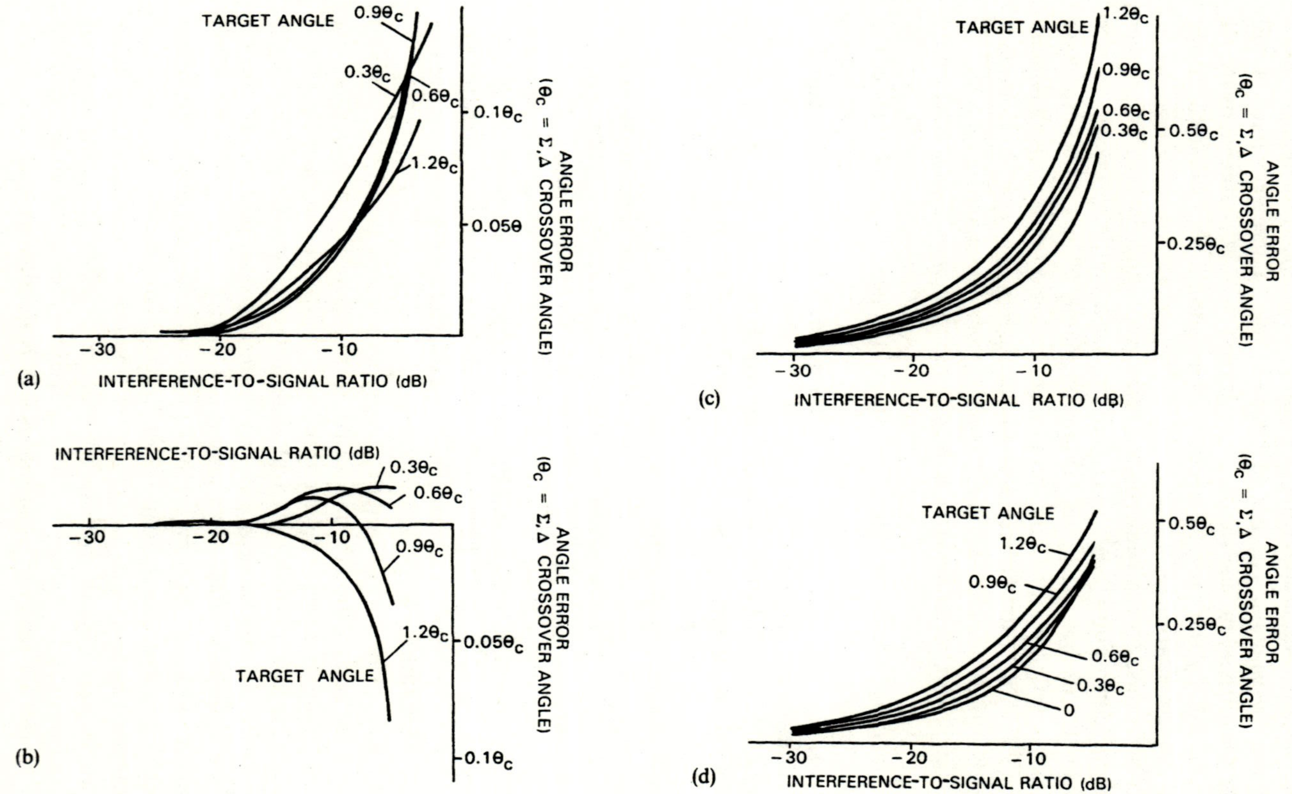




**Fig. 5.24** Standard deviation and bias errors due to interference: (a) L1 bias error; (b) P1 bias error; (c) L1 standard deviation error; (d) P1 standard deviation error (from ref. 5.13).

amplitude is measured. A frequency-dependent bearing measurement error would result from any lack of amplitude match between the two bandpass filters in the sum and difference channels. In practice these can be matched to within $\pm 0.3$ dB over at least $\pm 4$ MHz. In contrast, the P1 receiver uses a coherent detector which is sensitive to receiver mismatch errors in both phase and amplitude. For good performance the P1 receiver channels must track in amplitude and phase over the required frequency band and also over the signal dynamic range.

As is explained at the beginning of this section, the monopulse system that is used for SSR is based on an interferometer, but the signals are converted to sum and difference form for transmission from antenna to receiver. There they can be processed directly in amplitude form or converted back to phase and processed in that form. The results achieved are dominated by the antenna characteristics and are comparable whichever type of processing is utilized. Amplitude and phase processing are discussed in more detail below.

### 5.4.2 Amplitude processing

Given that the receiver will have a sum channel for the purpose of detecting and processing transponder reply data, a difference channel is needed to enable the direction of arrival of each reply pulse to be measured.

Figure 3.13 (Chapter 3) shows the characteristics of the sum and difference patterns of an antenna close to boresight. The requirement is to determine the ratio of the sum-channel signal to the difference-channel signal for each pulse detected. For pulse detection purposes it is convenient to provide the sum channel with a logarithmic characteristic. If this is also done for the difference channel then the division of the sum-signal amplitude by the difference amplitude can most easily be achieved by subtraction of the sum and difference video signals.

The sum–difference ratio characteristic has an ambiguity about the boresight, and to resolve this ambiguity it is necessary to make an additional measurement. The difference signal undergoes a phase reversal as the signal goes through zero at boresight, changing from $+90°$ to $-90°$ with respect to the sum-channel signal. This phase relationship is maintained at the intermediate frequency and the phase measurement can be made at the output of the i.f. amplifiers.

Since the reply signal angle from boresight is determined from the ratio of the sum- and difference-signal amplitudes, it is clear that the two receiver channels involved must be designed to permit this process to be performed to the required accuracy. There are three different techniques that can be used to realize a receiver channel with a logarithmic characteristic.

One approach is to have a linear receiver with a wide dynamic range and convert the detected linear video to logarithmic form using a video shaping circuit that could, for example, be based on the characteristic of a semiconductor junction. However, this technique is limited in the dynamic range that is realizable from a linear receiver and in the transient response available from practical video shaping circuits.

A more useful approach is to use a circuit known as a "true logarithmic amplifier". This circuit has the property of two alternative gains. During small-signal conditions the gain is high, typically 12 dB. A signal greater than a given threshold drives the circuit into its unit gain mode. An i.f. amplifier constructed from a series of such circuits provides an amplifier with an i.f. output which approximates a logarithmic characteristic. A detector is required to convert the i.f. output into a video signal. Typically the "true logarithmic amplifier" shows good phase performance over its dynamic range, and this is useful since phase information is needed to resolve the side-of-boresight ambiguity. However, present commercially available integrated circuits of this type are limited in their dynamic range.

The most popular form of logarithmic receiver is based on the successive detection technique. It has some similarity with the "true logarithmic amplifier" type except that the low-gain mode is one of limiting or zero gain. Each stage in the receiver chain has its own detector, so that the number of detectors which contribute to the video output is a function of the input signal strength. Any frequency band shaping filters that are required in the receiver must be placed before the logarithmic section. A circuit diagram of a successive detection logarithmic amplifier is shown in Fig. 5.14. Another disadvantage of the successive detector type of logarithmic amplifier is that

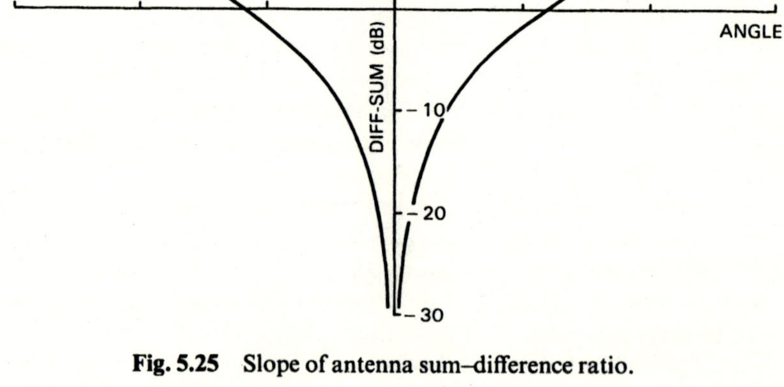


**Fig. 5.25** Slope of antenna sum–difference ratio.

the phase response is less easy to control as the stages successively enter hard limiting. However, these aspects of the successive detector logarithmic amplifier can be accommodated by careful design.

The accuracy of the monopulse measurement is influenced by the accuracy of the logarithmic amplifiers. Figure 5.25 shows a sum–difference ratio slope as it varies across the antenna beam. It follows that the effect of an amplitude error, in terms of the resulting measured angle error, is small near the beam center but is more significant towards the beam edge. A good

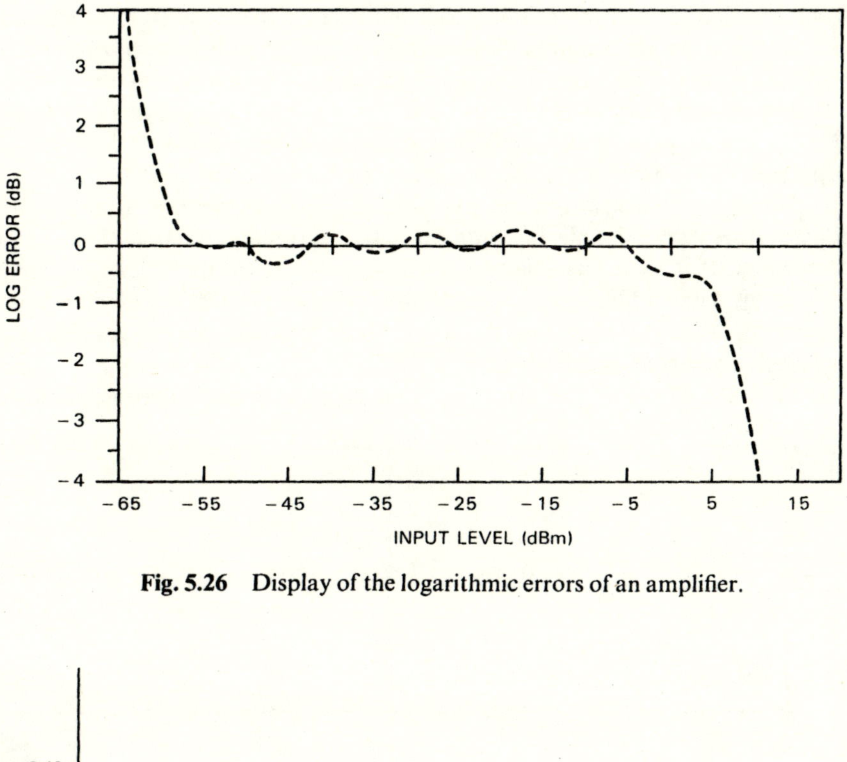


**Fig. 5.26** Display of the logarithmic errors of an amplifier.

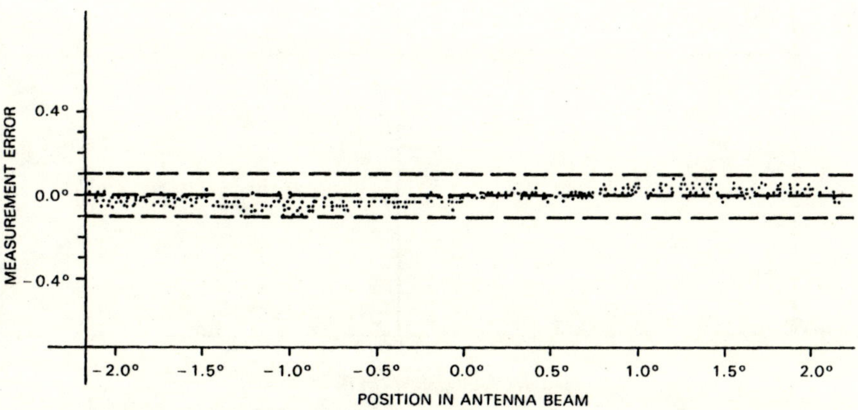


**Fig. 5.27** Monopulse measurement errors across the antenna beam.

successive detection logarithmic receiver can give a logarithmic accuracy of better than ±0.5 dB over a dynamic range of 80 dB or more. Figure 5.26 shows one method of displaying the errors in a logarithmic i.f. amplifier on an oscilloscope. The horizontal axis gives the input signal strength on a logarithmic scale. As the oscilloscope trace sweeps across the screen a high-*Q* resonant circuit decays exponentially to generate the corresponding input signal which varies logarithmically on a linear time-scale. The vertical axis gives the strength of the output video signal from the logarithmic amplifier under test. Alone, this technique would give an approximately linear trace from top right to bottom left and the small errors from the true logarithmic characteristic would be difficult to observe. By subtracting a proportion of the horizontal time-base waveform from the output video signal the vertical position of the trace is made constant for a perfect logarithmic response. Greater *y* axis gain of the oscilloscope can now be applied to indicate the departure of the amplifier under test from the perfect response.

Some implementations of monopulse receiver systems subsequently convert the video signals into digital form by means of analog-to-digital converters. It is then possible to use the output digital instantaneous video level as an entry address into a read-only memory (ROM) which contains a correction table. By this means the ripple in the logarithmic characteristic can be largely corrected or calibrated out.

Figure 5.27 shows an error diagram representing numerous measure-

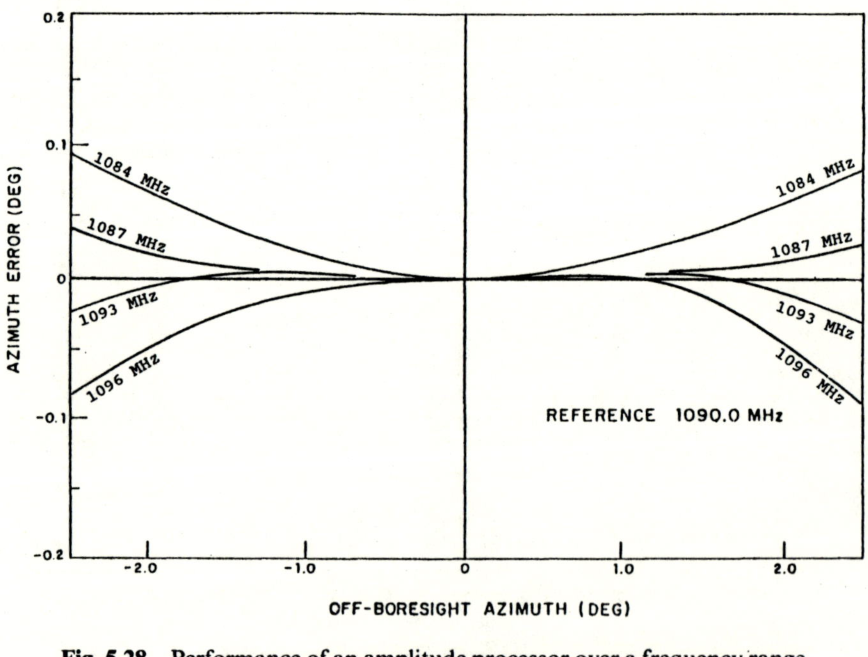


**Fig. 5.28** Performance of an amplitude processor over a frequency range.

ments made across the antenna beam. The horizontal axis shows the place within the antenna beam at which a reply was received. The vertical axis shows the error that accompanied that measurement. In a perfect error-free receiver there would be no vertical dispersion in the results and they would all lie on the horizontal axis. Figure 5.28 shows the bearing measurement error due to the reply frequency over the band 1090 $\pm 6$ MHz.

At weak signal levels receiver noise will limit the measurement accuracy, while at stronger levels other receiver processes, such as analog-to-digital conversion quantization, will become a limiting factor (see Chapter 10, Section 10.5.2). The variation in the standard deviation of the angle measurement over the receiver dynamic range is shown in Fig. 5.29 for the worst-case position in the antenna beam where the sum and difference beams are equal. The increasing influence of receiver noise with decreasing signal strength can be seen. At strong signal levels the measurements presented are influenced, in part, by the test equipment used.

### 5.4.3 Phase processing

The P1 phase-processing monopulse receiver (Fig. 5.20(b)) is also used for SSR. However, the receiver is impractical in this simple form as the angle over which the reply direction can be measured is limited. It can be seen from Fig. 5.18 that the signal direction is given by $\theta = \arcsin(\phi/2\pi d)$. The relative phase delay $\phi$ between the two half-antennas has an unambiguous range of $\pm 90°$ which limits the bearing $\theta$ to $\pm\arcsin(1/4d)$ or the 3 dB beam width of the antenna. The technique used to avoid this difficulty is to tap an output from the left-hand and right-hand signals and combine them to extract only the sum-channel component. This is shown in Fig. 5.30.

The phase of the left-hand and right-hand channels is now compared with that of the sum signal with each phase detector measuring the angle $\phi/2$. In effect the reconstituted sum signal represents the phase at the midpoint of

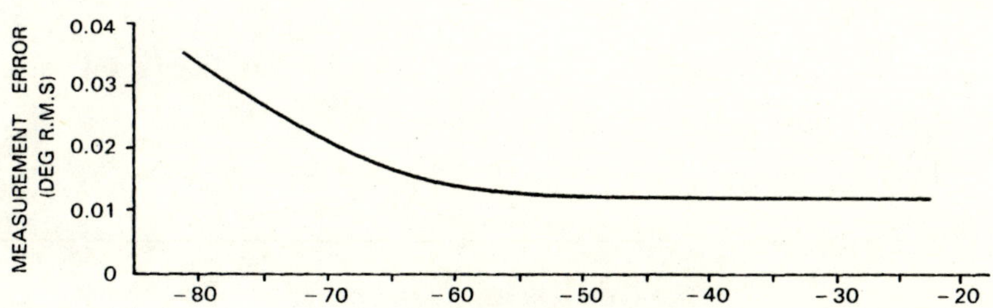


**Fig. 5.29** Variation in the standard deviation of angle measurement with signal strength (worst-case beam position).

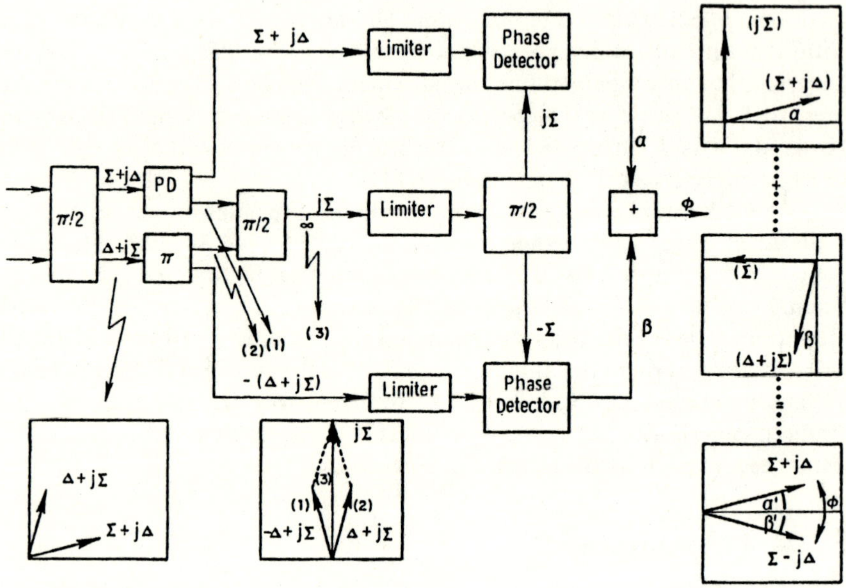


**Fig. 5.30** Half-angle monopulse receiver (from ref. 5.14). (Reprinted with permission of Lincoln Laboratory, Massachusetts Institute of Technology, Lexington, MA.)

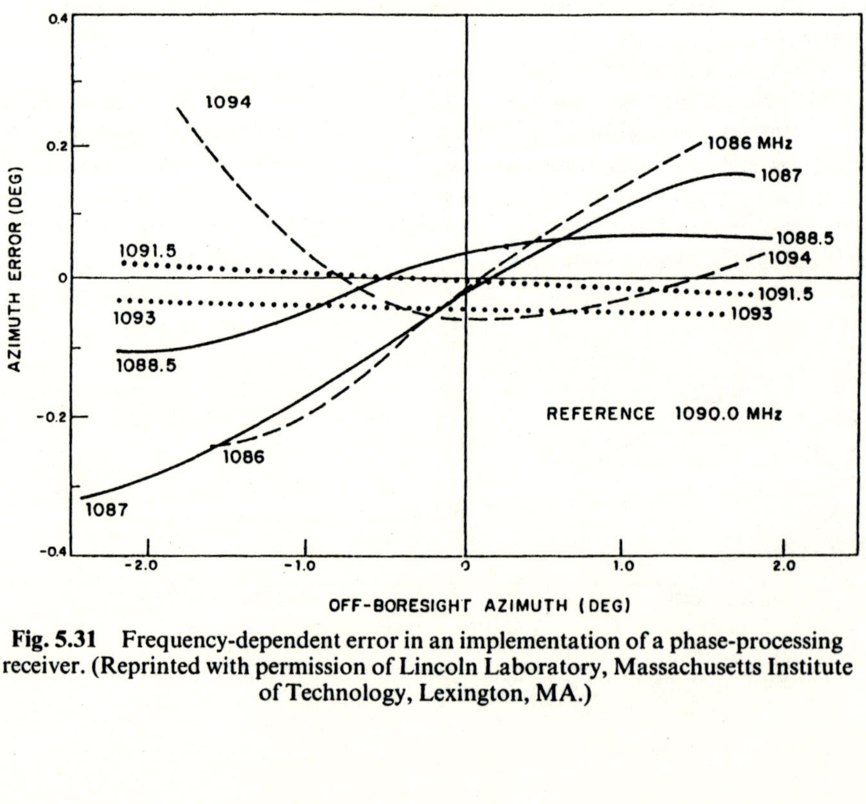


**Fig. 5.31** Frequency-dependent error in an implementation of a phase-processing receiver. (Reprinted with permission of Lincoln Laboratory, Massachusetts Institute of Technology, Lexington, MA.)

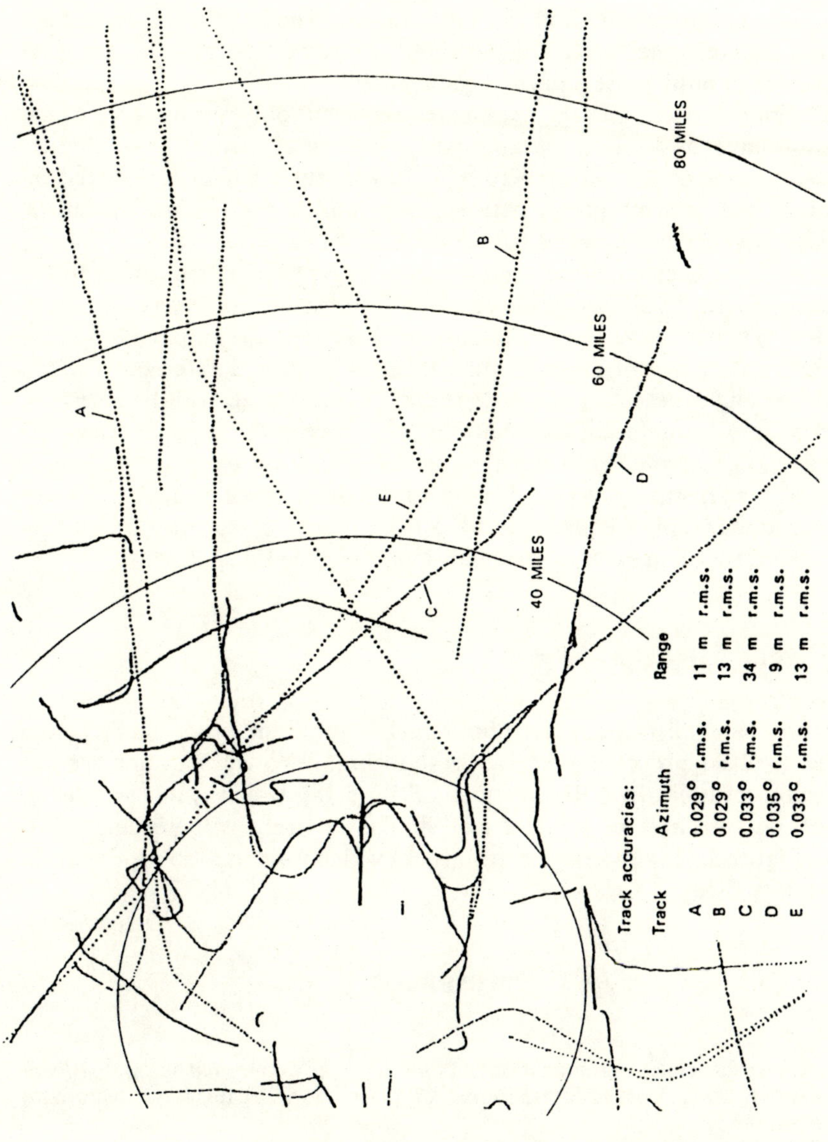


**Fig. 5.32** Aircraft tracks at Heathrow.

the two half-antennas in Fig. 5.18 and synthesizes two interferometers each with only half the original baseline, thereby doubling the angle over which the bearing can be measured.

It should be noted that the receiver in Fig. 5.30 does not have a suitable output whereby the reply data can be extracted. The system still requires a separate sum channel with a logarithmic characteristic for the accurate detection of framing pulse pairs and code pulses.

The modified P1 phase-processing receiver is the only receiver type which does not have a position of ambiguity. The equivalent L1 amplitude-processing receiver has an ambiguity at beam center which is resolved by measuring the relative phase between the sum and difference channels. Practical implementations of the modified P1 receiver (the half-angle processor) have been constructed and shown to give equivalent results to the L1 receiver [5.15, 5.16]. The designer of the P1 receiver should pay particular attention to two aspects. Firstly, unlike the L1 receiver, the antenna boresight (the electrical center of the antenna which is used as a reference for the monopulse angle measurement) is dependent on the stability of the receivers. It is necessary to maintain the phase match between the receivers over the dynamic range of the reply signals and the operating temperature range. Secondly, the sensitivity to phase errors can manifest itself as a sensitivity to the transponder reply frequency [5.14, 5.17]. Figure 5.31 shows a frequency-dependent error of an early implementation [5.14] (see Fig. 5.28).

### 5.4.4 Typical results

From the published results of both the L1 amplitude-processing receiver and the P1 phase-processing receiver both techniques offer bearing measurement errors of typically 0.03° rms [5.10, 5.15, 5.18], which is not surprising since all equipment uses antennas of similar aperture with similar beam widths. Figure 5.32 shows the aircraft tracks which were used to produce one set of results [5.10].

## References


5.1 Aeronautical telecommunications, *Annex 10 to the Convention on International Civil Aviation*, Vol. 1, International Civil Aviation Organization, Montreal, April 1985.

5.2 Technical characteristics of IFF Mk XA and Mk XII interrogators and transponders, *Standardization Agreement, STANAG 4193, Edition 3*, NATO, Brussels, 1986.

5.3 Zverev, A. I. *Handbook of Filter Synthesis*, Wiley, New York, 1967.

5.4 FAA Specification, Mode select beacon sensor, *FAA-E-2716*, 1982.

5.5 Spalding, D. J. Application of programmable GTC at Heathrow, *Cossor Internal Memo. 871/DJS/8962/MK*, 11 April 1984.

5.6 Jacovitti, G. Performance analysis of monopulse receivers for secondary surveillance radar, *IEEE Trans. Aerosp. Electron. Syst.*, **19** (6), 884–897, 1983.

5.7 Stevens, M. C. Precision secondary radar, *Proc. Inst. Electr. Eng.*, **118**, 1729–1735, 1971.

5.8 Majerus, J.-P. Comment on "Performance analysis of monopulse receivers for secondary surveillance radar", *IEEE Trans. Aerosp. Electron, Syst.*, **21** (2), 264–266, 1983.

5.9 Piccini, P., Benedetti, C., Grelli, A. and Pasquali, L. Theoretical evaluation of monopulse SSR and comparison with experimental measurements, *IEE Conf. Proc. RADAR 87*, pp. 227–231, IEE, London, October 1987.

5.10 Plant, R. J. and Stevens, M. C. A practical application of monopulse processing to the solution of SSR problems. *Conf. RADAR 84*, Institute Français de Navigation, Paris, May 1984.

5.11 Stevens, M. C. Monopulse secondary surveillance radar, *IEE Conf. Proc. RADAR 87*, pp. 160–164, IEE, London, October 1987.

5.12 Biffi Gentili, G., Giuli, D. and Jacovitti, G. Monopulse receivers for ATC beacon radar: analysis, comparison and selection criteria, *Alta Frequenza*, **52** (5), 323–334, 1983.

5.13 Biffi Gentili, G. and Guili, D. Analysis and comparison of monopulse receivers in the presence of SSR interference, *IEEE Int. Radar Conf. 1980*, pp. 445–450, IEEE, New York, 1980.

5.14 Karp, D. and Wood, M. L. DABS monopulse summary, *Lincoln Laboratory ATC-72, Rep. FAA-RD-76-219*, Lincoln Laboratory, MIT, Lexington, MA, February 1977.

5.15 Guerin, C. and Peyrat, A. Monopulse secondary radar: implementation and performance, *Conf. RADAR 84*, Institute Français de Navigation, Paris, May 1984.

5.16 Cole, E. L. and Enstrom, R. A. Mode S—a monopulse secondary surveillance radar for ATC, *IEE Conf. Proc. RADAR 87*, pp. 232–236, IEE, London, October 1987.

5.17 Alimenti, R. and Fox, D. P. DABS open array frequency sensitivity error measurements, *CT-81-100-5-LR*, FAA Technical Center, Atlantic City, NJ, July 1981.

5.18 Chapman, C. and Brady, J. J. Mode S system accuracy, *DOT/FAA/RD 81–90*, FAA, Washington, DC, February 1982.

## Chapter 6

# Sliding Window Plot Extraction

The SSR replies, transmitted by airborne transponders, are in the form of a series of pulses and are detected and amplified by the interrogator receiver. Some of the replies will be synchronized to the station's interrogations and others, called fruit, will be randomly timed since they are initiated by other interrogators.

In early implementations of SSR the receiver video was used to modulate the intensity of a PPI display. The human eye could recognize sequences of replies which were synchronized to the interrogator transmissions because they occurred at the same range on adjacent sweeps on the display. If the reply codes were limited to only the use of the A and B pulses then some attempt could be made to distinguish aircraft according to the pattern of their reply pulses. Further resolution could be obtained by requesting one aircraft to transmit the SPI pulse. Visual recognition became more difficult when the C and D code pulses were introduced to give 4096 different codes and if more than one interrogation mode was used. Recognition was also a problem if the replies from two nearby aircraft overlapped.

The provision of a reply decoder, which could gate replies at a given range and azimuth, enabled a numerical readout of the reply data from a single aircraft to be obtained. An alternative technique used a number of decoders to detect specified groups of identity codes or bands of flight level, so that when matching replies were detected special identification marks were placed on the display at the aircraft positions. Advances in display technology allowed alphanumerics to be written on the display using the spare time between the trace fly-back and the occurrence of the next interrogation transmission. This had the major advantage that each aircraft replying to SSR could be tagged with its identity number and flight level. With this type of display SSR became an extremely valuable ATC tool.

The new display—the labelled plan display—could not do the job by itself. In order to place the SSR information on the display in the correct position it needed the aircraft range and bearing in digital form together with the associated Mode A identity code and Mode C altitude information. The process of digitizing the Mode A and Mode C replies, associating synchronous replies and measuring the range and bearing of the aircraft responsible is known as plot extraction.

The availability of aircraft data in digital, instead of analog, form had a further advantage if the data had to be transmitted an appreciable distance from the radar site to the ATC display room. Narrowband links using telephone-type land lines could be employed instead of broadband links using microwaves; the consequent two or three orders of magnitude difference in the bandwidth required was reflected in the costs involved. The first step in plot-extraction is the conversion of the stream of pulses from the receiver into replies and the decoding of the data contained in those replies. This process is performed by a reply decoder.

## 6.1 Reply decoder

The function of the reply decoder is to detect individual reply pulses, recognize replies as indicated by pairs of pulses separated by 20.3 μs and determine the code data contained in the replies. In performing these functions garbled replies, phantom replies and military emergency and identification sequences must be identified.

### 6.1.1 Pulse detection

The transponder reply pulse specification requires the pulse width to lie between 0.35 and 0.55 μs as measured at half the peak voltage level of the detected pulse. The interrogator receiver is not a perfect device and has a limited bandwidth which marginally slows the rising and falling edges. The receiver adds noise which, for weak pulses, can modify the apparent pulse characteristics.

While, no doubt, transponders conform to their specification requirements when first delivered to the user and after subsequent recalibration, it is not unknown for them to drift from specification during service. Figure 8.5 (Chapter 8) shows the spread in measured pulse widths. Plot extractor designers are well advised to take the specification limits with some caution and attempt, as far as is practically possible, to tolerate marginally out-of-specification transponder performance.

Pulse width tolerances notwithstanding, a more usual reason for a reply pulse's being longer than the specification limit is that it is in fact an amalgam of two or more overlapping replies from different aircraft. It is necessary to detect that more than one pulse may be present in order that an attempt can be made to detect the two (or more) aircraft at a similar range. The occurrence of two replies which overlap and garble one another can cause an aircraft to be misdetected just when it is important to detect all aircraft.

Early implementations of reply decoders accepted raw video output from the interrogator receiver and applied their own processing to reject narrow pulses and noise spikes. More recent designs exploit the pulse detection described in Chapter 5, Section 5.3.3, where the half-voltage width of the received pulse is measured. The latter process permits the pulse trailing edge, as well as the leading edge, to be accurately determined.

### 6.1.2 Bracket detection

Figure 3.4 (Chapter 3) shows the format of an SSR reply. The framing pulses F1 and F2, which are also known as the "brackets" since they embrace the code pulses, are always present. The code pulses are those labelled A, B, C and D with the suffixes 1, 2 and 4 according to the binary value of the pulse if present. Figure 3.5 shows an example with a reply code of 4321. Shown in Fig. 3.4 is the X pulse which is reserved for future expansion but is not currently used. Most decoders either ignore the X pulse if it is present or declare the reply garbled and unusable. The SPI pulse, which can be set by pilot action, is also included in this figure.

Figure 6.1 shows the circuit of a bracket detector. The first pulse F1 is stored by the use of a delay line or digital shift register for sufficient time for the second bracket pulse F2 to be received. An AND gate, with inputs from both the input and output of the shift register, is used to detect the occurrence of the two pulses.

In practice some timing tolerance must be allowed in the separation between F1 and F2. ICAO *Annex 10* permits an F1–F2 tolerance of $\pm 0.1$ μs about a nominal spacing of 20.3 μs. The effects of receiver noise and its limited bandwidth requires this tolerance to be widened in a bracket detector so that a received reply can always be recognized. A detection tolerance of $\pm 0.15$ to $\pm 0.2$ μs is typical. In Fig. 6.1 this timing tolerance is realized by a multi-input OR gate for the position of the delayed F1.

Detecting brackets is quite straightforward, but false bracket pulse pairs can arise in two ways.

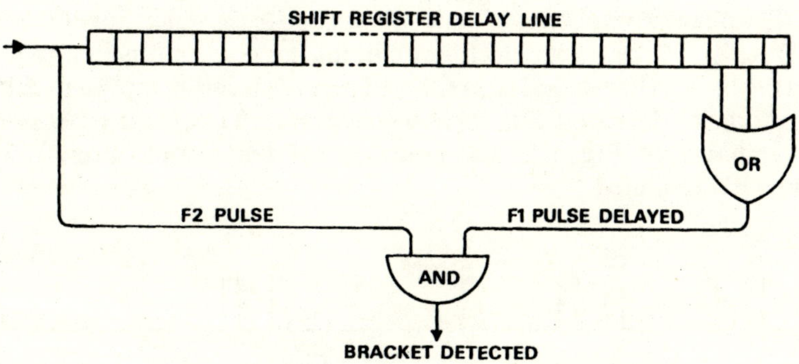


**Fig. 6.1** Bracket detector using a shift register.

Figure 3.11 (Chapter 3) shows an example of two replies. The chance timing of the replies provides two instances where a pulse in the first reply and a pulse in the second reply have a separation equal to the pulse separation of F1–F2 pairs so that four replies may be present. The bracket detector is unable to distinguish between these "phantom" brackets and true framing pulse pairs transmitted from aircraft. Unless these phantom replies are inhibited they will give rise to indications of aircraft that do not exist in reality. The phantom aircraft will appear very close to two real aircraft and the ATC controller will have difficulty in confirming the aircrafts' safety.

A remedy is to permit only two replies to be processed at a time. In the example shown in Fig. 3.11 the first reply will be detected. Subsequent replies during the following 21 μs are detected provisionally but the detection is cancelled if a further overlapping reply is found. This process continues, with each provisional reply being cancelled when a further reply is found, until the last reply. This last reply remains because there is no additional reply to cause it to be cancelled. The rationale for this process is that the first bracket detection must be a true reply since there are no earlier pulses which can have falsely produced it and, likewise, the last bracket detection must also be a true reply since there are no later pulses which can have falsely produced it. Intermediate detected brackets are doubtful replies and are assumed to be erroneous.

It can be seen that for the example shown in Fig. 3.11 the first and last detected bracket declarations survive and the intermediate "phantom" replies are ignored. That is the correct action to detect the two replies that actually exist in this example. Unfortunately if there are indeed three aircraft present the center aircraft will not be shown on the ATC display. While this situation does not occur often, it is most likely to occur when aircraft are at close range which is when the detection of all aircraft is most important.

The second problem that can arise in bracket detection is when both the SPI pulse and the code pulse C2 are present in one reply, as shown in Fig. 6.2. Together the C2 and SPI pulses form the correct timing for another bracket pair. The phantom detection circuit will not delete the detection of the second (false) reply. The bracket detector circuit must include a function to identify when the timing corresponds to a C2–SPI pair of an earlier reply and delete the second bracket detection. Should two actual aircraft exist with replies with the timing shown in Fig. 6.2, the detection of the second aircraft will be erroneously inhibited.

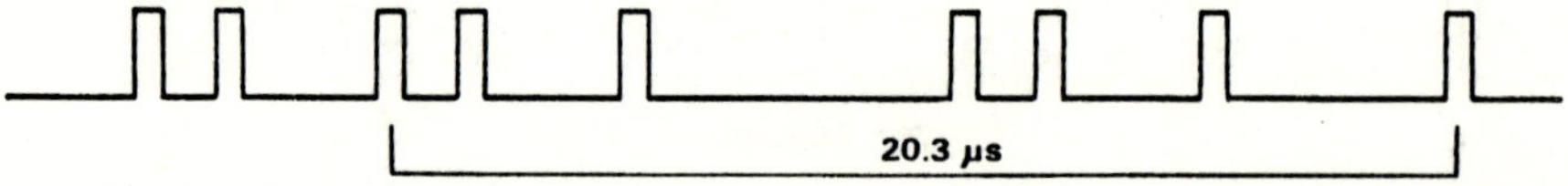


**Fig. 6.2** C2–SPI phantom.

Decoders usually incorporate circuits to determine when two separate F1–F2 reply brackets are detected with timing such that the two sets of reply pulses will interfere. Both replies are then flagged as garbled.

### 6.1.3 Code extraction

Having detected the existence of a reply by detecting the framing pulses as described above, the next task is to determine the code data contained in that reply. Figure 3.4 (Chapter 3) shows the reply format. When only a single reply is received, clear of any other replies, there is little difficulty in establishing the code content with accuracy. However, when more than one reply overlaps there can be a problem in detecting the presence of pulses and determining to which of the replies present the pulses belong. The problem is exacerbated by the need to allow a timing tolerance to the pulse position to cater for timing variations in transmission and the subsequent processing in the ground receiver.

A high clock frequency, such as 16 or 20 MHz, is frequently used in a decoder employing a digital shift register so that the timing of reply pulses can be performed accurately. By this means the probability of allocating a pulse to the correct reply is improved. However, if a pulse can be out of nominal position by up to 0.2 μs, for any number of reasons, it could be argued that a clock rate of faster than 5 MHz is unlikely to provide a reliable improvement in decoding efficiency in practical cases.

Replies on Mode C indicate the aircraft flight level as indicated by the aircraft's altimeter. Mode C replies are coded using a modified Gray code called a Gillham code. The coding used ensures that a transition from one flight level to the next does not require more than a single bit change so that data read at the time of a change cannot be more than 100 ft in error. An example of a Mode C reply is shown in Fig. 3.6 (Chapter 3).

### 6.1.4 Special military replies

Special SSR Mode A codes, as described in Chapter 3, Section 3.4, were developed for civil use and postdate the earlier use of IFF Modes 1, 2 and 3 for military purposes. The military emergency signal uses a normal reply code followed by three further replies, spaced 4.35 μs apart, consisting of empty brackets, i.e replies with no data pulses present. An example is given in Fig. 6.3. The military equivalent of the civil single SPI pulse is to duplicate the normal reply as shown in Fig. 6.4.

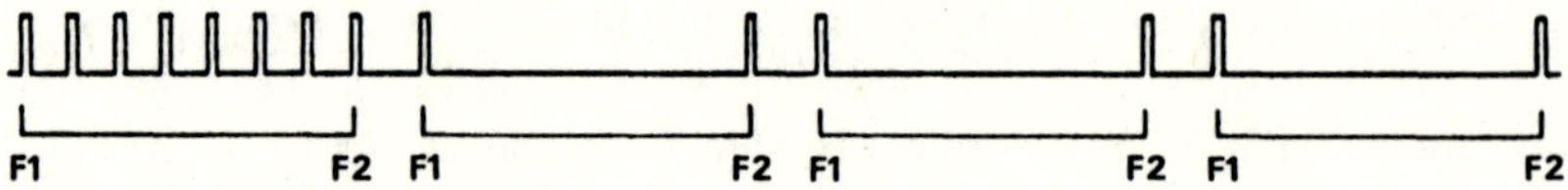


**Fig. 6.3** Military emergency reply.

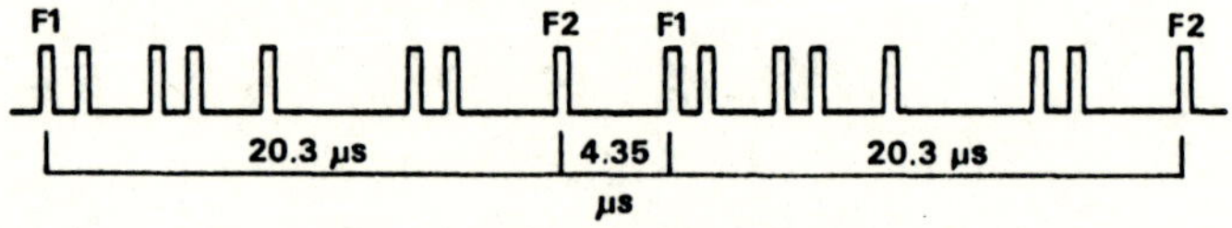


**Fig. 6.4** Military identity reply.

### 6.1.5 Mode determination and range measurement

Following the receipt of a system trigger from the interrogator (see Chapter 5, Section 5.1.5) the decoder will first expect the interrogation pulses P1 and P3 to appear imposed upon the receiver output. By this means the decoder is able to determine the interrogation mode and the type of replies to be expected. The detection of the transmission time of interrogation pulse P3 permits the aircraft range to be determined since the transponder reply is timed to P3. Failure to detect a valid P1–P3 spacing would cause the decoder to indicate a fault.

## 6.2 Defruiting

The introduction of labelled plan displays necessitated that the Mode A and Mode C replies from aircraft be correlated and associated with the range and bearing measurements made using those replies. A problem is that additional replies which are triggered by other interrogators will be received. These unwanted replies are given the acronym fruit. Fruit must be eliminated so as to avoid the chance of forming target reports for aircraft that do not exist in reality. The equipment that performs this task is, appropriately enough, termed a defruiter.

A common technique of defruiting makes use of the sliding window process. Replies are stored in the memory in range-ordered cells with a new area of memory used following every interrogation (Fig. 6.5). When a reply is received the contents of the range cells corresponding to the same range but storing replies to earlier interrogations are inspected to determine whether synchronous replies are being received.

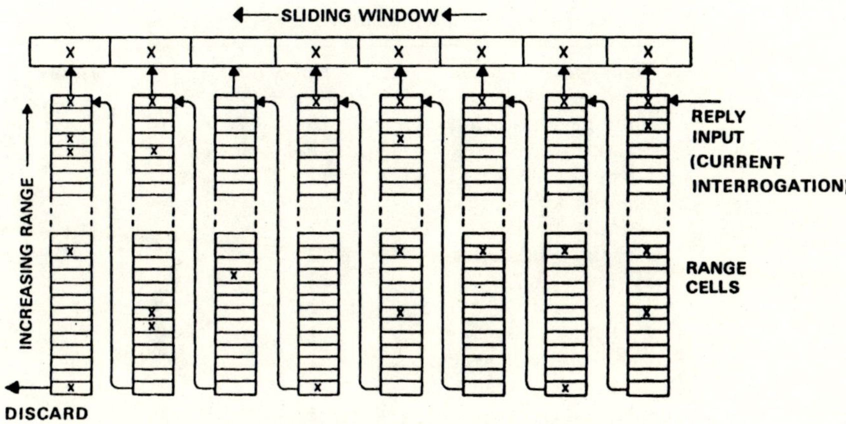


**Fig. 6.5** Range-ordered memory.

In general, fruit replies will not occur at consistent ranges or times following the local interrogations, and few fruit replies will appear in the "window" of cells of common range. A requirement for a minimum number of replies to be exceeded in order for a target to be recognized will cause false reports due to fruit replies to be rejected.

Separate defruiters are of value where raw video or bracket detection signals are presented on unlabelled displays. When a plot extractor is employed to provide labels on a display, the defruiting function is normally included in the plot extractor.

## 6.3 Target detection and position measurement

The output from the decoder is placed in a range-ordered store similar to that used in the defruiter described in Section 6.2. Typically, range cells with a width of 1/8 n miles are used, so that 2048 cells are required for a maximum range of 256 n miles. The contents of the range cell are given in Table 6.1.

The sliding window field is used to record whether a reply has been received in that range cell in the current or a previous interrogation. The field "slides" by shifting the oldest record out of the window at one side in order to make room for the most recent record at the other side (see Fig. 6.5). The functional width of the window can typically be set between eight and sixteen history positions.

In order to avoid declaring a target from the chance association of a few fruit replies, a minimum number of replies are needed in the window. The value of this minimum number is dependent upon the number of replies from an aircraft on each scan of the antenna, and this depends in turn upon the

**Table 6.1** Contents of the range cell

| *Data field* | *No. of bits* |
|---|---|
| Sliding window (hits) | 16 |
| Azimuth | 12 |
| Identity 1 | 12 |
| Identity 1 validation | 2 |
| Identity 2 | 12 |
| Identity 2 validation | 2 |
| Height | 12 |
| Height validation | 2 |
| Fine range | 1 |
| Leading edge detected | 1 |
| Trailing edge detected | 1 |
| Emergency | 1 |
| Communication failure | 1 |
| Hijack | 1 |
| Military emergency | 1 |
| SPI | 1 |
| Total | 78 |

interrogation rate, the antenna turning rate and the interrogation mode interlace program. Since not all aircraft reply to Mode C interrogations, it is common to use an interrogation interlace program of the form A, A, C, A, A, C, ... so as to increase the number of replies. A typical requirement is for five replies in a window eight interrogations long. Such a requirement is known as the "leading edge criterion".

When a leading edge situation is reached the antenna azimuth is placed in the appropriate field in the range cell together with the reply code for the latest reply. If the next interrogation is in a different mode the code of the reply is placed in a second identity field in the range cell. A further interrogation will obtain a reply for an interrogation mode for which a reply has already been received, and this reply will be compared with the earlier reply and the validation field set according to Table 6.2.

In due course the antenna beam will rotate past the aircraft and the aircraft will cease to reply to interrogations. The sliding window continues to be shifted to discard the oldest data and, since replies are no longer being received, no further entries are added at the input to the window. When the number of hits in the window falls below another preset number a "trailing edge" is declared. Typically the trailing edge criterion requires one less reply than for the leading edge criterion.

At trailing edge the antenna azimuth is combined with the stored value at leading edge to determine the center of the replies as the target bearing (see

**Table 6.2** Determination of the validation field

| *Validation level* | *Action* |
|---|---|
| Level 1 | Code has been received only once and was garbled |
| Level 2 | Code has been received only once and was ungarbled, or two identical codes have been received and at least one was garbled |
| Level 3 | Previous level was 2 and a subsequent identical code was received ungarbled |

Chapter 3, Fig. 3.10) according to the following equation:

$$\text{aircraft bearing} = \frac{\text{leading edge azimuth} - \text{trailing edge azimuth}}{2} - \text{offset}$$

The offset takes account of the delay in declaring the leading edge after the first reply is received and the delay in declaring the trailing edge after the last reply is received. It is a function of the leading edge criterion, the trailing edge criterion and the window length and is independent of transponder reply probability.

At the trailing edge the contents of the range cell can be assembled to form a target report for that aircraft. Reply code data are included if they have reached validation level 3. The aircraft range is given by the position of the range cell and can be improved in resolution by storing the fractional position of the reply in the range cell, i.e. whether the reply was received at the time of the first or the second half of the range cell.

The range cell width of 1/8 n mile corresponds to a time width of 1.55 μs. Some aircraft replies occurring near the edge of the range cell may appear in an adjacent range cell owing to movement of the aircraft while in the antenna beam or to jitter in the transponder replies. An overlap bit can be set in the range cell and in the adjacent cell if a reply range is within 0.5 μs of a cell outer boundary. When this bit is set the target report should contain an amalgam of the reply data in both range cells with the shorter range cell predominating if there is any disagreement between range cell data.

## 6.4 Shortcomings of the sliding window plot extractor

Although the application of plot extractors provided the air traffic controller with more complete code information to enable him to monitor the safe progress of aircraft, he also lost information in the process. The earlier presentation of unprocessed video replies on the display enabled him to judge the quality of the position data. Replies were normally obtained over the full antenna beam width, and aircraft separation could be assured by not allowing the returns from two aircraft to overlap. When a decreased number of replies were obtained, because the aircraft was flying through a fade region for example, this was apparent to the controller who could widen the aircraft separation if necessary.

In contrast, the plot-extracted aircraft position is presented as a graphic symbol on the display which suggests a measurement of consistently high accuracy. That this assumption of accuracy was not always valid became apparent when the course of the aircraft's plotted position jittered about a smooth track. Some plot positions, particularly in azimuth, suggested that the aircraft might be commencing an unannounced turn, and the original video was no longer available as an indicator of poor reply data.

The sliding window method of determining the aircraft bearing as the center of its replies (see Section 6.3 and Chapter 3, Fig. 3.10) is unreliable even when the maximum interrogation rate of 450 Hz is used. The bearing accuracy is dependent upon the system performance at the antenna beam edges which is a weak signal area most vulnerable to distortion due to multipath effects. In addition, loss of replies due to capture or extra replies due to fruit can cause the center of the replies to be offset. Aircraft unequipped to reply to some of the interrogation modes used will give rise to measurement errors.

The presentation of identity and flight level data on the display was a great improvement, but the occasional absence of these data or errors in their presentation, particularly when aircraft were in close proximity, gave rise to demands for still better performance. To avoid producing false replies due to phantoms the decoder was limited to processing only two overlapping replies even when three aircraft were present. The need for a minimum number of replies to confirm the existence of an aircraft meant that, when only a very few were received, the aircraft was not reported.

Ghost aircraft can be produced as a result of reflections and appear in the direction of the reflecting object. Often the replies from such reflections are weaker than those for a real aircraft and this is manifested as a reduced or erratic number of replies. The plot extractor, however, can convert the reflected replies into a target report indistinguishable from that of a real aircraft. The presentation of ghost aircraft as true targets became more of a problem with the increasing density of air traffic.

Improvements to SSR reply processing were required and these were realized by the use of better antennas, by better direction finding using monopulse techniques and by tracking. The features of a plot extractor which is able to exploit the additional data available from the employment of monopulse are described in Chapter 7.

Chapter 7

# Monopulse Plot Extraction

## 7.1 Summary

The general principles of video processing to detect SSR replies, extract the reply data, assemble the replies into synchronous groups to form target reports relating to real aircraft and discard the nonsynchronous fruit replies are described in Chapter 6. The monopulse plot extractor performs all these functions but with considerably improved precision.

The monopulse technique takes a direction-of-arrival measurement on every received pulse. This process is very useful for accurately determining the bearings of aircraft in the SSR system as is described in Chapter 5, Section 5.4. The same measurements also give powerful assistance in the decoding of SSR replies, particularly in overlapping reply, or garble, situations.

In the case of the standard nonmonopulse SSR plot extractor reply code validation is achieved by repetition, i.e. by requiring a number of replies from each aircraft for each interrogation mode. When overlapping replies are detected, the whole reply is labelled as suspect and, in general, not used. In contrast, a monopulse plot extractor assigns a garble bit to each pulse of a reply and this allows the ungarbled pulses to be used [7.1, 7.2]. Any garbled pulses can be resolved by reference to further replies from the aircraft, obtained from subsequent interrogations, or by the use of reply data from earlier antenna scans. This technique, together with the fact that monopulse provides good directional data, means that target reports can be reliably produced from as few as two replies. Table 7.1 highlights the differences between the two methods.

The processing technique used by the monopulse plot extractor can be understood by considering the reply signals shown in Figs 7.1–7.3.

Figure 7.1 shows a simple SSR reply as output from the sum and difference channels of a monopulse receiver. Only one reply is present and the plot extractor has no difficulty in detecting the reply and extracting its code.

**Table 7.1** Comparison of standard and monopulse SSR

| *Attribute* | *Standard SSR* | *Monopulse SSR* |
|---|---|---|
| Run length | 16–30 | 4–8 |
| Garble bits | 1 | 12 |
| Azimuth | Run center | Monopulse |

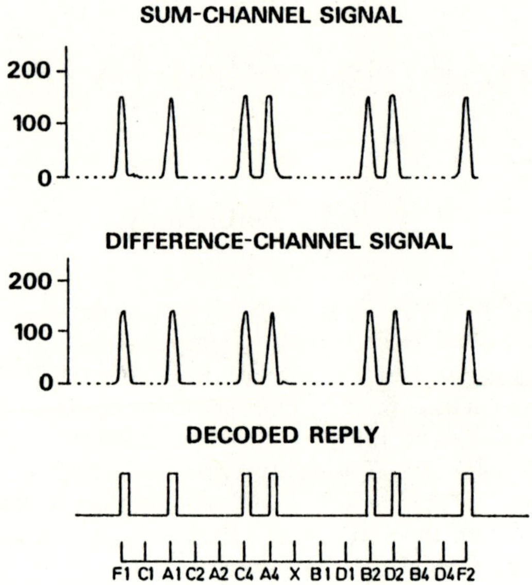


**Fig. 7.1** Video for a single reply (courtesy of the CAA).

The extracted data are shown on the bottom line as clean rectangular pulses.

Figure 7.2 shows a slightly more complex case. Considering first only the sum-channel signal, it is clear that the sequence of reply pulses lasts too long for them to be from just a single aircraft. Indeed it is possible to detect F1–F2 framing pulse spacings at four different positions, and so four aircraft could be present. Inspecting the difference-channel signal shows that two sum-difference pulse ratios exist and that signals from two distinct directions are involved. The presence of this additional direction information, produced by the monopulse process, now makes it relatively easy to determine that only two replies are present.

However, the reply decoding to extract the reply data from the two replies is not completely clear since it is not possible to be sure that only one pulse was received at each pulse position. The difference-channel signal at the position marked A in Fig. 7.2 is relatively weak, so that it is unlikely that two reply pulses received simultaneously from the two aircraft are present. The same cannot be said of the pulse at position B since it is strong and could be masking a weaker pulse from the other reply.

When a reply pulse timing is such that the pulse can belong to more than one reply then, initially, it should be regarded with some caution and decisions made on it flagged as low confidence. Later processing can be used to resolve any doubt. The two pulses in Fig. 7.2, where there is some doubt as to which

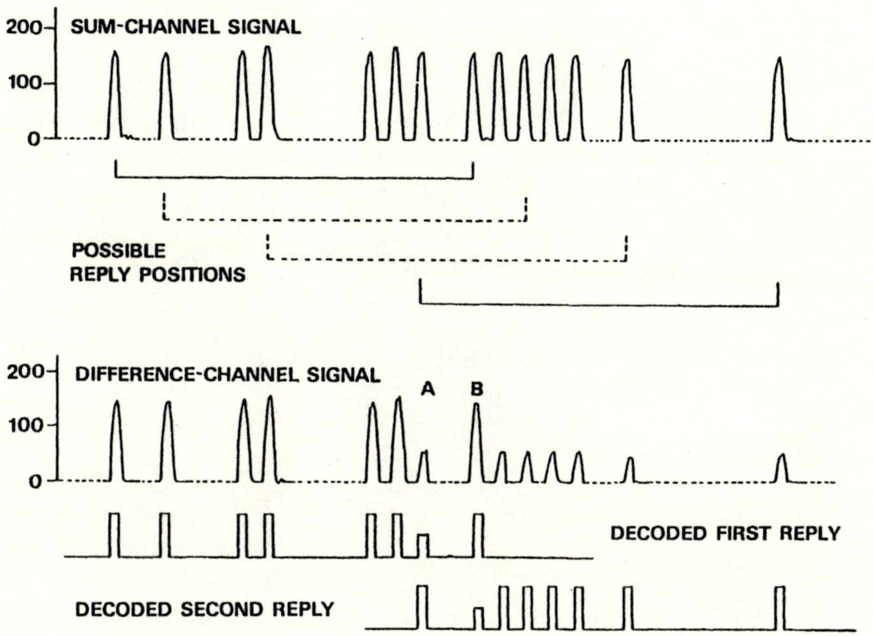


**Fig. 7.2** Video for multiple replies (courtesy of the CAA).

reply they belong, are indicated diagrammatically as half-amplitude in the bottom two lines. In the plot extractor each reply data pulse is stored together with a confidence flag. The framing pulses do not form part of the reply data and have no need to carry confidence flags.

Figure 7.3 shows an example of a much more complex reply. Four replies are present but the situation is more difficult for the human eye to resolve. Nevertheless the plot extractor provides a good estimation of the replies and their reply data.

The decoding process has been described on the basis of decisions made on the direction of arrival of pulses, but other data can also be used to similar effect. Useful results based on sum-channel signal strength can also be obtained. Indeed, results presented below will show that the use of signal amplitude is more powerful than the use of direction of arrival. Pulse decisions based on both monopulse direction of arrival and signal strength can be included in the decoding process with benefit.

The examples shown in Figs 7.1–7.3 are in response to a single interrogation. Other interrogations within the same antenna beam dwell-time could produce different results. Some of the replies may be due to fruit and will not repeat with the same relative timing. Other replies may be from aircraft which will quickly move out of the antenna beam. The results from subsequent

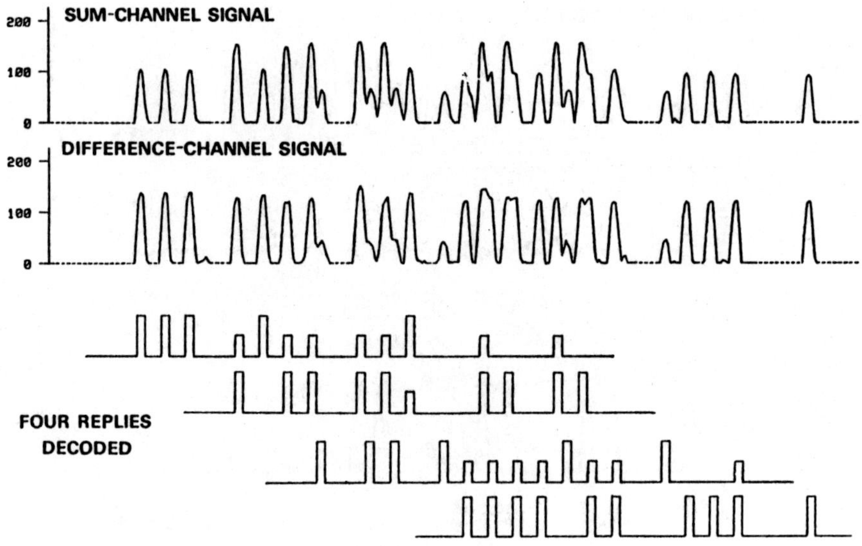


**Fig. 7.3** Video for complex multiple replies (courtesy of the CAA).

interrogations can be used to resolve pulse decisions earlier declared as low confidence. In addition, aircraft tracking will permit low confidence pulse decisions to be resolved using data obtained on earlier scans of the antenna.

Tracking, or surveillance processing, plays an important part in achieving the impressive performance obtained from monopulse SSR. It permits the resolution of incomplete reply codes, allows processing errors to be recognized and corrected, and also provides the basis of detecting and eliminating false reports that have arisen from reflections. These processes are discussed in turn in this chapter.

## 7.2 Reply processing

Modern decoders invariably use clocked digital circuitry, such as shift registers, to store the full reply so that measurements and decisions on it can be made once all the data are available. The circuit designer has a choice of clock frequency for this circuitry: a high frequency gives rise to fine time discrimination but needs a long shift register for the required time delay.

The pulse position tolerance allowed for a transponder reply is $\pm 0.1$ μs. This uncertainty in pulse position can be increased by receiver noise and by the presence of multipath conditions such that the pulse arrives by several different routes. A further difficulty is that transponders can drift out of specification without this fact being noticed. It is the task of the designer of the ground equipment to detect aircraft with marginally out-of-specification transponders if reasonably possible.

The decoder clock frequency should be matched to the reply pulse timing tolerance. The resulting pulse decoding time windows should be sufficiently wide to embrace all valid pulses of a reply but not so wide that garbling pulses are also accepted as belonging to the reply. Such a requirement cannot be met simultaneously and a compromise position must be adopted. Unpublished work at the Lincoln Laboratory shows that dividing the pulse spacing of 1.45 μs by 12, requiring a decoder clock frequency of 8.276 MHz, produces an optimum design. The use of a higher frequency necessitates inspection of more clock periods to ensure identification of a reply pulse that may be at the limit of its permitted tolerance, and this increases the probability of accepting a garbling pulse as a member of the reply. A higher clock frequency will not necessarily improve the measurement of aircraft range since, in the limit, this is determined by the jitter in the transponder reply delay, the difference in reply delay between interrogation modes, and the effects of receiver noise, bandwidth and pulse rise-time.

Figure 3.4 (Chapter 3) shows the format of an SSR reply. The framing, or bracket, pulses F1 and F2 are always present and the other pulses may or may not be present depending upon the data that the transponder is sending. The X pulse is not currently used.

### 7.2.1 Interrogation mode detection

The interrogation pulses P1 and P3, as shown in Fig. 3.3 (Chapter 3), are generated in the transmitter and radiated by the antenna. These pulses are frequently superimposed on the receiver reply video that is passed to the plot extractor. The reply decoder measures the spacing of P1 and P3 in order to determine the current interrogation mode and thereby correctly interpret the reply code data. Upon detection of the P3 pulse of a valid P1–P3 mode pulse pair a range counter is started which, for convenience, can be based on the 8.276 MHz clock.

### 7.2.2 Pulse leading edge detection

The time of the leading edge of a reply pulse is defined in the transponder specification in ICAO *Annex 10* and is a powerful discriminant in the process of recognizing the existence of a reply pulse and associating it with a given reply.

The time of the leading edge is defined at the half-voltage point of the modulation of the transponder reply. In the receiver this half-voltage point can be determined by first measuring the pulse peak amplitude and then inspecting the leading edge of the same pulse delayed in a delay line. This process is described in more detail in Chapter 5, Section 5.3.3. The detection of the pulse leading edge gives rise to the generation of an actual leading edge (ALE) signal at the following clock rising edge (Fig. 7.4). Although overlapping pulses may be of different amplitudes it is possible to detect their separate leading edges if the weaker pulse is first, as discussed in Chapter 5, Section 5.3.3. All such detected leading edges generate ALE signals. When the two leading edges of overlapping pulses are not separately determinable, use must be made of data provided by the timing of the trailing edge.

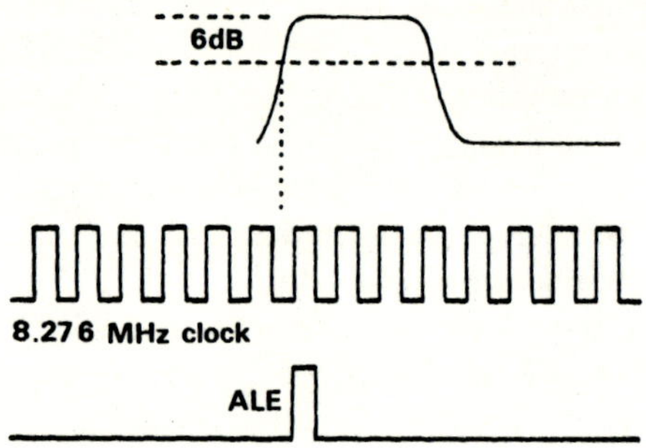


**Fig. 7.4** Generation of the actual leading edge.

### 7.2.3 Pulse trailing edge detection

The ability to detect reply pulses is of particular importance when two aircraft are at close range as, for safety reasons, the accurate detection of aircraft, as well as their reply data, is critical. Overlapping reply pulses caused by the reception of two or more aircraft replies at a similar time are a common occurrence, and improvement in detection has been given particular emphasis in recent years. When two reply pulses overlap, the problem becomes one of determining that two pulses are present since it is frequently not possible to detect two leading edges.

An indication that more than one pulse may be present is obtained if a pulse has an excessive duration. ICAO *Annex 10* permits the duration of a transponder reply pulse to lie between 0.35 and 0.55 μs, and any excess duration over the maximum of 0.55 μs may indicate that two pulses are present.

The position of the pulse trailing edge can be determined in a similar way to that of the leading edge, i.e. by measuring the time of the half-voltage point on the back edge. If a decoder clock frequency of 8.276 MHz is used, for reasons stated above, the time resolution available is a multiple of 0.121 μs. A pulse width between the measured positions of the leading and trailing edges in excess of five clock periods, or 0.605 μs, can be taken to indicate that two pulses are present. The position of the second hidden leading edge can be assumed by measuring backwards in time from the trailing edge by four clock periods or 0.484 μs. This assumed leading edge location is called a pseudo leading edge (PLE) (Fig. 7.5).

Occasionally more than two pulses overlap so that at least three pulses are present. The first leading edge is evident and an estimation of the position of the final leading edge can be made by measurement from the trailing edge position. Intermediate pulse leading edges can only be guessed at. The extra leading edges (XLEs) are assumed by measuring nominal pulse widths of 0.484 μs from the first (actual) leading edge (Fig. 7.6).

Interference, which consists of very long pulses or continuous waves, can produce a very large number of XLEs, which in turn can generate a large number of spurious replies. Means should be provided to limit the number of consecutive XLEs which are allowed, and replies should not be declared if their bracket pulses F1 and F2 are both XLEs.

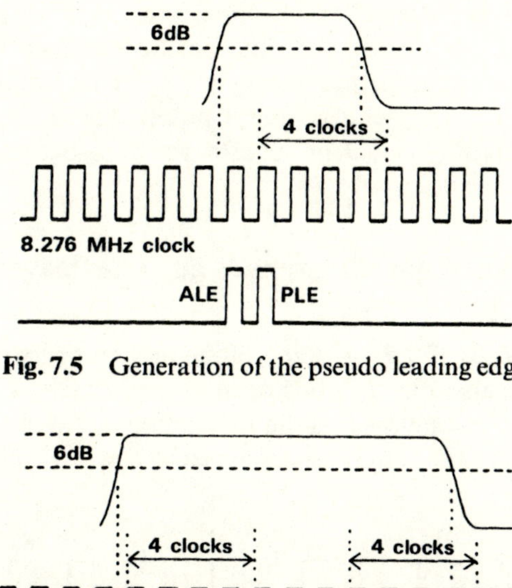


**Fig. 7.5** Generation of the pseudo leading edge.

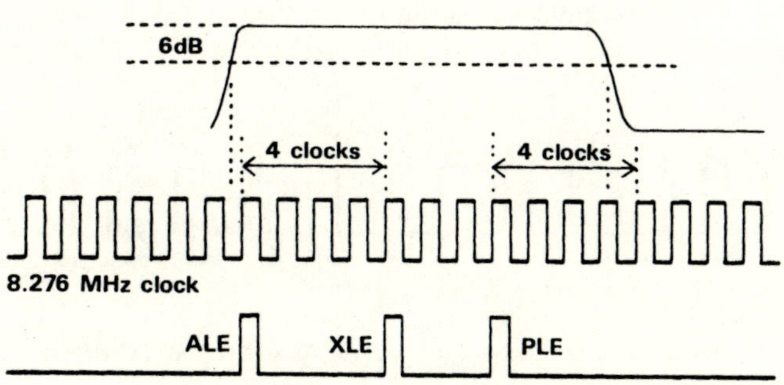


**Fig. 7.6** Generation of an extra leading edge.

## 7.2.4 Bracket detection

The process of detecting the existence of an SSR reply involves the detection of the framing pulses F1 and F2 which are always present in the reply. The process of framing pulse detection is one of measuring a pair of pulses with a timing between leading edges of 20.3 μs, or the 167th, 168th and 169th clock pulse periods of the 8.276 MHz clock. The F1 and F2 pulse leading edges can be ALEs, PLEs or XLEs.

Detection of a framing pulse pair is inhibited if one or more of the following apply:

(a) both the F1 and F2 pulse RSLS flags are set, indicating that they were both sidelobe pulses and therefore that the whole reply is presumably a sidelobe response;
(b) the reply is indicated as a phantom as defined in Section 7.2.7;
(c) a Mode S preamble has been detected (see Section 7.2.9).

## 7.2.5 Code pulse position timing

For a reply code pulse to be considered as a candidate pulse in a reply its leading edge must conform to the time position defined as $n \times 12^{+1}_{-\frac{1}{2}}$ clock pulses after the clock pulse period during which the F1 leading edge occurred, where $n = 1, 2, 3, \ldots, 13$ (including the X pulse). The timing for an SPI code pulse leading edge is $36 \pm 1$ clock pulses after the clock period during which the F2 pulse leading edge occurred.

## 7.2.6 Definition of garbling replies and interleaved replies

For the purposes of the operation of the reply processor, garbling replies are defined as (see also Fig. 7.7) two or more replies whose F1 leading edges occur within 170 clock pulses and the F1 leading edges of succeeding replies occur in one of the 14 time intervals described by $n \times 12 \pm 2$ clock pulses, where

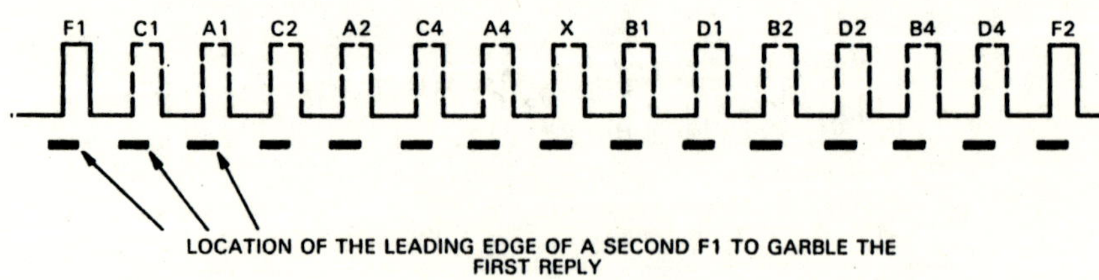


**Fig. 7.7** Location of a second F1 to qualify as garbling.

$n = 1, 2, 3, \ldots, 14$ after the clock period containing the F1 leading edge of the earlier reply.

Interleaved replies are defined as two or more replies whose F1 leading edges occur within 170 clock pulses provided that the F1 leading edges of succeeding replies do not occur in one of the 14 time intervals that would constitute a garbling reply.

### 7.2.7 Phantom replies

A phantom reply is defined as a reply whose F1 pulse is a code pulse or framing pulse of one reply and whose F2 pulse is a code pulse or framing pulse of another reply. The two replies can be overlapping (Fig. 7.8(a)) or nonoverlapping with a gap of less than 170 clock pulses (Fig. 7.8(b)) or only a single reply but with pulses C2 and SPI both present (Fig. 7.8(c)). Phantom replies are removed from further processing.

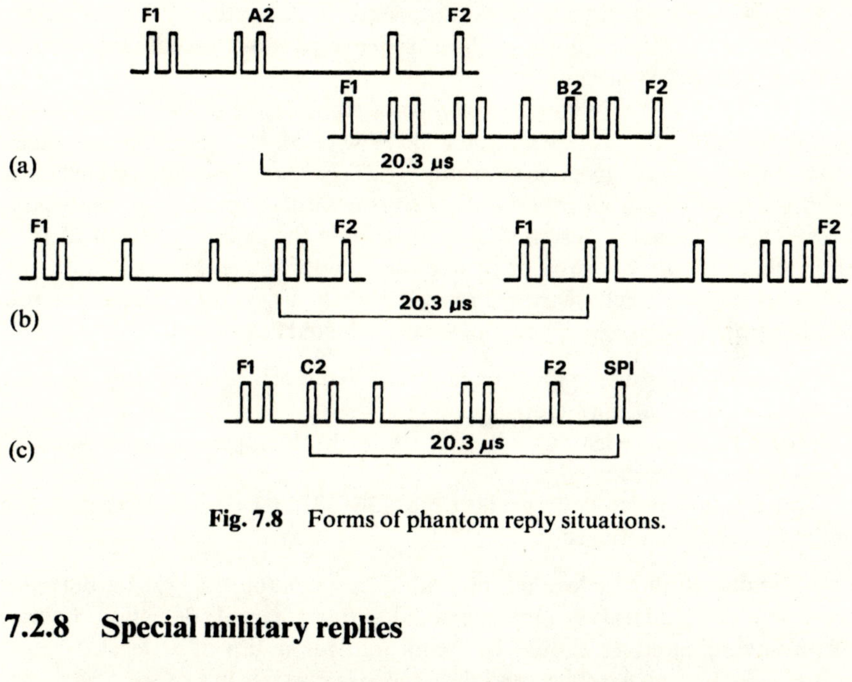


**Fig. 7.8** Forms of phantom reply situations.

### 7.2.8 Special military replies

Two forms of multiple replies can be transmitted to indicate an emergency condition or an identification of position. The military emergency reply is used on Modes 1, 2 and 3. The identification of position response is only used on Mode 1 with the SPI pulse used on Modes 2 and 3.

(1) Military emergency (Chapter 6, Fig. 6.3) is indicated as four closely spaced reply pulse trains where the F1 pulse leading edges of the succeeding replies are positioned at $36 \pm 1$ clock periods relative to the F2 pulse leading edges of the preceding replies. The first reply contains code information (which may or may not be 7700) and the succeeding reply trains have zero code content.

(2) Military SPI (Chapter 6, Fig. 6.4) is indicated as two closely spaced pulse trains with identical code content where the F1 pulse leading edge of the second reply is positioned $36 \pm 1$ clock periods after the F2 pulse leading edge of the first reply.

### 7.2.9 Mode S preamble detection

The monopulse SSR plot extractor discussed in this section is concerned with the processing of SSR replies in response to interrogation Modes 1, 2, 3/A, B, C and D. Mode S is discussed in Chapter 12. Nevertheless consideration should be given to the possible impact of Mode S replies on the SSR reply decoding process.

The Mode S reply consists of a long train of 0.5 μs and 1.0 μs pulses lasting for up to 130 μs (see Chapter 12, Fig. 12.6). It is possible that such a train of pulses would give rise to multiple SSR bracket detections which could overload the plot extractor and affect its proper operation. It is advisable that an SSR plot extractor design should anticipate the introduction of Mode S and include circuitry to provide protection against Mode S replies.

In order to detect a Mode S reply it is only necessary to detect the four Mode S preamble pulses. These pulses are detected as follows:

(a) a second pulse leading edge occurs 7–9 clock periods after detection of the first pulse leading edge;

(b) a third pulse leading edge occurs 28–30 clock periods after detection of the first pulse leading edge;

(c) a fourth pulse leading edge occurs 36–38 clock periods after detection of the first pulse leading edge.

The detection of a Mode S preamble should suppress bracket detection for a period of 120 μs (994 clock periods). After preamble detection the train of pulse leading edges should be monitored for a gap of at least 22 clock periods. Such a gap would exceed that which is possible in a Mode S reply and would indicate that the preamble detection had been spurious, possibly owing to chance overlapping fruit replies. If a gap of that size is detected, the bracket detection suppression should be released so that SSR reply processing can resume.

### 7.2.10 Pulse sampling

Decisions are made on various qualities of reply pulses and in order for this to be done these qualities must first be measured. The principal pulse quality in the processing of replies in a monopulse system is the monopulse angle of arrival in the antenna beam or, as it is usually known, the off-boresight angle (OBA). Boresight is the name given to the center of the antenna beam.

The video signals obtained from the monopulse receiver are not initially in angle units but must be converted to these units at some place in the subsequent processing. The video signals may be in the form of a phase difference, if a phase-processing monopulse system is employed, or an amplitude difference, if an amplitude-processing system is used. The various forms of monopulse are described in Chapter 5, Section 5.4. For the purposes of this discussion it will be assumed that the monopulse system is of the amplitude-processing kind, although the principles are very similar for the alternative form.

The amplitude-processing monopulse system uses two receiver channels—the sum channel and the difference channel—to produce the angle information which is given by the sum–difference amplitude ratio. This sum–difference ratio can be calculated before or after reply processing. Similarly, the sum–difference ratio can be converted to angle units before or after reply processing. These alternatives offer little in the way of relative benefits or disadvantages and the choice is usually made for reasons of implementation convenience rather than for system performance. For the remainder of this discussion it will be assumed that the sum and difference video signals are each converted into 8-bit digital amplitude measurements and processed separately by the reply decoder. After reply processing the sum–difference ratio is converted to the OBA.

Associated with the OBA measurement is the sign of that measurement which indicates whether the angle is to the left or right of the antenna center. The receiver can make a decision on the basis of the angle measurement as to whether the signals being processed are of sufficient quality to provide an accurate sign indication. When this decision process is used there is a confidence signal to accompany the sign indication.

Sidelobe replies are determined separately by comparing the relative amplitudes of the receiver signals in the sum and control channels.

Table 7.2 summarizes the reply signals available for participation in reply processing. The processed sum video is the output from the receiver video pulse detection process described in Chapter 5, Section 5.3.3, and the ALEs, PLEs and XLEs are obtained from this signal as described earlier. The quantized sum video, the quantized difference video, the sign data and the sidelobe flag are sampled from these leading edge signals. The sampling times

**Table 7.2** Typical monopulse receiver output signals

| | |
|---|---|
| Processed sum video | 1 bit |
| Quantized sum video | 8 bits |
| Quantized difference video | 8 bits |
| Sign data | 2 bits |
| Sidelobe flag | 1 bit |

following the leading edges are as follows: ALE, 2 clock pulse periods after the ALE; PLE, 2 clock pulse periods after the PLE; XLE, no samples are taken.

### 7.2.11 Reply reference selection

In order to test each reply pulse for consistency of direction of arrival with other pulses in the reply it is necessary to select a pulse as reference against which to measure the other pulses. If only a single reply is present then any pulse can be used but, equally, no problem in decoding the single reply normally exists. The decoding process is taxed more when more than one reply is present, and in this case the selection of a pulse to act as a reference for the other pulses in the reply must be performed with care.

The framing pulse F1 is taken as the first choice for the reference pulse. The choice is somewhat arbitrary, but there is a small possibility that F2 could be disturbed by a multipath reflection from an earlier code pulse. However, if F1 is garbled by being close to the reply pulse position of another reply (see Section 7.2.6), pulse F2 is taken. If F2 is also garbled, the reply is probably a phantom (see Section 7.2.7) and decoding is not possible. A further restriction on the selection of a pulse as reference is that it should not be flagged as a sidelobe since there is a likelihood that it has been corrupted.

It should be noted that pulses are only considered garbled if they are close to the time of a possible pulse of an overlapping reply. It is quite possible for framing pulses to fall in the spaces between the pulse positions of overlapping replies, in which case they can be used as the reference. It is usual to provide for the simultaneous processing of four overlapping replies. More than four is unproductive since the possibility of more replies overlapping but still having ungarbled framing pulses available for use as references is remote.

### 7.2.12 Reply decoding

Having detected a pulse pair with the correct F1–F2 spacing the next task is to identify those data pulses which may belong to that reply. In the first instance this identification is performed on the basis of pulse timing, making due allowance for timing errors in transmission from the transponder and subsequent errors introduced in processing on the ground. An ALE, PLE or XLE is sought in each of the 14 possible code pulse positions including the X pulse and the SPI pulse.

A time window, which allows the leading edge to be detected in the nominal pulse position or 1 clock period (0.121 μs) later or 1 or 2 clock periods earlier, is placed around each pulse position. The time window is asymmetric since, in weak signal conditions, receiver noise is more likely to advance than to retard the apparent leading edge. If a pulse is present in the time window, as indicated by the existence of an ALE, PLE or XLE, it is compared with the chosen reference pulse in terms of its monopulse value and, independently, its amplitude. Each possible code pulse is declared a 1 or a 0 and assigned high or low confidence according to the following rules.

H0: a high confidence 0 is declared if no pulse is detected in the time window or, in the case of decisions made on amplitude measurement only, a weak pulse is detected (see Section 7.2.13).

H1: a high confidence 1 is declared if a main beam pulse, which correlates in its monopulse value or its amplitude value and does not also correlate with the reference of another suitably timed reply, is detected in the time window.

L0: a low confidence 0 is declared if the pulse is flagged as sidelobe (RSLS) or fails to correlate with the reference but does correlate with the reference of another suitably timed reply.

L1: a low confidence 1 is declared if a main beam pulse exists in the time window but fails to correlate with the reference or with the reference of another suitably timed reply, or does correlate with the reference and also with the reference of at least one other suitably timed reply.

It should be noted that the definition of H0 has a weakness. If a processing error, due to receiver noise for example, displaces a leading edge outside the time window, a high confidence zero will be declared even though the pulse exists. The high confidence declaration will prevent the reply from being merged with other similar replies from the aircraft in which the pulse has been correctly detected. The problem is most likely to occur with very weak signals when the signal-to-noise ratio is low, and alternative merging rules could be employed if amplitude data are available. A better test for the existence of a code pulse may be to detect the presence of r.f. energy at the center of the possible pulse position which would avoid the problem. As far as the author is aware this alternative has not been fully explored.

Where a high confidence 1 has been declared, the measured monopulse and amplitude value can be used. One approach is to add the monopulse value to that of the reference to form an accumulated value. The total number of pulses accumulated in this way is also recorded so that an average monopulse value can be calculated. By averaging the monopulse values in this way some improvement in the effective signal-to-noise ratio of the measurement can be obtained. An alternative approach which only requires a division-by-2 function uses the reference pulse value, either monopulse or amplitude, and updates it by forming the running average, i.e. adding the code pulse value to the reference value and dividing the result by 2 (a binary shift to the left) to form a new reference. This has the benefit that a poor reference measurement is improved by adding measurements made on subsequent pulses. Using either form of averaging permits an azimuth or amplitude estimate to be improved by averaging with later correlating measurements.

The monopulse and amplitude code and confidence decisions thus made separately are combined according to Table 7.3. In general, when a decision based on either the monopulse value or the amplitude value is high confidence, that decision dominates. For the monopulse value test, a high confidence zero can only occur if no leading edge appears in the pulse position time window. By definition, the amplitude test cannot produce a different decision. However, the converse is not true since the amplitude test can produce a high confidence zero decision when a pulse is present if that pulse is weak (Section 7.2.13).

**Table 7.3** Combining monopulse and amplitude decisions for code and confidence

| *Monopulse decision* | *Combined decision* | | | |
|---|---|---|---|---|
| | *Amplitude decision* | | | |
| | *H0* | *L0* | *L1* | *H1* |
| H0 | H0 | ←Not possible→ | | |
| L0 | H0 | L0 | L1 | H1 |
| L1 | H0 | L0 | L1 | H1 |
| H1 | L0 | L0 | H1 | H1 |

If no valid monopulse reference is available, the amplitude decisions alone are used. The reasons for a possible lack of a valid monopulse reference are as follows.

(1) No valid monopulse value was associated with the reference.

(2) The monopulse reference value was not confirmed by a correlating code pulse.

(3) The monopulse reference value was greater than a preset value.

Both amplitude and monopulse angle data are used to assist the proper association of pulses with replies. In practice it is found that amplitude data are more effective (see the measured results presented in Section 7.8). When two aircraft are on the same bearing and at a similar range, angle data may not be sufficiently discriminating to distinguish the separate pulses of the two replies, and this situation can apply to all the replies obtained from the two aircraft. Although aircraft at similar range tend to present reply signal strengths within about 10 dB of one another, finer discrimination can be applied to the amplitude measurements compared with the angle measurements to perform high confidence pulse detection.

Figure 7.2 gives an example of overlapping replies where there is no difference in sum-channel signal strength. Clearly, decisions based on amplitude data could not discriminate between the replies. In contrast, Fig. 7.9 shows two replies from aircraft at a similar bearing (similar sum–difference ratio). The decoded reply data indicate that angle decisions are of no assistance and that amplitude-based decisions provide the best result. A similar signal situation can arise with delayed multipath signals as described in Section 7.2.13. If the ground is flat the bearing of the reflected signal will be the same as that of the target, although the signal strengths can be very different.

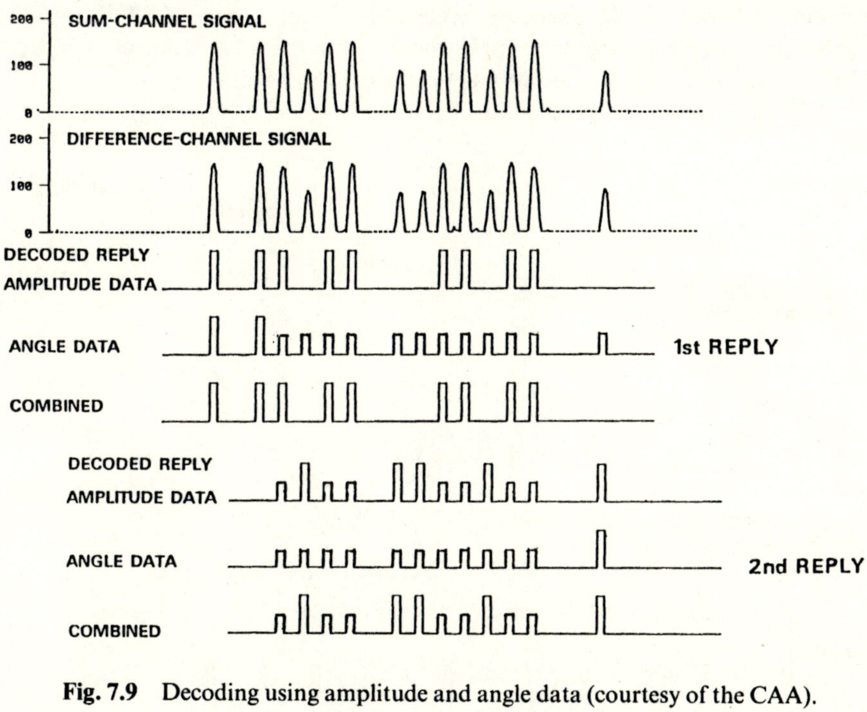


**Fig. 7.9** Decoding using amplitude and angle data (courtesy of the CAA).

Figure 7.9 also illustrates some other points of interest. The timing of the second reply places its F2 pulse at the SPI pulse position of the first reply and the angle decision includes this as a possible first reply pulse. The pulse is shown half-height to indicate that the decision is low confidence since it correlates with the references of both replies. The first two pulses of the first reply are shown as high confidence by the angle decision since they are too early to belong to the second reply. As regards the second reply, the amplitude decision finds five pulses as high confidence and the remaining pulses as low confidence. The latter decision is reasonable since the pulses in both channels are large and could be masking weaker pulses belonging to the second reply. The angle decision can do no better than regard all pulses but the last as low confidence.

### 7.2.13 Low amplitude pulses

SSR replies can be received via reflections as well as on the direct path from aircraft to ground station. False targets produced by reflections are discussed in Chapter 9, Section 9.1.3, and their detection and elimination is discussed in Section 7.5. However, one class of reflected reply, known as ground bounce, is of particular relevance here. Figure 7.10 shows the geometry of replies being received by the direct path and by a reflected path off the ground, and Fig. 7.11 shows the resulting reply video.

Replies received by reflection from the ground are common where the

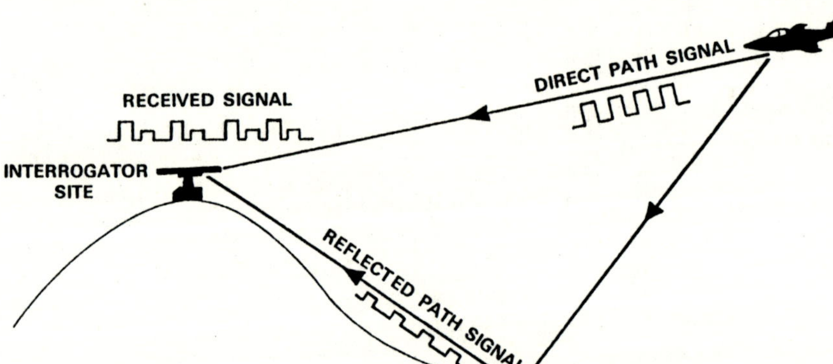


**Fig. 7.10** Geometry of direct and reflected reply paths.

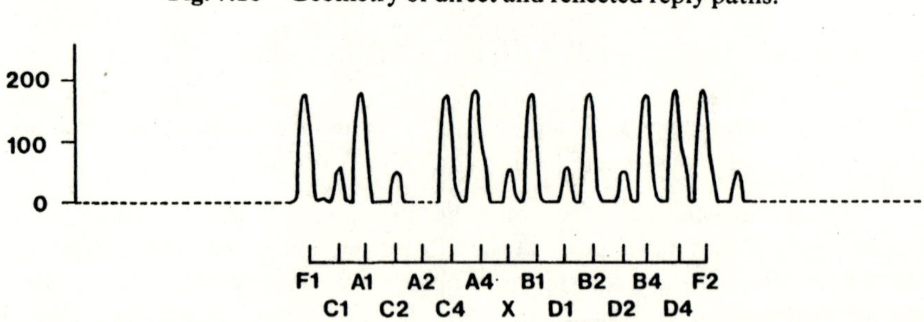


**Fig. 7.11** Reply video with echo.

ground station is situated on a hill. There can be measurable delay between the direct signal and the reflected signal. The improved detection performance of monopulse SSR has demonstrated that many more of these reflected replies are present than may have previously been realized.

At a number of sites, of which Clee Hill in England is one, the geometry of the terrain is such that the delay in the reflected pulse can equal the spacing between pulses in a reply. When this happens it may be difficult to recognize that the reflected pulse is not a code pulse of the direct reply. If reply pulse decisions are made on the basis of monopulse values, the reflected pulse may correlate with the direct reply reference since it is arriving at substantially the same azimuth. This process would produce a low confidence decision for all pulses. Since the ground-reflected reply is usually weaker than the direct reply, particularly if an antenna which discriminates against signals arriving from negative elevation angles is used, amplitude decisions can prevent the decoding error.

Figure 7.12 shows the fall in signal strength that can be expected from a ground bounce pulse as a function of delay. These values can be used to reject pulses from a reply although their timing may suggest that they belong to that reply. The same data can be used to reject the reflected reply, which because of the short time difference is included as a class of wide pulse reply (see Section 7.3.1). To be deleted the delayed reply must be weaker than the curve in Fig. 7.12 with respect to the earlier reply and exhibit the same code as the earlier reply.

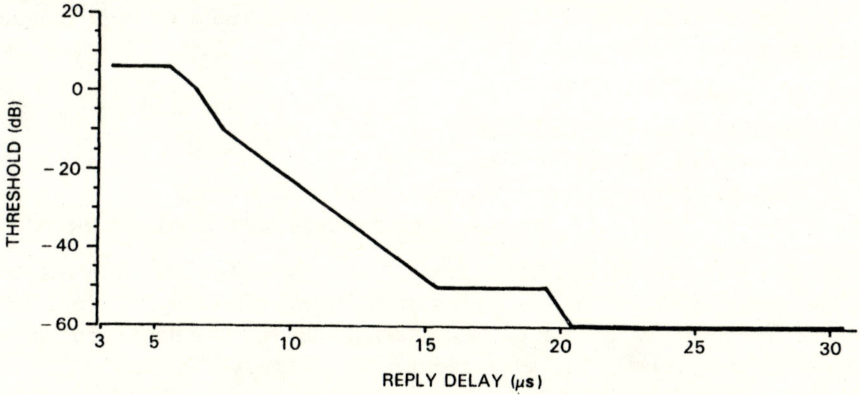


**Fig. 7.12** Variation in ground bounce reflection with pulse delay.

### 7.2.14 Monopulse azimuth data

The reply azimuth relative to the antenna beam center is determined from the values of every monopulse-correlating reply pulse. The separate monopulse measurements are added, taking note of the sign of the value, and an average value is calculated. This average reply azimuth value can be calculated either before or after the monopulse measurements have been converted to angle units.

It is possible that a reply has been decoded using an amplitude reference but that the monopulse reference did not correlate with any other pulse. The monopulse value is unsafe and cannot be used. In this case the antenna boresight direction is taken as the reply azimuth and a "boresight" flag is set to indicate the situation. If a pulse decision is a high confidence 1, it is then changed to a low confidence 1.

### 7.2.15 C2–SPI phantom detection and rejection

The C2–SPI phantom will be recognized by the process described in Section 7.2.7 and confirmed as a phantom if the monopulse or amplitude measurement of the SPI pulse correlates with the reply reference. This double check on a C2–SPI phantom ensures that two overlapping replies, which have a chance timing that corresponds to a phantom, are not rejected.

## 7.3 Reply-to-reply processing

The reply reports that are produced by the reply decoder include extraneous replies due to the following:

fruit replies
residual phantoms and hybrids
false replies due to reflections
spurious replies due to transponder errors (wide pulses)
multiple replies due to a military emergency or identification response

The first task is to identify the extraneous replies and reject them. Following this the remaining replies are grouped according to range and azimuth to form raw target reports and then passed to surveillance processing to refine the reports further before they are sent to the ATC center.

### 7.3.1 Initial processing

**Residual phantoms and hybrids**

A large proportion, about 70%, of fruit replies are detected in the antenna sidelobes. Most of these are eliminated by the reply detector which will reject replies with both the F1 and F2 pulses flagged as a sidelobe response. However, although eliminated, the sidelobe fruit reply can interact with an overlapping main beam reply and give rise to a number of undesirable effects. The principles of these effects are illustrated in Fig. 7.13.

If the timing of the sidelobe reply is such that some of its pulses align with pulse positions in the main beam reply, additional hybrid replies can be generated. Specific examples are given in Fig. 7.14 which shows various possible overlapping conditions for sidelobe and main beam replies. The resulting replies output by the reply decoder are indicated with the attributes that can assist in the recognition of false and real replies.

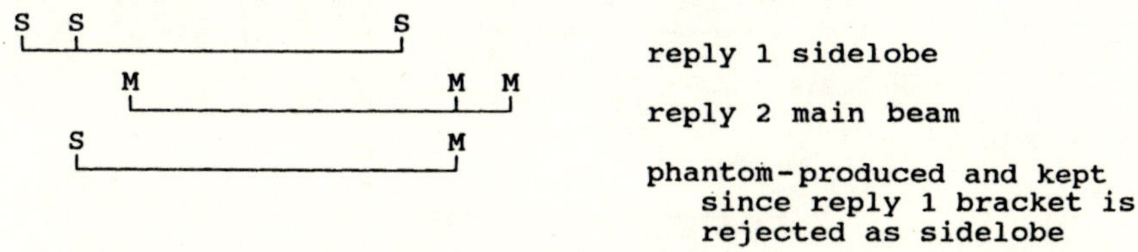


**Fig. 7.13** Formation of hybrid replies. (Reprinted with permission of Lincoln Laboratory, Massachusetts Institute of Technology, Lexington, MA.)

In some cases one of the bracket pulses (F1 or F2) is declared garbled as defined in Section 7.2.6. In some other cases one of the bracket pulses is flagged as a sidelobe response. All hybrid replies which are false (phantoms) have one framing pulse which is garbled and one which is a sidelobe. Since neither garbled pulses nor sidelobe pulses can be used as a reference for reply decoding, it follows that the false replies present have no valid reference and can be eliminated on those grounds. On this basis the hybrid reply cases 1, 2, 4 and 6 in Fig. 7.14 are rejected except for consideration as part of a military emergency or special identification sequence.

**Military emergency and identification replies**

Special military replies are composed of two or four framing pairs transmitted in response to a single interrogation. If unrecognized as such these multiple responses will give rise to multiple target reports and false aircraft indication. The multiple replies are characterized by a following F1 pulse occupying the SPI pulse position of an earlier reply. The SPI pulse will be declared low confidence by the reply decoder.

A military special identification reply can only result from a Mode 1 interrogation and can be recognized as follows.

(a) The range difference between the two replies should be $204 \pm 2$ RU (1 range unit (RU) = 0.121 μs).

(b) The azimuth difference between the two replies should be less than 11 AU (1 angle unit (AU) = 0.022°).

(c) The code of the second reply should not be 0000, ignoring confidence flags.

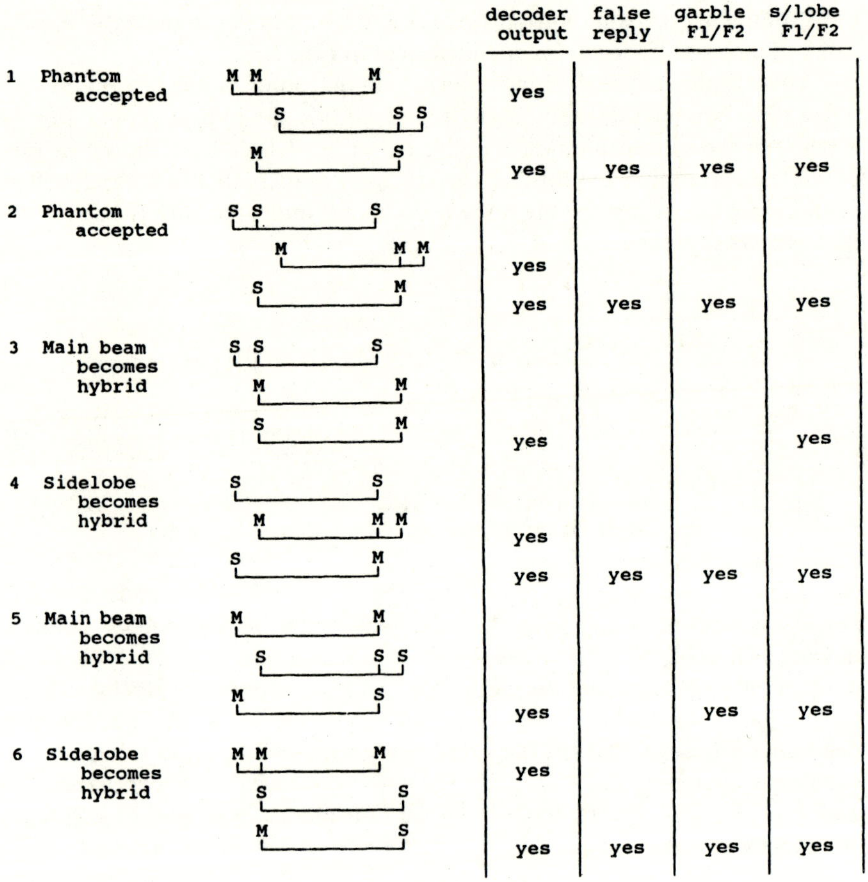


**Fig. 7.14** Types of hybrid reply.

(d) The codes of the two replies should not disagree for any high confidence bits.

A second reply which fails the above tests is assumed to be from a second aircraft unless its code is 0000 in which case it is subjected to the test for a military emergency. A second reply which passes the above tests is deleted and the first reply is marked as being a military special identification.

A military emergency reply can only result from a Mode 1, 2 or 3/A interrogation and can be recognized as follows.

(i) There are four replies each with a range difference between successive replies of $204 \pm 2$ RU.

(ii) Although the four replies should agree in azimuth and the last three of the replies should have a code content of 0000, these criteria are not applied to avoid a chance fruit situation causing the test to be failed. It is better to declare an emergency reply situation erroneously than to miss a true one.

Second, third and fourth replies which fail the above test are assumed to be from additional aircraft. The second, third and fourth replies which pass the above test are deleted and the first reply is marked as a military emergency.

**Wide pulse replies**

Two mechanisms exist which may cause a reply from a single aircraft to be seen as more than one reply.

(1) The transponder may be out of calibration and transmitting wide pulses in the reply, although not all pulses will necessarily be of the same width.

(2) Multipath can produce delayed echoes of each pulse which may be seen as one or more additional replies.

When wide pulse or multipath replies are detected, all but the first reply can be eliminated as spurious. In each case a later reply is compared with an earlier reply for evidence of a spurious situation. For the later reply to be eliminated all the following tests must be passed.

(a) The range difference between the earlier and later reply is less than a parameter which can be varied to suit local multipath conditions.

(b) The amplitude (signal strength) difference between the first reply and the later reply is more than the values given in Fig. 7.12.

(c) A high confidence 1 code pulse in the later reply does not correspond to a high confidence 0 in the earlier reply since this would rule out the later reply's being a delayed version of the earlier reply.

## 7.3.2 Reply-to-reply correlation

After the extraneous replies have been filtered out according to the above procedure, the range, azimuth and high confidence code pulses (for replies to the same interrogation mode) of the remaining replies are compared so as to locate groups of replies which have originated from the same aircraft. The principles of the process are as follows.

(1) All replies received from an aircraft are used in the formation of the target report. Specifically, replies after a miss or those first detected in the antenna beam are included, unlike the process used for defruiting in the earlier nonmonopulse plot extractor.

(2) A new reply, which lies at a similar range and azimuth to an earlier reply or reply group and which agrees in its high confidence code decisions

with any earlier replies to that mode, is combined with the group. By this means low confidence code decisions present in earlier replies can be replaced by high confidence decisions in later replies.

(3) Rule (2) is relaxed for Mode C replies to permit a single high confidence code pulse disagreement. This caters for a possible incremental altitude change of 100 ft during the antenna beam dwell-time. The conflicting high confidence code pulse is converted to low confidence to permit the disagreement to be resolved by a subsequent reply.

(4) Replies which lie at a similar range and azimuth but disagree in their high confidence code decisions with a group are retained. Both the single reply and the correlating group are flagged as code-swap candidates. This caters for the case where the wrong initial reply was associated with the group. Marking the groups as code-swap candidates permits association errors to be resolved later.

(5) Groups of two or more replies are passed forward with the single-reply code-swap candidates for further processing. A minimum number of two correlating replies are needed in order to detect the presence of an aircraft.

(6) Remaining single replies which do not correlate with any other replies are discarded as fruit. In this way minimum processing time is wasted on fruit replies.

Groups of replies, determined as described above, which result from the same interrogation mode are combined into a single reply code. By definition the replies will not differ with respect to high confidence code pulses, but a low confidence pulse in one reply can be resolved if another reply has a high confidence decision in that pulse position. In this way a complete high confidence code reply can be assembled even though no single reply was received with all high confidence pulses. The code update rules are shown in Table 7.4.

The discussion so far has concerned only Mode A and Mode C replies. Mode A, which is similar to military Mode 3, is the principal identity mode that is used in the plot extractor as a reference to the target. The plot extractor may be required to operate with Mode 1 and/or Mode 2 replies as well. A high confidence code pulse disagreement on one of the latter modes is not allowed to start a new group; instead it has the effect of lowering the confidence level for the disputed code pulse in a similar way as for Mode C, thereby permitting further replies to resolve the conflict within the one group.

Each reply that is combined carries a measurement of target range, target azimuth and, where employed, reply signal strength. With the exception of reply signal strength, it would be possible to average all these quantities obtained from the replies, but it is found that good results are obtained if just the two replies closest to the beam center are used for this purpose. These same two replies can also be used to determine the reply signal strength since they will be closest to the peak of the beam.

Section 7.2.14 refers to the case where a reply is obtained without a valid monopulse value and in this case the antenna boresight direction is used instead. If all replies from an aircraft lack valid monopulse values the azimuth used in the target report is found from the average of all boresight values from all of the replies forming the group. In this way, without valid monopulse measurements, the azimuth determination defaults to a form of sliding window or beam-split value. Of course if one reply in the group possesses a valid monopulse measurement, that value is used.

**Table 7.4** Code update rules

| *Old group code* | *New group code* | | | |
|---|---|---|---|---|
| | *Reply code* | | | |
| | *H0* | *L0* | *L1* | *H1* |
| H0 | H0 | H0 | H0 | X |
| L0 | H0 | L0 | L0 | H1 |
| L1 | H0 | L0 | L1 | H1 |
| H1 | X | H1 | H1 | H1 |

X cannot, by definition, occur with a Mode A reply. X is taken as L1 for a Mode C reply. See (2) and (3) in the text.

### 7.3.3 Raw target report declaration

When the antenna has passed on by about one beam width beyond the target azimuth, the raw target report can be transferred to the surveillance processor. The antenna beam width corresponds to the angle over which the transponder replies are received and is about 5° with commonly used antennas with a 3 dB beam width of 2.5°.

In cases of ring-around, or severe antenna beam distortion due to multipath, replies may still be being received from the aircraft at the time that the raw target report is produced. In this case the continuing replies will generate further target reports from the one aircraft. These false reports are eliminated by later processing. Typically the raw target report will contain the data fields given in Table 7.5.

The raw target reports produced by the reply-to-reply correlation process can contain a number of errors. For example, the code data may be incomplete or may be erroneous owing to a decoding error such as a code swap, or the target report itself may be false and be due to a reflection or multipath effect. These problems can be resolved by the use of past knowledge of the target through the establishment of track files for each aircraft.

**Table 7.5** Raw target report data fields

| | |
|---|---|
| 1 | Range (15 bits) |
| 2 | Azimuth (14 bits) |
| 3 | Mode A code (12 bits) |
| 4 | Mode A confidence (12 bits) |
| 5 | Mode C code (11 bits) |
| 6 | Mode C confidence (11 bits) |
| 7 | Third Mode code (12 bits) |
| 8 | Third Mode confidence (12 bits) |
| 9 | Third Mode type (2 bits) |
| 10 | Number of correlating Mode A replies (4 bits) |
| 11 | Number of correlating Mode C replies (4 bits) |
| 12 | Number of correlating third Mode replies (4 bits) |
| 13 | SPI flag (1 bit) |
| 14 | SPI flag confidence (1 bit) |
| 15 | Military identification flag (1 bit) |
| 16 | Military emergency flag (1 bit) |
| 17 | Number of replies flagged as a potential reflector (5 bits) |
| 18 | Signal amplitude (9 bits) |
| 19 | Code-swap candidate indicator (1 bit) |
| 20 | Antenna revolution number (16 bits) |
| 21 | Boresight flag (1 bit) |

The third Mode refers to Modes 1, 2, B or D when a three-mode interrogation program is used.

# 7.4 Surveillance processing

## 7.4.1 Summary

Reply-to-reply correlation processing, as described in Section 7.3, converts individual replies into raw target reports. The majority of these target reports will have been obtained in interference-free conditions and contain high quality data on the current status of aircraft. However, a proportion of the target reports will possess one or more of a variety of errors. In most cases reference to a track history, maintained for each aircraft, will not only identify that an error exists but also enable it to be corrected. This is the function of surveillance processing. False target elimination, described in Section 7.5, is a related activity although it is treated separately. However, some reference to this function is made in this section where it assists understanding. An SSR target report results from the merging of two or more replies which are related in range and azimuth. The code content of these replies may be incomplete or inconsistent.

The first step in surveillance processing is to match a target report with the track established for that aircraft. This process is straightforward if the target and track Mode A codes are fully known, match and are unique in that area and if the target and track position and altitudes agree. In this case the association of target with track is on the basis of the Mode A code and is known as discrete code correlation.

If the Mode A code of either the target report or the track is incomplete because some bits are at low confidence or more than one track of that Mode A code exists in that area, the association of target with track is less obvious. This is an example of general association. The term "association" is used to mean the identification of the report(s) and track(s) which could relate to one another. The term "correlation" is the confirmation that a particular report does relate to a particular track.

### 7.4.2 Discrete code correlation

The prerequisites for discrete code correlation are as follows.

(a) Only one "real" (not false) track with that Mode A code exists.

(b) The Mode A codes of both the target report and the track are complete, i.e. they both contain only high confidence code bits.

(c) The Mode A codes of the target and track agree.

(d) The target position approximates to the predicted track position.

(e) The target and track agree in altitude within close limits.

If the range of the target is beyond the distance where ring-around can occur, a credibility check is applied to ensure that the target report position lies close in range and bearing to the position predicted in the track file. Failure of this test could indicate that the target report is the first report of a new aircraft which, by chance, has the same code. The target report is then subjected to general association rules (Section 7.4.3) where such conditions can be resolved. Ring-around reports are a form of false target and are discussed in Section 7.5.

If the range of the target is within the distance where ring-around can occur, only the range-credibility test described above is applied. If ring-around has occurred, several target reports, all at similar range and all originating from the one aircraft, will be present. The report with the strongest signal strength will correspond to the replies received in the antenna main beam and should be selected. Weaker target reports will correspond to replies received in the antenna sidelobes and should be discarded.

If discrete code correlation is not possible for one of the above reasons or because no track is found, the target report is passed for general association and correlation.

### 7.4.3 General association

It is not always possible to correlate a target report with a track through the use of the Mode A codes. This failure may be because the Mode A codes are not fully known, because there is more than one aircraft with that Mode A code or because there is an unacceptable difference in the two altitudes. Agreement in Mode A code is desirable but the possibility that it has been changed in flight must be allowed.

The matching of target and track is performed using more extensive tests of position, height and Mode A code.

**Association zones**

Three association zones are defined around each track predicted position with dimensions of range and angle [7.2] as shown in Fig. 7.15. Zone 1 corresponds to the expected future position of an aircraft flying a straight course with an allowance for small errors in the measurement of range and azimuth. The zone size should be increased if there is a failure to detect the aircraft on the previous antenna scans. Zone 2 allows for some aircraft acceleration, including the performance of a circular turn. Zone 3 is a catch-all to cater for abnormal maneuvers by military aircraft or prediction errors caused by a previous plot error. Detailed equations for the three zones are given in ref. 7.2.

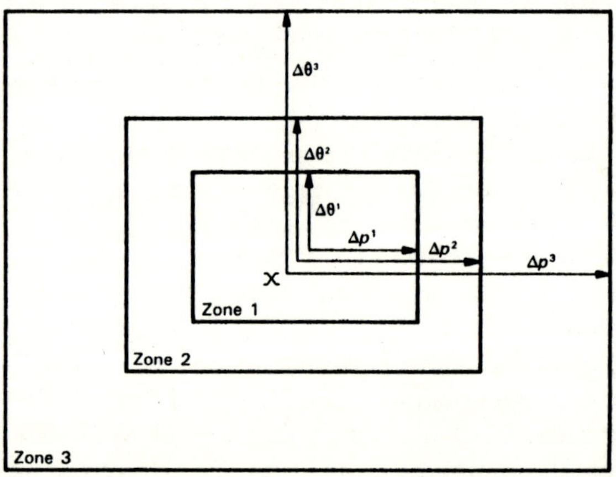


**Fig. 7.15** Association zones (from ref. 7.2). (Reprinted with permission of Lincoln Laboratory, Massachusetts Institute of Technology, Lexington, MA.)

At close range, particularly in the case of zone 3, the angle dimension of the association zone can become very large and impracticable and it is normal for it to be limited to about 90°. The first step in associating a report with a track is for the report to fall in zones 1, 2 or 3 of the predicted position of the track. No report associates with a track if it falls outside zone 3.

The next test for association is to compare the Mode A and Mode C codes of the report and track.

**Association categories**
Six categories of association can be defined as follows.

(1) Perfect association—all attributes (position, Mode A and C codes) match fully.

(2) Acceptable association—the Mode A or C attribute, or both, is suspect but positional agreement is excellent.

(3) Potential association—the combination of suspect Mode A or C, or both, with suspect position requires the application of the velocity reasonableness test (see below).

(4) Potential Mode A code-swap association—the Mode C code but not the Mode A code matches the track. The code-swap candidate flag is set so that code swapping with another report could improve the situation.

(5) Potential Mode C code-swap association—the Mode A code but not the Mode C code matches the track. The code-swap candidate flag is set so that code swapping with another report could improve the situation.

(6) No association—none of the above applies and the association is rejected.

**Velocity reasonableness test**
The purpose of the velocity reasonableness test is to determine the likelihood that a current target report is part of the sequence of reports from the same aircraft as that represented by the track.

Two velocity vectors are calculated. One vector $\boldsymbol{w}$ describes the direction and speed necessary to travel from the last reported position to the predicted position. The second vector $\boldsymbol{v}$ describes the direction and speed necessary to travel from the last reported position to the target report position. If both the angle between the two vectors $\boldsymbol{w}$ and $\boldsymbol{v}$ and the difference in their magnitudes are within reasonable limits, potential association is accepted.

**Code swapping**
In the identification of associating targets and tracks, consideration should be given to the possibility that two nearby reports may have their Mode A and Mode C codes transposed. This code-swap situation could have happened for one of two reasons.

The most frequent effect occurs when a fruit reply has been inadvertently included in the target report. This can happen when an earlier fruit reply, detected from a Mode A interrogation say, lies sufficiently close in range and azimuth to a first Mode C reply to become associated with it. The next Mode A interrogation elicits a Mode A reply from the aircraft but cannot be associated with the report because it conflicts with the code of the earlier (fruit) Mode A reply. The result is that a second report is formed which has only Mode A replies whilst the first report contains only Mode C replies plus one (wrong) Mode A reply (Fig. 7.16).

```
Mode A fruit reply

Valid Mode C reply - forms group with earlier Mode A reply

Valid Mode A reply - conflicts with earlier Mode A reply

Valid Mode C reply - associates with group

Valid Mode A reply - associates with earlier valid Mode A
                         reply to form second group
Valid Mode C reply - associates with first group

Valid Mode A reply - associates with second group

ETC.

RESULT:

First group forms a target report  - 1 Mode A reply (fruit)
                                     several Mode C replies

Second group forms a target report - several Mode A replies
                                     0 Mode C replies
```

**Fig. 7.16** Example of a code swap.

Code swaps can also be caused when two aircraft are very close in position so that all replies fall in the same reply correlation limits. Slight differences in Mode A and Mode C transponder reply delays can cause the Mode A code from one aircraft to become associated with the Mode C code from the other aircraft and vice versa.

Better association of the two target reports with a track can be achieved if the Mode A codes are swapped between the two reports. If the error was the result of a chance single fruit reply then, following the exchange of Mode A codes, one report will have several Mode A replies and several Mode C replies and the other report will have the single fruit reply. The latter report is discarded and the revised association is accepted.

## 7.4.4 General correlation

Once all associations for each track have been determined it is possible to decide which target report, if any, should be used to update each track. The following four association situations might exist.

1 on 1: only one target report associated with only one track
*m* on 1: *m* target reports associated with only one track
1 on *n*: only one target report associated with *n* tracks
*m* on *n*: *m* target reports associated with *n* tracks

The 1-on-1 case is the most common and no further tests need be applied. For the other cases it is necessary to determine the best association of target with track where a multiple choice exists. The decision of which to choose is determined by calculating a quality score for each association. The quality score is determined by giving a value to the degree of match achieved in the following tests:

(a) Mode A code agreement
(b) association zone
(c) number of replies in the report
(d) altitude agreement
(e) track maturity

Table 7.6 shows a typical quality score content which comprises a five-digit octal number with the contributions as indicated.

When the quality scores have been calculated for all the competitive report-to-track associations, those with the lowest value are chosen. When multiple reports have the same low quality score for the same track or multiple tracks have the same low quality score for the same report, the candidate chosen is the one with the greatest signal amplitude.

Should a target report fail to be associated with an existing track, it is considered for the initiation of a new track.

### 7.4.5 Track initiation

Raw target reports which remain uncorrelated after target-to-track correlation are considered for the initiation of new tracks. Two types of report are eliminated from this process. These are residues from code swapping, which are usually single reply reports originating from fruit, and reports with fewer than two replies on either of Mode A or Mode C. A large proportion of the latter will be the chance association of two fruit replies, one in Mode A and one in Mode C.

A new track is initiated only when two reports from the same aircraft are detected on consecutive scans with reasonable Mode A and Mode C code match and with a credible distance separation. When two such reports exist the data of the earlier report are considered as the track data, while the data of the subsequent report are the data of the target to be correlated with the "track".

### 7.4.6 Target report update

Track data are used to improve a variety of imperfections that may exist in the target reports produced by the reply-to-reply correlation process (Section 7.3). Normally, a mature track will have full Mode A code and Mode C code data available, i.e. with no low confidence code pulses.

**Table 7.6** Values for the five-digit score (digit 5 is the most significant)

| *Digit* | *Item* | *Value* |
|---|---|---|
| 5 | *Code agreement and zone* | |
| | Good code agreement | |
| | Zone 1 | 0 |
| | Zone 2 | 1 |
| | Zone 3 | 4 |
| | Poor code agreement | |
| | Zone 1 | 2 |
| | Zone 2 | 3 |
| 4 | *Number of replies in report* | |
| | Three or more | 0 |
| | Two of the same mode | 0 |
| | Two of different mode | 1 |
| | Single reply | 2 |
| 3 | *Code agreement* | |
| | Good code agreement | |
| | All report code pulses high confidence | 0 |
| | One but not all report code pulses high confidence | 1 |
| | Maximum permitted code difference | 3 |
| | Poor code agreement | |
| | Report code unknown | 2 |
| | All report code pulses low confidence | 2 |
| | All report code pulses not high confidence but code-in-transition flag set | 4 |
| | All report code pulses high confidence but code-in-transition flag set | 5 |
| | All report code pulses not high confidence but code-in-transition flag not set | 6 |
| | All report code pulses high confidence but code-in-transition flag not set | 7 |
| 2 | *Altitude agreement* | |
| | Agreement better than 600 ft | 0 |
| | Agreement better than 700 ft | 1 |
| | Agreement better than 800 ft | 2 |
| | Agreement better than 900 ft | 3 |
| | Agreement better than 1000 ft | 4 |
| | Agreement better than 1100 ft | 5 |
| | Otherwise | 6 |
| 1 | *Track maturity* | |
| | Established track | 0 |
| | New track | 2 |

### Mode A code data

The target report Mode A code and confidence data are compared with the track Mode A code and confidence data. If there is no conflict between the two sets of code pulses declared with high confidence, any low confidence pulses in the target report can be resolved using the track data.

Should there be a high confidence pulse code difference between target and track, either a decoding error has occurred or the code has been deliberately changed by pilot action. The track file contains the previous target report code and a count of the number of scans over which that code was stable. While a high confidence difference between track and target data exists, the target data cannot be confirmed and the pulse(s) in disagreement in the target report are reduced to a low confidence 1.

When the new target report code has remained stable for a minimum number of antenna scans, typically three, the new code is accepted by the track and the target data are confirmed. A pilot, responding to a ground ATC request to change code, may take several scans to complete the change during which time the code goes through a number of transitions of false code.

**Mode C code data**

The processing of the target report Mode C data is similar to that of the Mode A data except that Mode C changes are more expected and do not necessarily indicate a decoding error. High confidence code pulse differences are accepted if the consequent difference in height does not exceed 1000 ft. If the latter condition is met, the target report data will not be changed and will also replace the track data.

**Target position report**

Although a smoothed position can be derived from the tracking process, it is important that the target report contains the range and bearing of the aircraft as measured on that scan. Monopulse bearing measurement is sufficiently accurate for track jitter not to require smoothing. Smoothing can mask the start of a turn and mislead the air traffic controller as to the true position and intention of the aircraft.

### 7.4.7 Typical performance results

The use of track data permits target reports to be reliably produced with the minimum of replies, thereby achieving a high target detection probability. Table 7.7 gives the measured target detection efficiency in southeast England, the busiest part of the UK, for a typical en route radar with a range of 180 n miles. Both the overall performance and the performance in the more exacting crossing-target situation when the aircraft replies are overlapping are given.

**Table 7.7** Measured target detection efficiency

| | |
|---|---|
| Detection efficiency (overall) | 98.7% |
| Detection efficiency (crossing track) | 98.0% |

Once the presence of the aircraft is detected, the Mode A and Mode C codes are required. Table 7.8 shows the code validity performance at the same site as for Table 7.7. The data show the results obtained before and after code improvement by surveillance processing. The Mode C performance is slightly poorer than that of Mode A since allowance must be made for legitimate aircraft height changes and this inhibits the resolution of some low confidence situations.

Table 7.9 presents similar data to Table 7.8 for a different site where severe multipath phenomena are experienced. The code data are further subdivided to show the performance difference obtained by the use of monopulse angle-of-arrival data and of signal amplitude data in reply decoding, as described in Section 7.2.12.

**Table 7.8** Code validity performance

| | *Before surveillance processing* | *After surveillance processing* |
|---|---|---|
| Overall performance | | |
| Mode A code validity (%) | 98.5 | 99.7 |
| Mode C code validity (%) | 97.6 | 98.7 |
| Crossing-track performance | | |
| Mode A code validity (%) | 95.3 | 99.5 |
| Mode C code validity (%) | 92.0 | 97.0 |

**Table 7.9** Code validity performance

| | *Before surveillance processing* | | *After surveillance processing* | |
|---|---|---|---|---|
| | *Mode A* | *Mode C* | *Mode A* | *Mode C* |
| Overall performance | | | | |
| Angle only (%) | 91.8 | 93.1 | 99.3 | 94.6 |
| Amplitude only (%) | 97.0 | 95.6 | 99.7 | 97.9 |
| Combined (%) | 97.1 | 95.8 | 99.7 | 98.0 |
| Crossing-track performance | | | | |
| Angle only (%) | 84.1 | 84.6 | 98.2 | 89.2 |
| Amplitude only (%) | 89.4 | 87.1 | 99.4 | 93.3 |
| Combined (%) | 89.6 | 87.3 | 99.7 | 93.7 |

### 7.4.8 Track update

Following the correlation of a target with a track, the track data are updated so as to be available for repeating the process on the subsequent scan. If no target report correlates with the track, the track must still be updated but in this case using the predicted data obtained from the previous scan. Track updating consists of updating the code data and predicting the future target position. The target report parameters used to update the track data are as follows:

(a) measured range
(b) measured azimuth
(c) Mode A code
(d) Mode A code confidence
(e) Mode C code
(f) Mode C code confidence

The track parameters updated are as follows:

(a) predicted range
(b) predicted azimuth
(c) predicted range rate
(d) predicted azimuth rate
(e) Mode A code
(f) Mode A code confidence
(g) unique/nonunique Mode A code status
(h) Mode C code
(i) Mode C code confidence
(j) turn state
(k) track firmness
(l) history firmness
(m) pointer to false/real companion track
(n) track false/real status
(o) count of Mode A code transition scans
(p) transition Mode A code
(q) transition Mode A code confidence
(r) count of Mode C code coasts

**Mode A code update**

In the great majority of cases the Mode A code of the track will be the same as the Mode A code of the latest correlating target report since Mode A codes rarely change.

It is possible that a recently established track has not yet had a correlating report with a full high confidence Mode A code. In this case the new target report code can be used to improve the track code using a similar process to that employed in reply-to-reply correlation, i.e. by allowing high confidence pulse decisions in the target report to replace low confidence pulse decisions in the track file. This improvement process can only be used if there is no disagreement between code pulses declared high confidence in both the report and the track. If such a high confidence disagreement occurs, that is taken as possible evidence of an intentional code change by the aircraft pilot.

In order to safeguard against a possible high confidence decoding error appearing as an intentional change the new code must be stable for a minimum number of scans, usually set at three, before the new code is accepted and replaces the earlier code. The pilot will often have the four code switches ABCD to change and this may take several scans to complete, depending upon other demands on his attention. It is quite possible, therefore, for the Mode A code to go through some transitional code values before the final code is established.

**Mode C code update**

The Mode C code update process is very similar to that for Mode A. However, since the aircraft flight level can be expected to change during flight some relaxation of the rules must be allowed.

If the Mode C code is fully known, i.e. pulses are all high confidence, it is more convenient to hold the code in its flight level form. A disagreement between the flight level of the recent target report and that of the correlating track by, typically, a maximum of 1000 ft is taken as acceptable, and the flight level of the report is used to replace that of the track.

If the Mode C data of the target report are unacceptable, either because they differ from the track flight level by more than the permitted distance when they are translated into flight levels or because the code contains some low confidence pulses so that it is not fully known, they cannot be used to update the track. A counter is started which indicates the number of successive scans during which the Mode C data of the report are not valid. While the counter is active the last valid flight level is retained in the track file as the aircraft altitude. When a preset count is reached, typically three, the flight level in the track file is deleted and the flight level is "not known".

**Position update**

The positions of the targets are described in $(R, \theta)$ form, and target-to-track correlation is carried out using the same coordinate system. It is obviously convenient, and computer efficient, if tracking and prediction are also performed in $(R, \theta)$ terms. For target ranges over about 10 n miles the $(R, \theta)$ system is sufficiently linear for it to be used. However, the accuracy necessary at closer ranges requires conversion to and from $(x, y)$ Cartesian coordinates with the actual tracking performed in $(x, y)$.

Target prediction sufficient to enable target-to-track correlation to be performed on the following scan is all that is required, and the accuracy of the monopulse direction-finding system is sufficient to permit the use of a simple two-point extrapolator as a prediction function is most instances. Simple two-point extrapolation should not be used when there is suspicion of a bad measurement or the start of a turn. If an aircraft has been flying a reasonably straight course and has exhibited several successive good correlations but then correlates with a report some distance from its predicted position, caution should be applied. For this purpose a poor correlation is one where the report falls outside the zone 1 prediction area.

Figure 7.17 [7.2] shows how a bad measurement could take the track astray if it were used without modification. The technique adopted is to smooth the target position to the nearest edge of the zone 1 area and to use that value in the prediction.

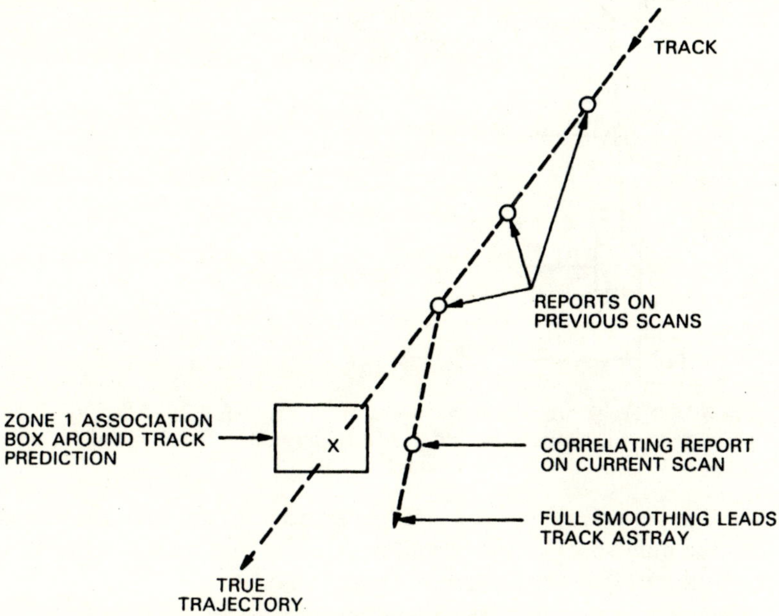


**Fig. 7.17** Track update with a suspect position measurement (from ref. 7.2). (Reprinted with permission of Lincoln Laboratory, Massachusetts Institute of Technology, Lexington, MA.)

Of course the so-called bad plot may be accurate and a result of the start of an aircraft turn. Figure 7.18 [7.2] shows an example of this and illustrates how the smoothing is varied in a logical manner through the turn.

### 7.4.9 Track coasting and dropping

If a track fails to correlate with a target report, it must be coasted for

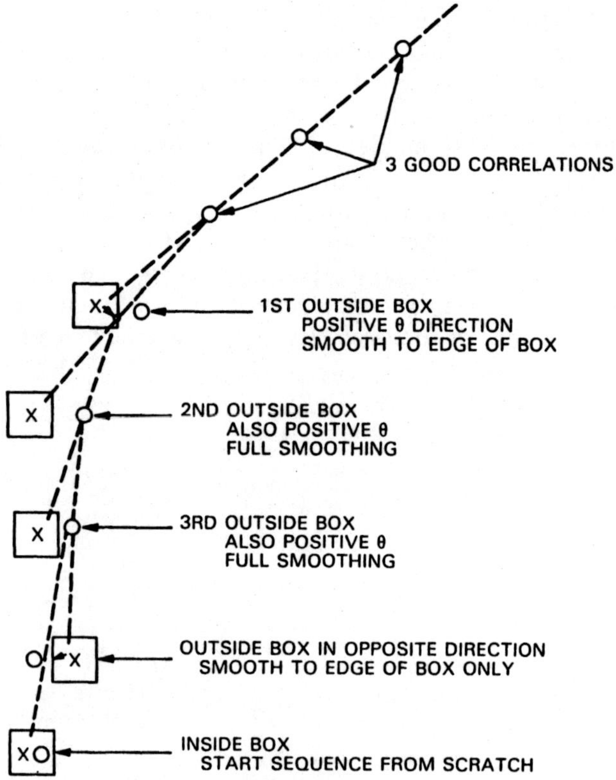


**Fig. 7.18** Track prediction through a turn (from ref. 7.2). (Reprinted with permission of Lincoln Laboratory, Massachusetts Institute of Technology, Lexington, MA.)

further scan periods so that correlation can occur at the next opportunity. A limit to the number of coasts permitted is set so that tracks can be dropped when an aircraft lands or flies out of range.

When the number of coasted scans exceeds some value, typically six, the track is dropped and all reference to the aircraft is deleted. An exception to this operation is when the target overflies the SSR antenna during which time the aircraft may be out of contact for 10 or more scans depending upon the size of the cone of silence exhibited by the antenna and the altitude of the aircraft. Since the aircraft will eventually re-emerge into the antenna beam, the track should be maintained in this case.

Care should be taken when coasting aircraft at airports. A landing aircraft will switch off its transponder, and if a second aircraft takes off shortly after it may correlate with the "predicted" position of the earlier aircraft. A different Mode A code will initiate the "code-in-transition" action. If this false correlation of a new aircraft with an earlier track is permitted, other attributes of the earlier aircraft may also be associated with the new aircraft.

## 7.5 False target processing

Not every target report produced by the reply-to-reply correlation process will correspond to a real aircraft position since weaknesses in the SSR system can produce various types of false target reports. Surveillance processing generates tracks on all aircraft, and these track data permit false and real aircraft to be identified.

The false reports can be divided into four types [7.3]:

(1) false targets—produced by reflected replies;

(2) fruit targets—produced when fruit replies coincidentally correlate;

(3) split targets—produced by the failure of reply-to-reply correlation to group together all replies emanating from an aircraft;

(4) Ring-around targets—produced by sidelobe replies which were not suppressed.

False targets are generally caused by the reflection of aircraft responses from buildings, fences or other structures near the sensor. Depending on the size of the reflector, such false targets may persist for several antenna revolutions and initiate false tracks. This phenomenon causes an apparent aircraft position behind the reflector and is discussed further in Chapter 9.

Since the reflection mechanism is deterministic, it is possible to compute the position of the aircraft whose signal is responsible for the target provided that the parameters of the reflecting surface are known.

The majority of false target reports formed by fruit replies will consist of one Mode A reply and one Mode C reply. This is because replies in different modes need not agree with respect to code, while two replies in the same mode require code agreement. If such a two-hit report fails to correlate with a track, it is suspected of being a fruit report.

Split reports occur when the reply sequence from an aircraft is separated by the reply-to-reply correlation process into two or more target reports. The split could result from code or azimuth errors in the reply processor, from intermode delay variations in the transponder (which may result in range splits) or from various terrain effects. Many of the more common types of splits have easily recognizable characteristics which permit them to be identified.

Ring-around target reports occur when sidelobe interrogations are received successfully by an aircraft and its replies are not rejected as sidelobes. Ring-around is likely to occur when an aircraft with a faulty transponder is flying close to the radar or as a result of antenna imperfections at high elevation angles. The process for identifying ring-around targets is very similar to that for identifying reflection false targets. In the case of ring-around targets, the "reflector" is at the sensor itself and all azimuths are searched. Any high elevation angle target report not correlating with a real track is subjected to the ring-around test.

Of the four types of false reports listed at the beginning of this section, those due to fruit and splits can be eliminated either as part of reply-to-reply correlation or as a consequence of target-to-track correlation. False targets due to reflections or ring-around require further processing.

To determine whether a target report is false it is first necessary to identify the track of the real aircraft that may have given rise to the false plot. If the Mode A code contained in the target report is complete, i.e. the code contains only high confidence bits, inspection of the track files will reveal all tracks with the same code. Each of these tracks can be tested in turn. Should the Mode A code of the target report be incomplete, i.e. some of its code bits are low confidence, the real track cannot be found by reference to the Mode A code. In this case knowledge of reflectors at the site is required so that the reflection path can be calculated and the location of the real aircraft computed.

### 7.5.1 Initial conditions

When a track or tracks of the possibly real aircraft that could have given rise to the target report by reflection have been found, the following conditions must apply.

(a) The Mode A codes of the target and the track must agree with respect to at least the high confidence bits.

(b) The Mode C codes of the target and track must not disagree by more than a specified height difference (typically 500 ft).

(c) The range of the target is not less than the predicted range of the track (within a small tolerance limit to allow for prediction errors).

If all the conditions apply, the possibility that the target is false and results from a reflection of the aircraft of the track can be considered. No single test is sufficiently reliable to indicate whether a target report is false or real since circumstances can arise to cause the test to give a misleading indication.

### 7.5.2 False target tests

Six different tests can be applied, and each is given a weight depending on the power of the test to indicate a false target. The tests are as follows, with the weights lying in the range 0 to 100:

| | |
|---|---|
| range test | weight $W_r = 70$ |
| signal strength test | weight $W_s = 70$ |

| | |
|---|---|
| track history test | weight $W_h = 50$ |
| potential reflection test | weight $W_f = 30$ |
| reflection zone test | weight $W_b = 50$ |
| potential ring-around test | weight $W_o = 70$ |

An overall weight $W_{tot}$ can be calculated by accumulating the weights resulting from the following tests.

**Range test**

Since the reflected path must be longer than the direct path, the false target can be expected to be at a longer range. However, since there may be an appreciable angular distance between the real aircraft and its reflection, some time will have elapsed between the antenna's pointing at the aircraft and the antenna's pointing at the reflection and during this time the aircraft will have moved. The problem is resolved by calculating the predicted track position for the same time that the antenna saw the target. Unfortunately the calculation is subject to small prediction and rounding errors which could make the false target appear closer than the real one. If the range difference is small, it is not possible to distinguish the real aircraft from the false.

If the target range exceeds the track predicted range by more than this computation tolerance, that is taken as a possible false target indication and the weight $W_r$ is subtracted from the overall weight $W_{tot}$ being calculated. However, if the track predicted range exceeds the target range by more than the computation tolerance, that is taken as a possible real target indication and the weight $W_r$ is added to the overall weight $W_{tot}$ being calculated.

**Signal strength test**

Normally a reflected signal would be expected to suffer some reflection loss such that the real target would appear stronger than a false target and this indeed is often the case. However, some reflectors can exhibit quite low loss, and occasionally targets are detected only at the beam edge and thus appear weaker than they really are. A minimum signal strength difference is required, typically 5 dB, in order to allow for small measurement variations.

If the target signal strength exceeds the track signal strength by more than the measurement margin, this is taken as a possible false target indication and the weight $W_s$ is subtracted from the overall weight $W_{tot}$ being calculated. However, if the track signal strength exceeds the target signal strength by more than the measurement margin, that is taken as a possible real target indication and the weight $W_s$ is added to the overall weight $W_{tot}$ being calculated. If signal strength is not available, an alternative to this test is the use of run-length, or the number of replies received from the aircraft on the current antenna scan, since the weaker target will tend to exhibit fewer replies.

**Track life test**

If a track has been established for an appreciable time and a second version of the target appears, it is reasonable to suppose that the long-established track is the true one and that the new version is suspect. When the SSR is at an airport it is possible for the aircraft's SSR antenna to be hidden from the direct path by the fuselage or wings. If a signal path exists via a reflecting surface, the reflection may be detected first. The target may or may not have correlated with its own track. If its own track does not exist, then its track life is taken as unity.

If the target's track life is shorter than the comparison track life by more than a preset number, typically five scans, that is taken as a possible false target indication and the weight $W_h$ is subtracted from the overall weight $W_{tot}$ being calculated. However, if the target's track life exceeds the comparison track life by more than a preset number, typically five scans, that is taken as a possible real target indication and the weight $W_h$ is added to the overall weight $W_{tot}$ being calculated.

**Potential reflection test**

If a target report has been received via a reflecting object, it is possible that the direct path signal was also received at a different azimuth and detected in the antenna sidelobes. The reply processor described in Section 7.2 can be designed to detect main beam replies which are preceded by overlapping sidelobe replies, as could be the case if the main beam replies were delayed by coming from a reflecting object. It will be recalled that the reply processor discards sidelobe replies so that the above two reply situations are known only at the time of detection; however, the reply processor can mark the main beam reply at the time to indicate that a sidelobe reply preceded it.

Of course a chance fruit sidelobe reply could overlap the main beam reply in a similar way, but this is unlikely to be the case for every reply. To cater for this possibility of erroneous indication of a reflection, the proportion of replies in the beam that are so marked is used to scale the weight value.

The weight subtracted from the overall weight $W_{tot}$ being calculated is given by

$$\frac{W_f \times (\text{marked replies})}{\text{total replies in report}}$$

The potential reflection test is not given a very high weight since a sufficient range difference between the direct reply into the sidelobes and the reflected reply must exist for the two replies to be detected as such by the reply processor. This range difference would also score in the range test described above. However, if a target has just come into cover, say taking off at an airport, a track of the real aircraft may not have been established and the range test cannot be performed.

**Reflection zone test**

If a target report is received at a bearing at which a reflecting object is known to exist, that provides some circumstantial evidence that the target report may be false. The weight $W_b$ is subtracted from the overall weight $W_{tot}$ being calculated.

**Potential ring-around test**

Ring-around target reports occur when the sidelobe suppression action fails. This often occurs because the aircraft is very close to the interrogator site and at a high elevation angle (see also Section 7.4.2). If the target report is marked as a potential ring-around case and the report is weaker than the comparison track, the weight $W_o$ is subtracted from the overall weight $W_{tot}$ being calculated.

**Total weighting**

After the weights above have been calculated, a definite real target will have positive results for the range test and the signal strength test and possibly all tests will give positive results. Consequently, if the weight values given above are assumed, the final overall weight $W_{tot}$ will exceed a value of 139. Similarly, a definite false target will have a final overall weight $W_{tot}$ which is more negative than $-139$.

If a threshold parameter $W_{min}$ is set to 139, the status of targets can be defined as follows:

| | |
|---|---|
| Real | $W_{tot} > W_{min}$ |
| Possibly false | $-W_{min} \leqslant W_{tot} \leqslant W_{min}$ |
| False | $W_{tot} < -W_{min}$ |

## 7.5.3 Track status

A target can exist in one of the above three states but the status of a track can have four possible states:

| | |
|---|---|
| Real | state 0 |
| Possibly real | state 1 |
| Uncertain | state 2 |
| False | state 3 |

Once the target status has been determined according to the process described in Section 7.5.2, the status of the correlating track can be modified. If the target is designated Real, the track status is decremented by 1 (ending up at Real). If the target is designated Possibly false, the track status is unaltered. If the target is designated False, the track status is incremented by 1 (ending up at False). A newly initiated track is assumed to be Uncertain (state 2) and is immediately modified by the status of the correlating target. False target reports are not output to ATC displays.

### 7.5.4 Determination of reflecting surfaces

The majority of target reports correspond to aircraft replying with unique Mode A identity codes and in these cases the code can be used to find all tracks, real and false, with that code. If the target report code is not known with certainty, i.e. it contains some low confidence bits or is not unique, it is not possible to use the Mode A code to find all tracks with that code. Should the target report be due to a reflection, it is important to find the track of the real aircraft in order that the true nature of the target can be discovered by the application of the false target tests.

If the code cannot be used as a pointer to the track, the geometry of known reflectors can be used to predict where the real aircraft might lie. A site map can often identify obvious reflecting objects such as hangars and fences. It is not quite so easy to determine the radar-reflecting properties of a group of buildings or of round objects such as fuel tanks and antenna masts. Some reflecting objects, such as the fuselages and tailfins of large aircraft, are only temporary.

The existence of a target report with a unique Mode A code permits the identification of the real track of that target without prior knowledge of reflecting objects. Such targets and tracks allow the geometry of reflecting objects to be calculated. The following conditions must be met.

(a) The target has a unique Mode A code and has the status False.
(b) The target correlates with a track which also has the status False.
(c) The target is not a ring-around candidate.
(d) In the range test the range difference exceeds the minimum threshold.
(e) A related track exists with the status Real.
(f) The related track is not flagged as turning.
(g) The track range is sufficient for slant range correction to be unnecessary.

The range, azimuth and orientation of the reflecting surface $s$ are calculated as follows.

*Range:*

$$p(s) = \frac{p(i)^2 - p(j)^2}{2[p(i) - p(j)\cos\{\theta(j) - \theta(i)\}]}$$

*Azimuth:* start azimuth $\theta_S(s) = \theta(i) - d\theta$

finish azimuth $\theta_E(s) = \theta(i) + d\theta$

*Orientation:* if $|\theta(j) - \theta(i)| < 90°$ and $p(j) < p(s)$, then

$$\phi(s) = \theta(i) - \tfrac{1}{2}\arcsin\left[\frac{p(j)\sin\{\theta(j) - \theta(i)\}}{p(i) - p(s)}\right] - 90°$$

Otherwise, let

$$a = \theta(i) + 180°$$

and

$$b = \theta(j) + \arcsin\left[\frac{p(s)\sin\{\theta(i)-\theta(j)\}}{p(i)-p(s)}\right]$$

If $a-b<180°$,

$$\theta(s) = \frac{a+b}{2} + 90°$$

Otherwise

$$\theta(s) = \frac{a+b}{2} - 90°$$

In the above expressions $p(i)$, $p(j)$ and $p(s)$ are the ranges of the target, track and surface respectively, $\theta(i)$ and $\theta(j)$ are the azimuths of the target and track respectively, $\theta_S(s)$ and $\theta_E(s)$ are the start and finish azimuths respectively of the reflecting surface, $\phi$ is the orientation of the reflecting surface, and $a$ and $b$ are in the range $-180°$ to $+180°$. The acute angle result of arcsin[ ] is used where it occurs.

This determination of reflecting surfaces can be used to identify temporary reflectors, such as parked aircraft, or for the location of permanent reflectors at a new site. In the latter case it may be desired to store the data in the permanent reflector nonvolatile memory once they have been determined. Figure 7.19 [7.4] shows the permanent reflecting surfaces at Heathrow determined using this method.

## 7.6 Plot combination with primary radar target reports

When target reports extracted from primary radar plots are available at the site of the SSR, it is often desired to combine these reports where they relate to the same aircraft. This reduces the capacity required in the communication channels from the radar site to the ATC control center.

The simplest combining process that can be employed is to place a range–azimuth zone around the SSR target position and mark the SSR target report with a "primary reinforced bit" if a primary radar target appears in that zone. When two reports are combined in this way, the primary target report is deleted. A primary report that does not correlate with an SSR report is output unchanged. The correlation zone can be elliptically shaped to conform to the range and azimuth errors of the primary and secondary radars.

If the primary radar signal processor is also tracking targets and track data, such as track number, speed and heading, are available, these data can be used in the combining process to avoid erroneous combinations. In many

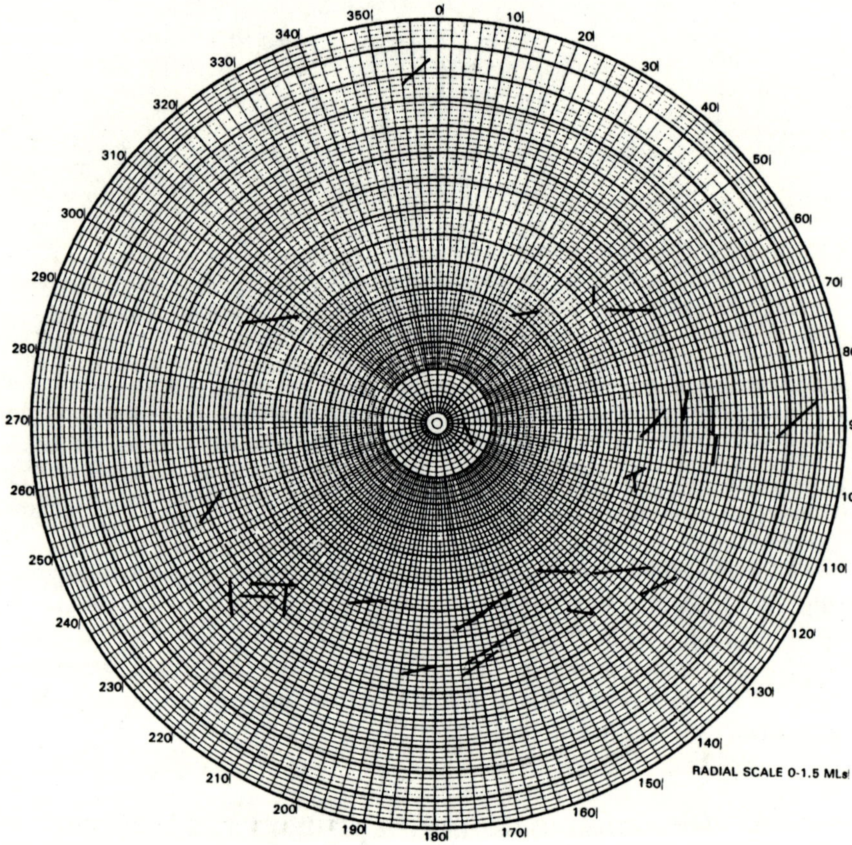


**Fig. 7.19** Radar-determined reflecting surfaces at Heathrow.

instances this additional processing adds little in the way of improved performance in return for the complexity involved. There is a likelihood that if the combination rules are too exacting then some reports will fail to combine, with the result that two targets are reported where only one exists.

In some implementations a primary radar target report is used to update an SSR track if the SSR report is missed on that scan. This process enables a continuous output of target report data to the ATC center during SSR fades. Care must be taken, however, to ensure that the correct primary radar data are used with the SSR track. Examples exist where the track of a landing aircraft, after its transponder is switched off, becomes associated with a primary radar return from a land vehicle on a nearby access road. The "aircraft" plot, complete with call-sign label, can then be observed departing the airport via the approach road.

Some equipment uses the absence of a primary radar report as an indication of a possible false SSR target. Since primary radar detects its own reflected transmissions, a false primary radar target would have to overcome the two-way path losses of the reflecting object in order to be detected. Consequently primary radar suffers from fewer problems of false targets due to reflections than does SSR. However, primary radar does suffer from problems due to clutter in the area where SSR experiences most of the reflections, so that there is a possibility that a primary clutter return could combine with a false SSR target report to cause that report to appear real. The false target processing described in Section 7.5 is very effective and there is little benefit to be gained by using primary radar as a further false target discriminant.

## References

7.1 Nelson, R. G. and Nuckols, J. H. A hardware implementation of the ATCRBS reply processor used in DABS, *Lincoln Laboratory ATC-78, Rep. FAA-RD-77-92*, Lincoln Laboratory, MIT, Lexington, MA, 19 September 1977.

7.2 Gertz, J. L. The ATCRBS mode of DABS, *Lincoln Laboratory ATC-65, Rep. FAA-RD-77-39*, Lincoln Laboratory, MIT, Lexington, MA, 31 January 1977.

7.3 Stevens, M. C. Multipath and interference effects in secondary surveillance radar systems, *Proc. Inst. Electr. Eng., Part F*, **128** (1), 43–53, 1981.

7.4 Spalding, D. J. Fixed reflector file generation, *Cossor Internal Memo. 871/DJS/8658/JB*, 21 March 1984.

# Chapter 8

# Transponders

The operation of the airborne element of the SSR system—the transponder installation—is discussed in this chapter. The information provided is not intended to guide the design of the airborne equipment but to allow the designer of the ground equipment to understand his task better. The basic parameters of the transponder are discussed and measurements are given which show how these parameters are found to vary in practice. Often a drift in a transponder parameter can appear as a reduction in the performance of the ground equipment. The ground equipment designer is well advised to accommodate such transponder performance variations and not rely too much on transponders' conforming to their specifications. In doing this the designer must, of course, take care not to degrade the performance of his equipment when processing replies from the majority of aircraft which are within specification.

## 8.1 Transponder specification documents

*Annex 10 to the Convention on International Civil Aviation* [8.1] is one of the most important documents regulating the international application of SSR. It defines the SSR signals in space and the performance criteria for the transponder. Some guidance is given with respect to the ground equipment, but this topic is generally left to national administrations to determine in accordance with their local needs. *Annex 10* attempts to ensure that the specification of airborne transponders is standardized so that the ground equipment can give a consistent performance with any aircraft that may fly into local airspace.

*STANAG 4193, Part 1* [8.2] describes the performance of military transponders. However, since these transponders are required to meet the relevant sections of *Annex 10*, *STANAG* contains little in the way of additional requirements as far as the civil system is concerned. Table 8.1 summarizes the transponder parameters which are specified in these publications.

*ARINC Characteristic 718* is produced by Aeronautical Radio Incorporated [8.3]. One of the purposes of the document is to specify the requirements of the airlines with respect to the mechanical dimensions and electrical interfaces of the transponder. RTCA and Eurocae [8.4, 8.5] produce minimum operational performance standards for the airborne equipment to meet national and ICAO requirements. In particular the documents include statements on how the transponders should be tested in order to demonstrate that the requirements are met. The ARINC and RTCA specifications include Mode S operation.

**Table 8.1** Summary of transponder parameters

| | *Minimum* | *Typical* | *Maximum* |
|---|---|---|---|
| Antenna gain (dB) | 0 | 0 | 3 |
| Sensitivity[a] (dBm) | −69 | −71 | −77 |
| Power output[a] (dBW) | 21 | 24 | 27 |
| Reply delay (μs) | 2.5 | 3.0 | 3.5 |
| Transmitter recovery (μs) | 25 | 50 | 125 |
| Sidelobe suppression (μs) | 25 | 35 | 45 |
| Squitter replies ($s^{-1}$) | 0 | 0 | 30 |
| DME suppression time (μs) | 7 | 45 | 60 |
| Reply rate threshold ($s^{-1}$) | 500 | 1200 | 2000 |

[a]Receiver sensitivity and transmitter power are measured at the antenna end of the r.f. cable. General aviation transponders can be operated with a reduced output power specification if they are to be used in aircraft limited to flying below 15 000 ft. For this class of transponder the output power is 18.5 dBW minimum.

## 8.1.1 Receiver sensitivity

Receiver sensitivity is defined as the interrogation signal strength at the antenna necessary to produce a reply at least 90% of the time. The nominal sensitivity requirement is −71 dBm with a minimum value of −69 dBm and a maximum value of −77 dBm. If the sensitivity is defined at the antenna terminal, the r.f. feeder loss to the transponder need not be known although *Annex 10* recommends the assumption of a feeder loss of 3 dB when designing the transponder itself. *Annex 10* also recommends the assumption of a simple quarter-wave antenna which has a gain of 1.5 dB although 0 dB is often used in system design. The transponder sensitivity level should not differ by more than 1 dB between different interrogation modes.

The transponder transmitter is protected from possible overload by reducing its sensitivity during high interrogation rates. This process maintains replies to the closer interrogators but denies them to the more distant. Typically sensitivity is reduced by 3 dB when the reply rate reaches 1100 replies per second and by 30 dB at 1800 replies per second.

Some protection against the generation of false responses in the presence of weak signal echoes is realized by desensitizing the receiver by 9 dB following a pulse. The receiver sensitivity then recovers, at about 3 dB $\mu s^{-1}$, until full sensitivity is restored.

### 8.1.2 Reply delay and jitter

The leading edge of interrogation pulse P3 is taken as the time reference and the leading edge of the first pulse F1 of the transponder reply should follow $3.0 \pm 0.5$ μs later. The jitter between successive replies to the same interrogation mode should not exceed $\pm 0.1$ μs. Delay variations between different interrogation modes should not exceed $\pm 0.2$ μs.

### 8.1.3 Reply suppression

The transponder should not reply if the amplitude of the P2 pulse in an interrogation exceeds the amplitude of the P1 pulse. If the amplitude of P2 is weaker than 9 dB below the detected P1, the transponder should reply. If P2 is between 0 and 9 dB below P1, the transponder may or may not reply. After sidelobe suppression the transponder remains insensitive to interrogations for $35 \pm 10$ μs. The *ARINC Characteristic* [8.3] and the *MOPS* [8.4] recognize the limits of practical measurement and interpret the "should reply" requirement as meaning at least 90% replies and the "should not reply" requirement as meaning no more than 1% replies, with both measurements applying to signals at least 3 dB above minimum trigger level. The transponder is also permitted to fail to reply to an interrogation shortly after it has replied to an earlier interrogation. This suppression period is typically 50 μs but can be a maximum of 125 μs.

If the aircraft is also fitted with distance-measuring equipment (DME), an external suppression signal can be applied from that equipment to prevent mutual interference between the two systems. The duration of the DME suppression pulse can vary between 7 and 60 μs with an average value of 45 μs. The rate of DME suppression can vary between 30 and 150 per second. In the case of military aircraft, external suppression signals can also be generated by onboard u.h.f. equipment, but the impact of this feature is usually small and is ignored.

### 8.1.4 Reply transmitter power

The transmitter power available at the antenna end of the r.f. feeder cable should be $24 \pm 3$ dBW. As with the specification of receiver sensitivity, an antenna gain equivalent to a quarter-wave stub is recommended although an antenna of 0 dB gain is often assumed in system design. Aircraft which never exceed an altitude of 15 000 ft are permitted a minimum transmitter power of 18.5 dBW. The rationale for this relaxation is that the radar range for low-flying aircraft is limited and so there is no requirement for a higher transponder transmitter power.

### 8.1.5 Random trigger rate

It is difficult to design transponders that will not occasionally transmit a reply in response to noise spikes or interference received from other onboard equipment. A limit of 30 random replies per second (squitters) is allowed for this.

Occasionally a transponder can develop a fault which has the effect of causing the transmitter to produce a high rate of random replies which can saturate the ground SSR equipment while the delinquent aircraft is in the antenna beam. Equipment designers are advised to ensure that the ground equipment can recover from such overload conditions.

### 8.1.6 Replies to Mode C interrogations

Not all small aircraft are equipped with altitude encoders suitable for supplying Mode C reply data. In these cases it is recommended that the transponders reply to a Mode C interrogation with framing pulses only.

### 8.1.7 Dual-antenna installations

Some modern aircraft, particularly those equipped for Mode S, are fitted with dual antennas, one on the top of the fuselage in addition to the one conventionally underneath the fuselage. In these installations the antenna receiving the stronger signal is used to transmit the reply. This process results in improved detection, particularly when the aircraft is banking in a turn and in climb-out.

Certain military aircraft have a similar dual-antenna installation, but with the difference that the transponder is switched between the upper and

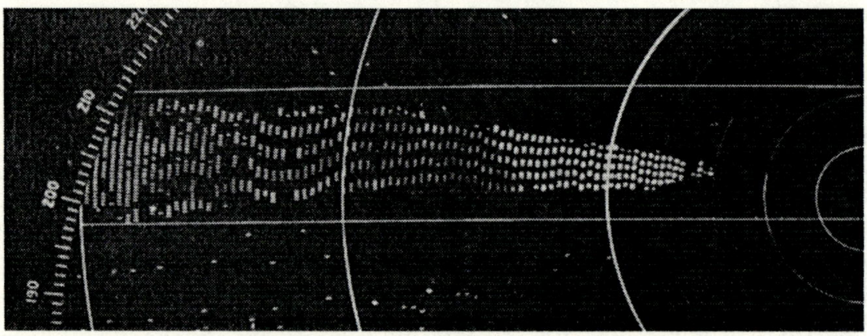


**Fig. 8.1** Elapsed-time photograph of replies from an aircraft equipped with switching antennas.

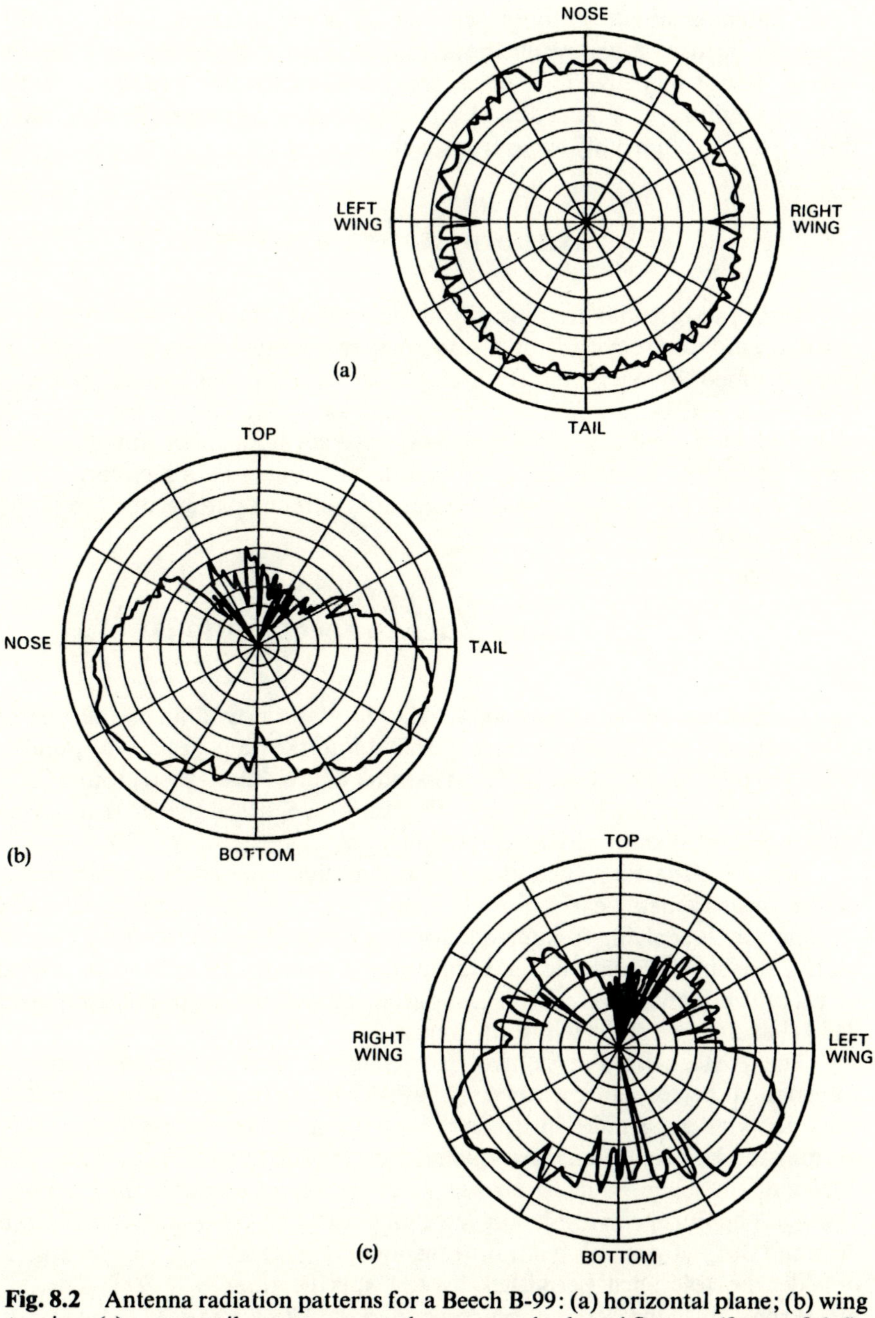


**Fig. 8.2** Antenna radiation patterns for a Beech B-99: (a) horizontal plane; (b) wing to wing; (c) nose to tail, center-mounted antenna, wheels and flaps up (from ref. 8.6). (Reprinted with permission of Lincoln Laboratory, Massachusetts Institute of Technology, Lexington, MA.)

lower antennas about 40 times per second. While this system also permits detection in turns it can result in the loss of about 50% of replies when the aircraft is not maneuvering, which can impose detection problems on the ground equipment. Figure 8.1 is an elapsed-time photograph of a radar display showing the reply video from such an aircraft.

## 8.2 Airborne antenna patterns

Airborne antennas are generally in the form of a quarter-wavelength stub or a flush-mounted cavity. Both are designed to operate in an extensive ground plane formed by the aircraft fuselage. In practice the ground plane is noninfinite and is curved in at least one dimension. The antenna may be obscured by engines or wings and may interact with other antennas also mounted nearby. It is to be expected that the antenna radiation pattern will be far from ideal. Figure 8.2 shows the measured antenna pattern obtained from a scale model [8.6].

## 8.3 Conformance to specifications

Considerations of size, weight, reliability and cost limit the degree of automatic self-test that can be incorporated into the design of a transponder. Often the pilot is only aware of a transponder malfunction when ground ATC complains that either it is not detecting him on SSR or that it is receiving incorrect identity or height data from his transponder.

Simple self-testing, such as automatic receiver sensitivity or transmitter power measurement, can be applied and the pilot can monitor that the transponder is replying to interrogations by observing a lamp on the control unit. More subtle faults such as transmitter frequency error are less easy to detect but can cause severe problems for the designer of the ground equipment which has the task of detecting all aircraft.

The FAA, with Eurocontrol participation, have developed a mobile transponder performance analyzer (MTPA) which can be taken to airfields to test transponders installed in aircraft. The equipment determines a wide range of transponder performance parameters by interrogating taxi-ing aircraft and analyzing the replies. It has been used at a number of airfields in the USA and Europe. The European results suggest that about 15% of air-carrier transponders and 30% of general aviation transponders have at least one parameter outside the tolerance permitted by the specification [8.7, 8.8]. The US measurements concentrated on general aviation aircraft [8.9–8.11] and showed results which varied with location. At one location only 7% of aircraft met all specification limits.

### 8.3.1 Reply frequency

The European measurement of the distribution of the air-carrier transponder reply frequency is shown in Fig. 8.3(a). About 3% of transponders are outside the ICAO limits. Typical US measurements on general aviation transponders are indicated in Fig. 8.3(b). About 9% are outside the limits. Significant frequency error will result in a loss of detected signal at ground interrogator receivers which could result in a failure to detect the aircraft. Some monopulse systems could convert the frequency error to a bearing error.

### 8.3.2 Transmitter power and receiver sensitivity

The transmitter power and receiver sensitivity of a transponder are difficult to measure accurately owing to the distortion of the results by the presence of signals reflected from the ground. ICAO *Annex 10* specifies the requirements at the antenna terminals and does not include the gain of the antenna, which is also an unknown in any measurement using radiated interrogations and replies.

After allowing 3 dB for measurement error, the European measurements of transmitter power for air-carrier aircraft showed only 1.5% falling outside the specification limits. The US measurements showed about 25% of powers outside specification. Nearly all the failures were for too high a power, and this included most of the military aircraft in the test. The higher power would give a better signal-to-noise ratio at the ground receiver but would also result in a slightly high fruit rate for other ground stations.

Receiver sensitivity was not measured in the European tests and Fig. 8.4 shows the US results for general aviation aircraft only [8.10]. Allowing for the possible measurement error of 3 dB, 13% of aircraft had sensitivities of less than the specification limit of −69 dBm with 7% exceeding the upper limit of −77 dBm. The average sensitivity was −71 dBm.

Aircraft with inadequate receiver sensitivity would show poor probability of detection at long range and in the lobing minima of the ground interrogator antenna. Too high a sensitivity could result in extra sidelobe suppressions, which would marginally degrade reply probability, or replies to distant interrogators to give a small increase in the fruit rate.

### 8.3.3 Pulse widths

Figure 8.5 shows a histogram of measured reply pulse widths [8.8, 8.10]. About 6% of both samples exceeded the maximum width limit of 0.55 μs with

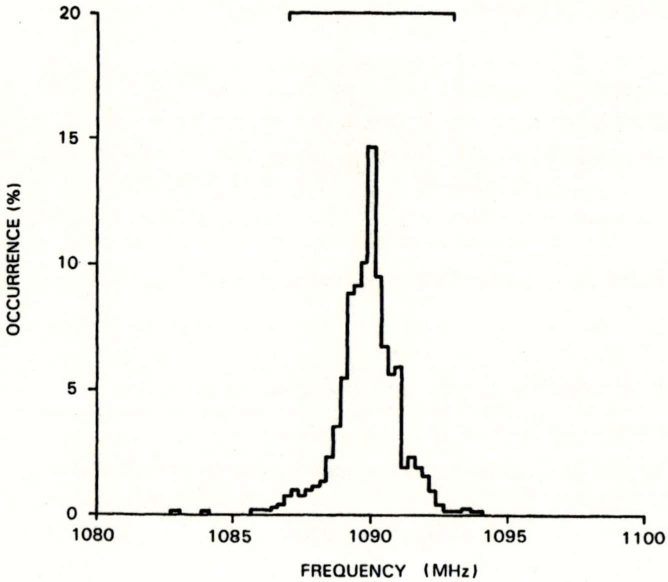


(a)

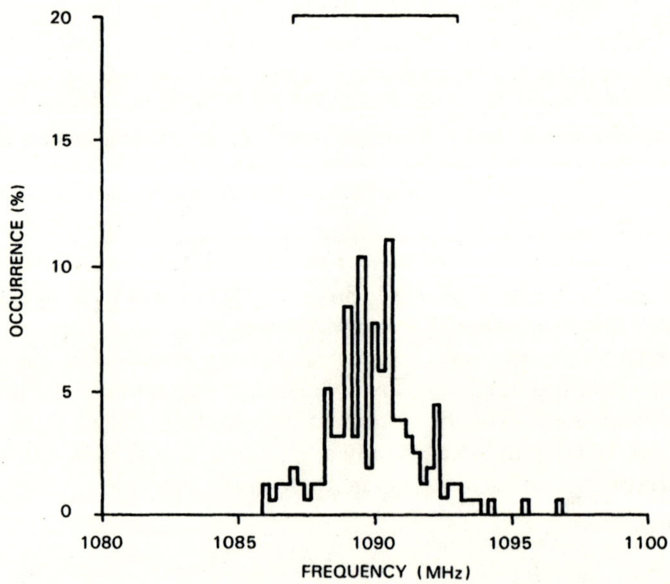


(b)

**Fig. 8.3** Measured transponder reply frequency distribution: (a) air-carrier aircraft (b) general aviation aircraft.

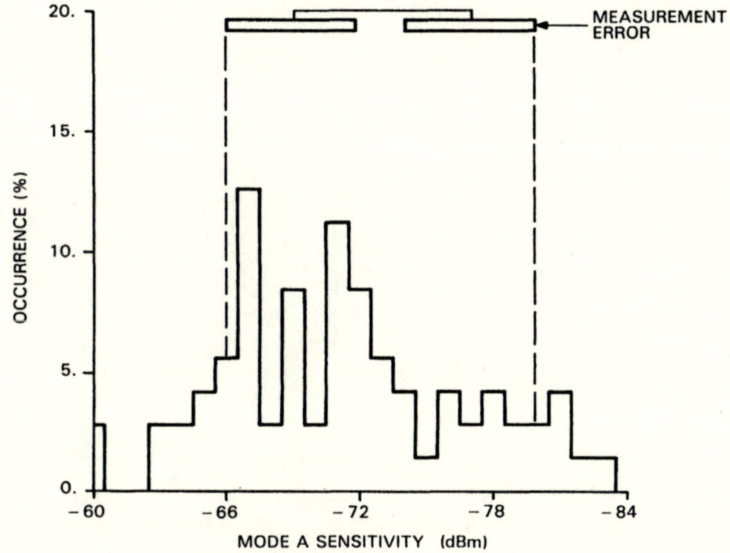


**Fig. 8.4** Sensitivity distribution for the transponder receiver.

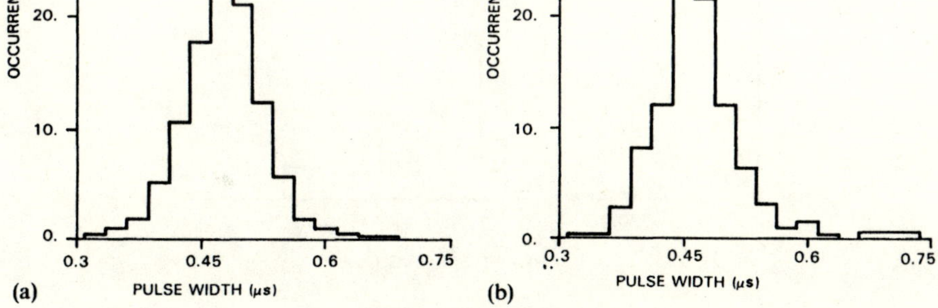


**Fig. 8.5** Transponder reply pulse width distribution: (a) air-carrier aircraft; (b) general aviation aircraft.

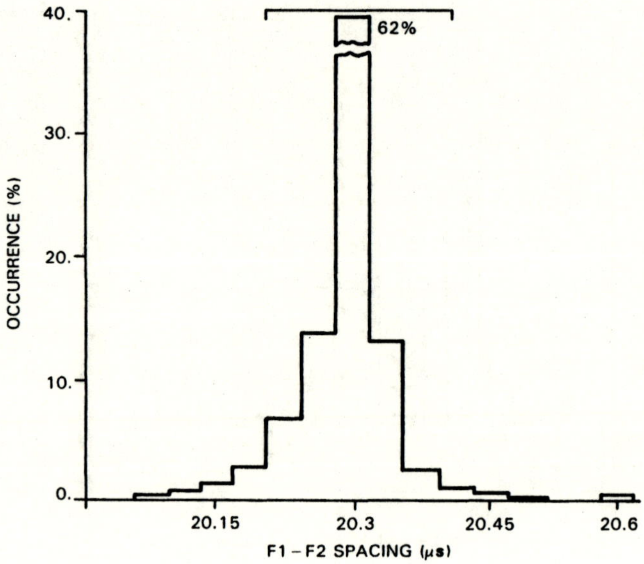


(a)

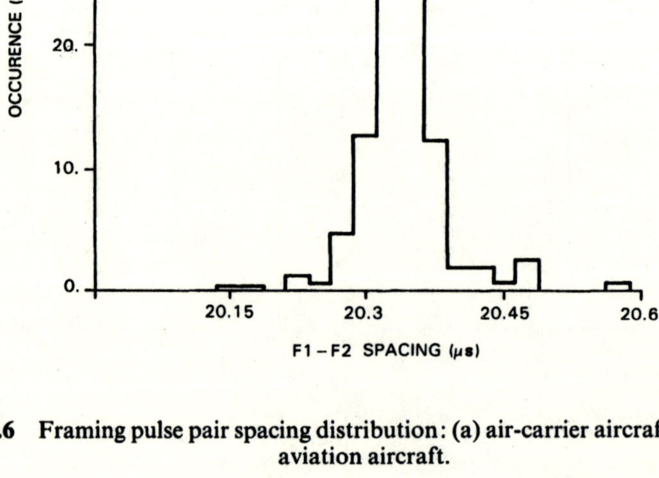


(b)

**Fig. 8.6** Framing pulse pair spacing distribution: (a) air-carrier aircraft; (b) gei aviation aircraft.

a very small proportion being narrower than the lower limit of 0.35 μs. Both departures from the specification can be accommodated in the plot extractor design (see Section 7.3.1). At least one example is known of a transponder fault which produced pulses of different widths in the same reply and needed special processing to avoid the generation of two separate target reports from one aircraft.

### 8.3.4 Reply pulse errors

Figure 8.6 shows the distribution of the measured F1–F2 framing pulse spacing from the European (air-carrier) and the US (general aviation) measurements. Both show that a fraction of aircraft produce replies outside the specification tolerance, 6% in the general aviation case. However, widening the plot extractor detection tolerance to $20.3 \pm 0.2$ μs (the ICAO *Annex 10* limit is $\pm 0.1$ μs) would result in the acceptance of most replies. The spacing of the bracket pulses F1 and F2 is a critical parameter as the detection of pulse pairs with this separation is the first and essential step to reply detection. However, too lax an acceptance tolerance would degrade the decoding of overlapping replies and increase the false declaration of phantoms.

About 1.5% of air-carrier transponders and 6% of general aviation transponders had data code pulses displaced more than the 0.1 μs allowed from the true position. These transponders could cause decoding errors which produce false or erratic identity codes or height indications.

On occasion, a fault develops in a transponder which causes a data pulse to be permanently on or off. If the error occurs in a Mode A reply then it will be evident to the air traffic controller, assuming that the error resulted in an identity code error. In the case of a Mode C reply the problem may not be immediately apparent.

Mode C uses a modified Gray code, or Gillham code, such that there is not a linear relationship between the code value and the corresponding flight level. Figure 8.7 [8.12] shows the reported flight level of a steadily climbing aircraft which had the B1 pulse permanently missing. When no B1 pulse was required the flight level data were accurate; however, when it was needed the reported level was wrong and the aircraft was indicated as descending.

Measured data [8.12, 8.13] suggest that the incidence of reply data errors due to pulses which are stuck either on or off is about 1 in 3000 movements.

### 8.3.5 Suppression characteristics

ICAO *Annex 10* requires transponders to be designed so that they reply to a valid interrogation if pulse P2 is weaker than pulse P1 by more than 9 dB

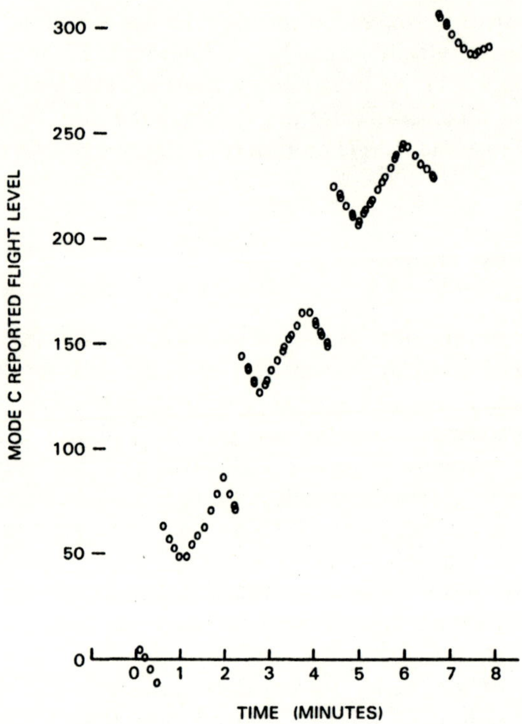


**Fig. 8.7** Reported flight level with pulse B1 permanently missing (courtesy of the CAA).

and that they do not reply if P2 is stronger than P1. The test is generally applied for signals 3 dB or more stronger than the minimum trigger level. Figure 8.8 shows the measured data [8.11]. The number of transponders replying at a rate of greater than 90% rises significantly when P1 exceeds P2 by more than 6 dB. Nearly all transponders reply at a rate of less than 1% when P2 is equal to or stronger than P1; however, about 3% continue to reply at full rate when they should have suppressed.

Transponders are also required to deny a reply to an interrogation following an earlier interrogation or a sidelobe suppression. The measurements show that transponders generally conform to specification in this respect.

## 8.3.6 Reply delay

Following an interrogation a transponder delays its reply by 3 μs timed from the interrogation pulse P3. The tolerance on the delay is ±0.5 μs with a

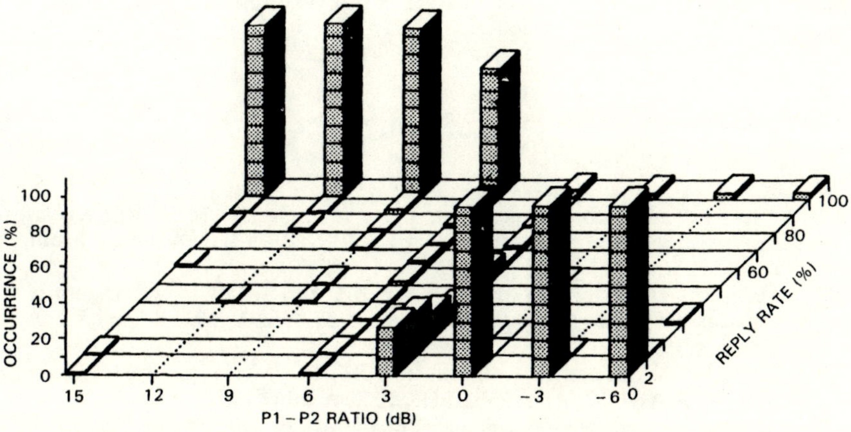


**Fig. 8.8** Measured transponder suppression characteristics (minimum trigger level + 3 dB).

jitter between successive replies of $\pm 0.1\ \mu s$ and a permitted differential delay between Mode A and Mode C interrogations of a further $\pm 0.2\ \mu s$.

The measurements of reply delay show that about 5% of transponders exceed the permitted tolerance but not by a wide margin. In the case of one make of transponder [8.11] it was found that the reply time was related to P1 instead of P3, as required in all recognized specifications, so that the reply time was influenced by errors in the P1–P3 spacing. No aircraft were found which exceeded the allowed jitter tolerance.

The European measurements did not include the differential reply delay between Mode A and Mode C interrogations but the US measurements on general aviation transponders indicate that about 7% exceed the allowed tolerance. Excessive differential delay is a condition that can be anticipated in the design of the ground plot extractor.

## 8.4 Need for transponder performance reporting

A transponder operating out of specification will not be apparent to the pilot unless the error is so significant that a complaint is made by ATC. Before the drift in performance reaches this extreme situation the performance of the SSR can degrade by producing increased position jitter, decoding errors or reduction in detection—all of which may be incorrectly attributed to the ground equipment and initiate unnecessary maintenance action.

Tests could be devised in the ground equipment which could indicate the possibility of some of the transponder errors. Target reports could then be marked so that the aircraft responsible could be identified and the operator alerted.

# References


8.1 Aeronautical telecommunications, *Annex 10 to the Convention on International Civil Aviation*, Vol. 1, International Civil Aviation Organization, Montreal, April 1985.

8.2 Technical characteristics of IFF Mk XA and Mk XII interrogators and transponders, *Standardization Agreement, STANAG 4193, Edition 3*, NATO, Brussels, 1986.

8.3 Mark 3 air traffic control transponder (ATCRBS/Mode S), *ARINC Characteristic 718-3*, ARINC, Annapolis, MD, 10 September 1981.

8.4 *Minimum Operating Performance Standards for Air Traffic Control Radar Beacon System/Mode Select (ATCRBS/Mode S) Airborne Equipment, DO-181*, RTCA, March 1983.

8.5 *Minimum Operating Performance Requirements for SSR Transponders and Altitude Measurement Encoding Systems, ED-42*, Eurocae, Paris, October 1983.

8.6 Keeping, K. J. and Sureau, J. C. Scale model pattern measurements of aircraft L-band beacon antennas, *Lincoln Laboratory ATC-47, Rep. FAA-RD-75-23*, Lincoln Laboratory, MIT, Lexington, MA, 4 April 1975.

8.7 Transponder Central File—Yearly Report 1986, *EEC Note 6/87*, Eurocontrol Experimental Centre, April 1987.

8.8 Eymann, C. Histograms of measured transponder performance, *Private communication CE/612/B3*, Eurocontrol Experimental Centre, 22 June 1987.

8.9 Wapelhorst, L., Zvanya, J. and Schoenthal, M. ATCRBS transponder data from initial field test of MTPA, *DOT/FAA/CT-TN83/04*, FAA Technical Center, Atlantic City, NJ, October 1983.

8.10 Wapelhorst, L. and Schoenthal, M. ATCRBS transponder data from Dayton, Ohio Airshow—1983, *DOT/FAA/CT-TN83/53*, FAA Technical Center, Atlantic City, NJ, March 1984.

8.11 Wapelhorst, L. ATCRBS data from Atlanta, Georgia, Spring 1985, *DOT/FAA/CT-TN85/38*, FAA Technical Center, Atlantic City, NJ, December 1985.

8.12 Gent, H. Secondary radar and ATC, II—The reliability of height and identity data, *J. Navig.*, **35** (2), 204–219, 1982.

8.13 Jenkins, D. B., Wyndham, B. A. and Banks, P. Errors in aircraft height information telemetered by secondary surveillance radar systems, *IEE Conf. Proc. RADAR 87*, pp. 200–202, IEE, London, November 1987.

## Chapter 9

# Multipath and Interference

The invention of radar followed research into the effects of reflections on the propagation of radio signals. Reflections, or multiple paths (multipath), are still a dominant influence and limitation on the performance achieved by radar including SSR. The term "multipath" describes the phenomenon where there is more than one signal path between the transmitter and the receiver. Often, but not always, one of these paths is the intended direct line between the transmitter and the receiver. Other paths may exist which involve reflections from natural and man-made objects such as the ground and buildings.

In the case of a continuous radio signal, multiple paths from transmitter to receiver result in signal fluctuation and fading. Fading can also occur in radar including SSR. However, owing to the short pulses normally used, direct and reflected signals can differ sufficiently in time for the multiple signals to arrive separately. Time separation between the direct and reflected path signals can alter the apparent number of reply pulses and cause decoding errors or be falsely interpreted as indicating that more than one aircraft is present. The problems caused by multipath can be alleviated by careful siting of the ground antenna, although the radar designer rarely has complete freedom in the choice of site.

SSR can also suffer from self-inflicted interference due to the use of common frequencies (1030 MHz on the uplink and 1090 MHz on the downlink). Transponder replies are received by other interrogators within range in addition to the one initiating the replies. These superfluous replies, known as fruit, can cause processing errors at the receiving interrogators. Closely spaced aircraft will have their replies overlapped or garbled. Interrogations or sidelobe suppressions emitted by one ground station can prevent transponders from replying to other ground stations with a consequent reduction in detection probability.

The SSR frequencies lie in a part of the L band which is also shared by DME. Although guardbands exist around the SSR frequencies, DME signals can often be detected by both the transponder and interrogator receivers. SSR equipment must be designed such that it will not be confused by the wider DME pulses.

SSR is frequently operated in conjunction with a colocated primary radar which emits r.f. power measured in hundreds or thousands of kilowatts. Although primary radar uses different frequencies from those used by SSR (1215–1365 MHz in L band or 2700–3400 MHz in S band), careful design is required to prevent the primary radar emission from breaking through into the SSR ground receiver.

# 9.1 Multipath

The relationship between the direct and reflected paths can vary widely, and the various multipath configurations can be divided into three classes [9.1]:

(1) those where signals reflect off the ground but arrive in the same vertical plane as the direct path signal;

(2) those which reflect from objects or sloping ground having a small horizontal angular offset but are detected in the same beam dwell-time as the direct path signal;

(3) those where there is a large horizontal angular separation between the direct and reflected paths.

Each of these three classes of multipath can be divided into two subclasses:

(i) those for which the time difference between the two paths is small so that there is substantial overlap between the same pulses arriving by the two paths;

(ii) those for which there is a large time difference so that there is little or no overlap between the corresponding pulses arriving by the two paths.

Each type of reflection phenomenon produces different effects. These effects are summarized in Table 9.1.

**Table 9.1** Classes of multipath and resulting system errors

| | *Azimuth separation between direct and reflected paths* | | |
|---|---|---|---|
| | *Zero angle* | *Small angle* | *Large angle* |
| Short path difference | Vertical lobing<br>Fades<br>Main beam suppression<br>Sidelobe punchthrough | Beam shape distortion<br>Bearing errors<br>Split plots | Ghost targets |
| Long path difference | Code corruption | Code corruption<br>Bearing errors | Ghost targets |

## 9.1.1 Multipath signals in the same vertical plane

**Short path difference**

Short path difference multipath results in substantial overlapping of the direct and reflected signals. The r.f. energy arriving at the antenna via the two paths combines to add or cancel according to the relative phase between the two signals. The resultant signal is thereby strengthened if the two signals are in phase, weakened if the two signals are in antiphase or affected to an intermediate extent depending on the actual phase relationship.

If the multipath reflection is via horizontal ground, the direct path and the reflected path will lie in the same vertical plane. Figure 9.1 shows the reflection geometry for this case. The aircraft elevation angle, as well as the height of the ground antenna, determines the relative phase between the two signal paths. The aircraft range is usually very much greater than the height of the radar antenna so that the direct path and the reflected path are very nearly parallel. Figure 9.2 shows a typical range–height diagram including the variations in range caused by the effects of ground reflection.

At certain elevation angles, as can be seen in Fig. 9.2(a), the signal at the aircraft is enhanced as the direct signal and ground-reflected signal add; the effect is to increase the range of the transmitter. At other elevation angles the two signals largely cancel and the range of the SSR is severely reduced.

Figure 9.2(a) is plotted for a small vertical aperture antenna, of the type illustrated in Fig. 4.6 (Chapter 4), where much of the radiated energy is directed at the ground. The large vertical aperture antenna shown in Fig. 4.1 reduces this energy and the degree of lobing caused, as indicated in Fig. 9.2(b).

Figure 9.3 shows an elapsed-time photograph of individual replies from an aircraft. As the aircraft passed through minima in the range–height diagram of the antenna the number of replies decreased considerably. The weakened signal strength reduced the beam width over which the antenna had sufficient gain to overcome the losses due to the ground reflection. At some ranges the replies ceased completely.

When the relative phase between the two signals is 180° they will completely cancel only if they are of equal strength. The two signals are rarely of equal strength because there is usually some difference in the antenna gain

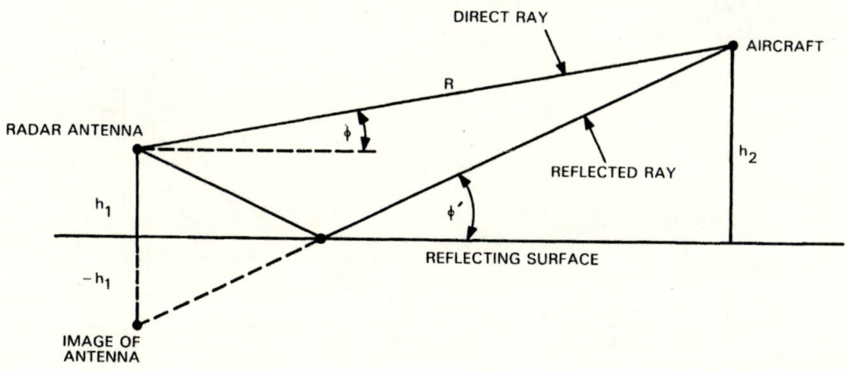


**Fig. 9.1** Signal path for ground reflection.

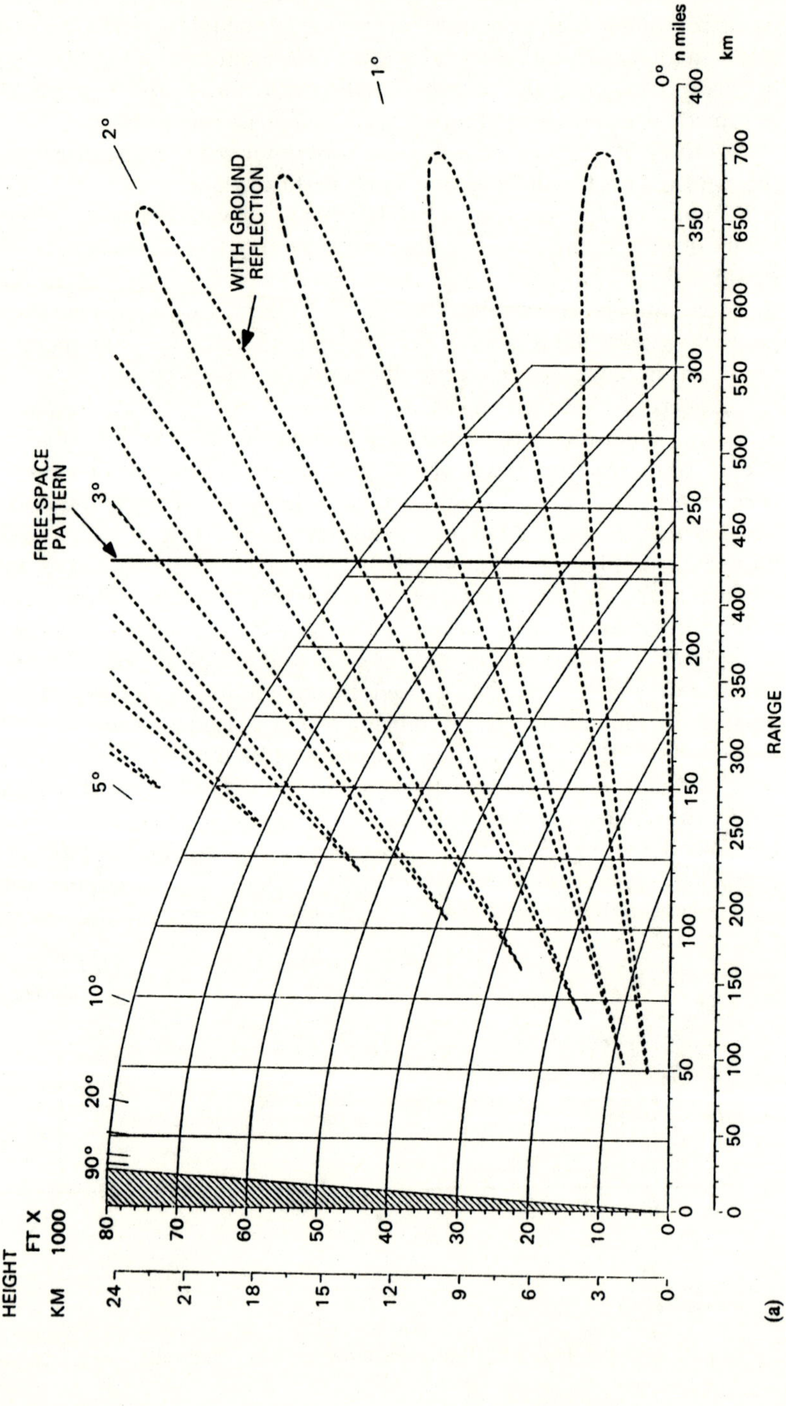
HEIGHT
KM
FT X 1000
FREE-SPACE PATTERN
WITH GROUND REFLECTION
90°
20°
10°
5°
3°
2°
1°
0°
n miles
km
RANGE

(a)

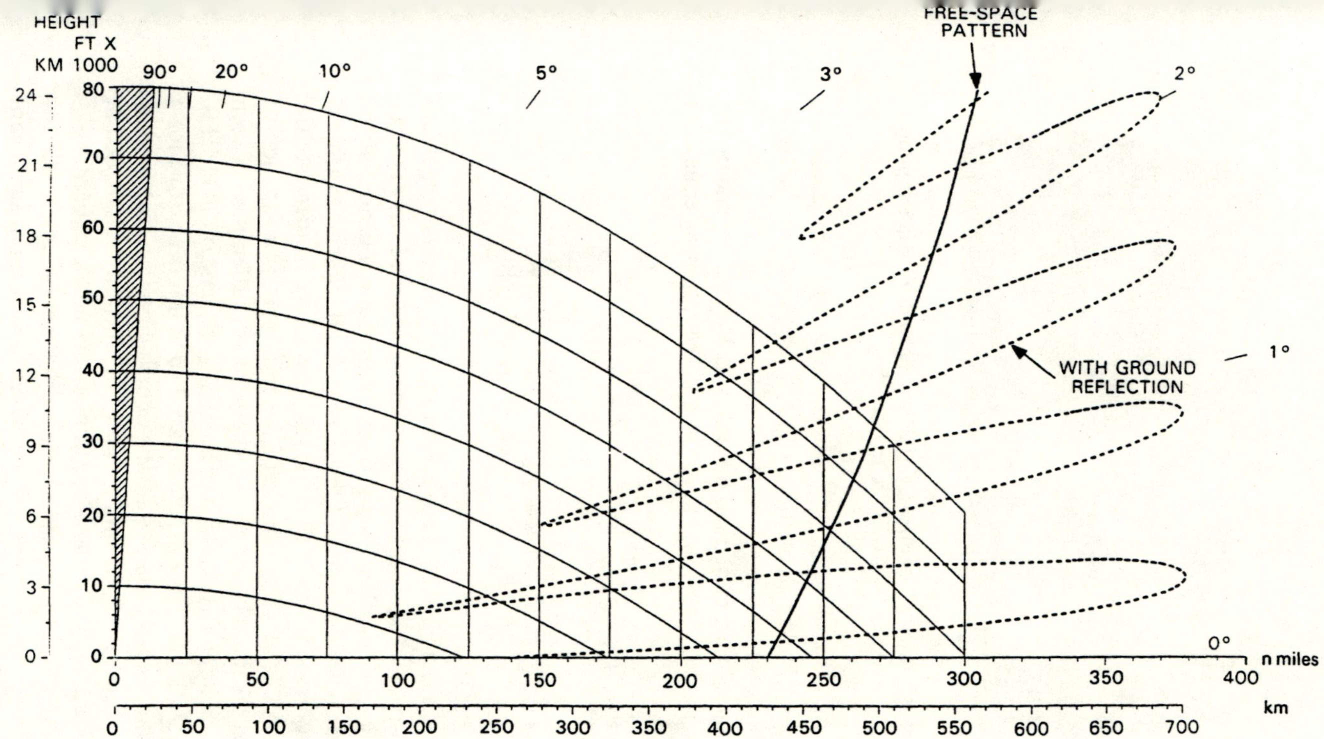


(b)

**Fig. 9.2** Range–height diagram showing the free-space range and the effects of ground reflection: (a) small vertical aperture antenna; (b) large vertical aperture antenna.

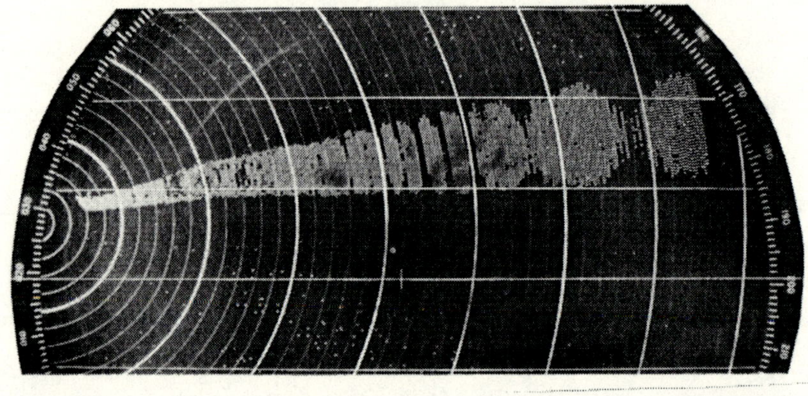

**Fig. 9.3** Elapsed-time photograph of an aircraft flying through antenna minima caused by ground reflection.

in the two directions and some attenuation upon reflection. Large vertical aperture antennas (see Chapter 4, Section 4.2) are designed specifically to increase the differential gain between the direct and reflected signals.

The elevation angles of the minima and maxima in Fig. 9.2 depend on the height of the antenna above the ground according to the following equations [9.2]:

$$\theta_{\text{min}} = \sin^{-1}\left(\frac{n\lambda}{2h}\right) = \sin^{-1}\left\{\frac{492n}{f(h\,\text{ft})}\right\}$$

$$\theta_{\text{max}} = \sin^{-1}\left\{\frac{(2n+1)\lambda}{4h}\right\} = \sin^{-1}\left\{\frac{246(2n+1)}{f(h\,\text{ft})}\right\}$$

where $n = 0, 1, 2, 3, \ldots$ and $f$ is in megahertz.

Note that there is a minimum at zero elevation angle. At other elevation angles a minimum occurs where the direct ray and the reflected ray arrive at the receiver 180° out of phase. The depth of the minimum depends on the difference between the antenna gains in the above-ground and below-ground directions and also upon the ground reflectivity at the angle of incidence of the ground reflecting ray. The elevation angle of the minimum is a function of signal frequency, and it should be noted that there is a 6% difference between the interrogation frequency (1030 MHz) and the reply frequency (1090 MHz). The uplink and downlink minima will be displaced so that it is possible for a reply to lie in a minimum even though a strong interrogation was received by the transponder. In practice the minima are only significantly misaligned at high elevation angles where the depths of the minima are much reduced.

Figure 9.4 shows the reflectivity of four different types of soil for incidence angles ranging from 0° to 90°. The soil reflectivity has a significant effect on the depth of the minima, and the data show that the reflectivity can vary widely with the seasons and during a single day as well as for different

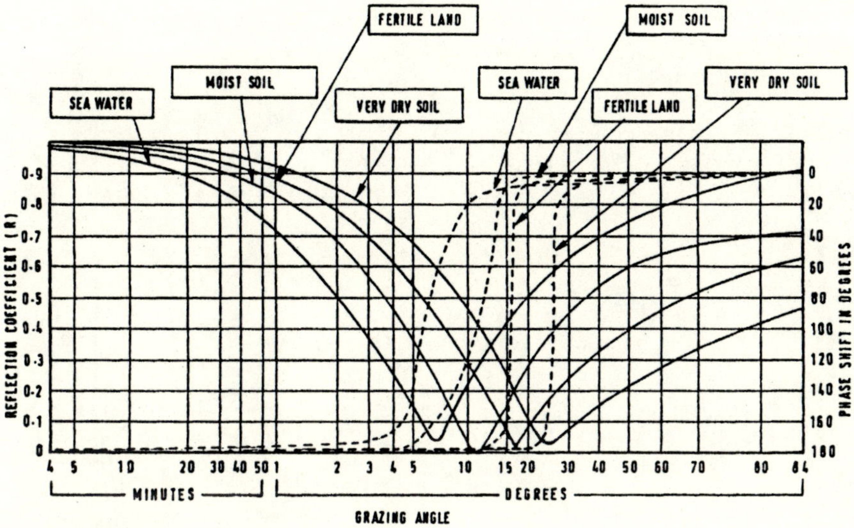


——— Reflection coefficient

---- Phase shift

Sea water: $\varepsilon = 80, \sigma = 4 \text{ mho m}^{-1}$
Moist soil: $\varepsilon = 30, \sigma = 2 \times 10^{-2} \text{ mho m}^{-1}$
Fertile land: $\varepsilon = 10, \sigma = 2 \times 10^{-3} \text{ mho m}^{-1}$
Very dry soil: $\varepsilon = 4, \sigma = 10^{-3} \text{ mho m}^{-1}$

**Fig. 9.4** Reflectivity of four soil types.

bearings from the antenna. Figure 9.5 shows the loci of the peaks of the maxima and the depths of the minima for moist soil for small and large vertical aperture antennas.

The data given in Figs 9.4 and 9.5 are pessimistic in that relatively smooth ground is assumed. In practice it is rarely smooth and some scattering occurs. Ground roughness can alter the path length of closely spaced rays so that the reflected ray does not have a clean phase front to provide a high degree of signal cancellation and a deep minimum. Minima are rarely as deep as calculated and a practical depth of two-thirds of the calculated range reduction can be used.

The first minimum is only a small angle above the ground with the corresponding reflected ray a similar small angle below the ground. Even in the case of the large vertical aperture antenna the gain difference between these two angles cannot be made very large within the constraints of a practical vertical aperture. The direction of the horizon will cut the underside of the antenna beam at a position of rapidly decreasing gain. The first

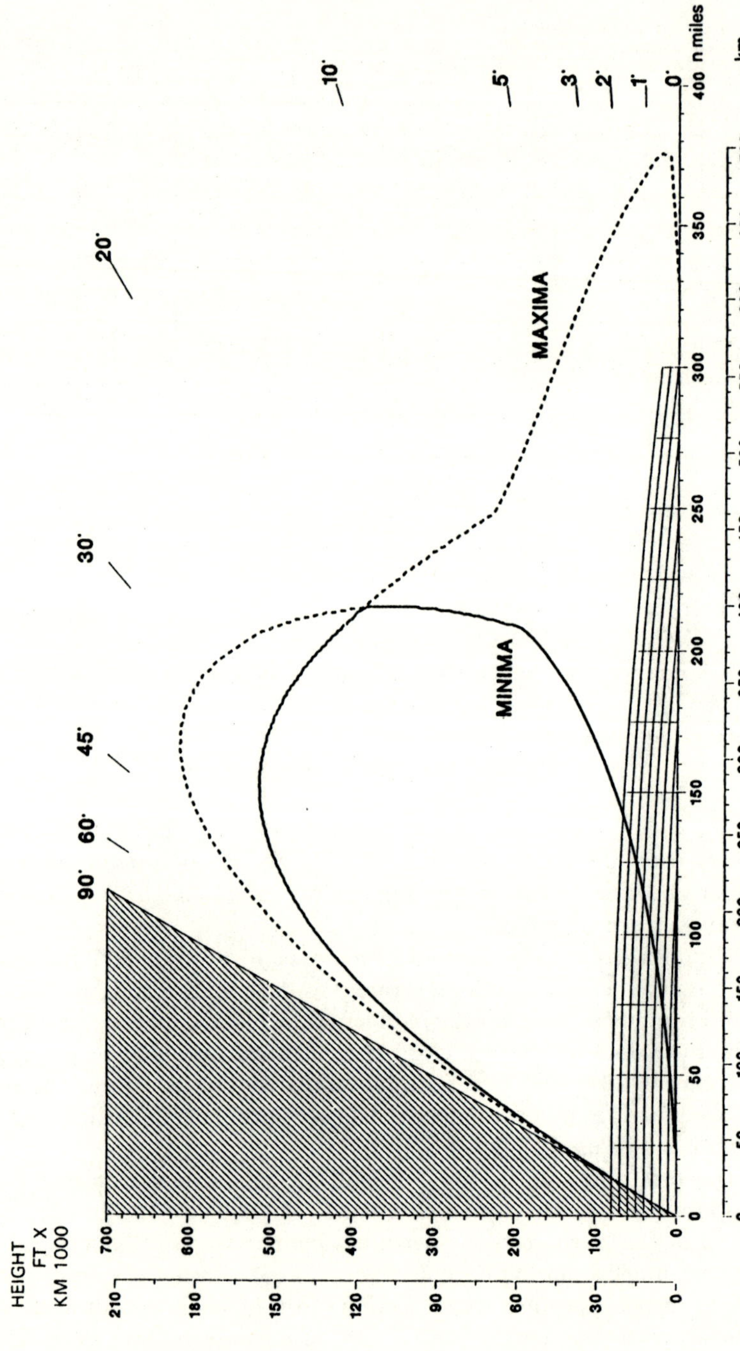
HEIGHT
FT X
KM 1000
MAXIMA
MINIMA
RANGE
n miles
km
90˚ 60˚ 45˚ 30˚ 20˚ 10˚ 5˚ 3˚ 2˚ 1˚ 0˚

(a)

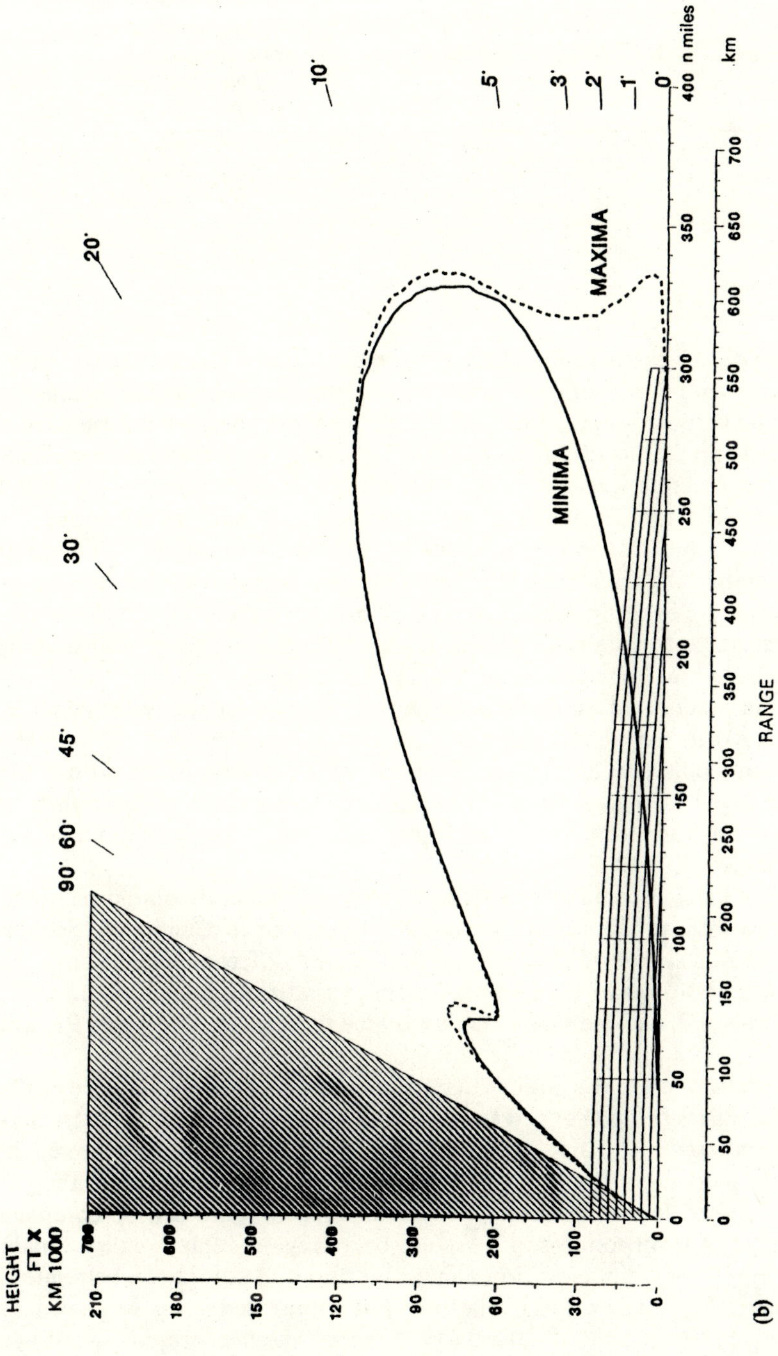


(b)

**Fig. 9.5** Loci of the depth of minima and peak of maxima of lobing: (a) small vertical aperture antenna; (b) large vertical aperture antenna.

**Table 9.2** Total gain reduction at the first minimum

| *Antenna tilt (free-space gain reduction)* (dB) | *Depth of first minimum (due to ground reflection)* (dB) | *Total gain reduction (sum of first two columns)* (dB) |
|---|---|---|
| −0 | −22.2 | −22.2 |
| −2 | −15.0 | −17.0 |
| −3 | −14.1 | −17.1 |
| −4 | −13.0 | −17.0 |
| −6 | −12.0 | −18.0 |

minimum produces the deepest fade and the question arises as to the optimum tilt of the antenna to minimize the fade. An increased downward tilt brings the peak of the vertical beam closer to the horizon. More energy is reflected from the ground and a deep minimum results. Conversely, an increased upward tilt reduces the antenna free-space gain at the angle of the minimum. Table 9.2 shows how the total antenna gain, free-space gain and reduction by the ground reflection change with antenna tilt for the first minimum. The total effective gain in the direction of the first minimum is practically constant over antenna tilt angles equivalent to free-space gain reductions of −2 to −4 dB. The latter angle is usually chosen as it also reduces the energy that may be reflected from buildings and cause false targets to appear.

The interaction of the same pulses arriving via direct and reflected paths is considered in the above discussion. A system problem can occur if the interrogation pulses P1 and P2 are affected in different ways. If the ratio of the pulses received at the transponder is different to the ratio of the pulses as transmitted by the interrogator, the transponder suppression action may not be as intended.

The practice used at some sites of employing a separate omnidirectional antenna to radiate the control pulse P2 is described in Chapter 4, Section 4.1.2. When this separate antenna is mounted at a different height above the ground, the positions of the minima for the omnidirectional antenna, which radiates pulse P2, and the main antenna, which radiates pulses P1 and P3, will differ.

This mismatch in the minimum positions can result in the balance of the P1 and P2 pulses being changed such that the sidelobe suppression characteristics are altered. At some elevation angles pulse P1 can be reduced in amplitude to narrow the effective beam width and reduce the number of replies obtained from the aircraft. In extreme cases there can be complete suppression of transponders in the main beam region—this is called "main beam killing". At other elevation angles the amplitude of P2 can be reduced with the result that some sidelobes are not suppressed—this is known as "sidelobe punchthrough". Figure 9.6 [9.3] shows this effect graphically. When

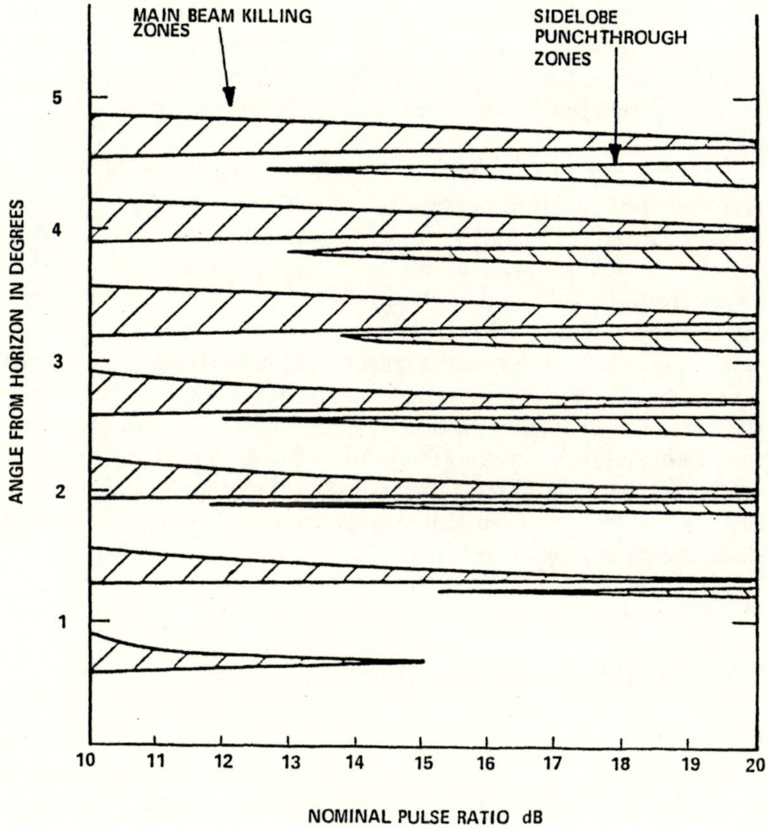


**Fig. 9.6** Main beam killing and sidelobe punchthrough (from ref. 9.3).

"improved" sidelobe suppression (IISLS) is used, careful phasing of the interrogate and control antenna beam feeds is required [9.4].

**Long path difference**

For the purposes of this discussion a long path difference between the direct and a reflected ray is taken to mean that the direct and reflected pulses overlap only slightly. Pulse amplitude and direction of arrival is measured in the first 25%–50% of the pulse, so that where a long path difference exists there is no corruption of the pulse in the region where the measurement is made.

Accurate measurement of the pulse arrival time is very important in decoding and degarbling reply signals, and use is made of either the leading edge or the trailing edge of the pulse for this purpose. Where a long path difference exists there is usually no difficulty in separately measuring the arrival time of pulses on the two reply paths. A particularly troublesome time delay between a direct and a reflected reply occurs when the difference is equal to 1.45 μs since this is the interval between successive pulses in a reply. Unless care is taken in the decoding of reply signals, such a multipath condition can alter the apparent reply code to change the aircraft's identity code or reported altitude.

The SSR site at Clee Hill in the UK was particularly prone to code corruption due to delayed multipath, and the phenomenon became known as the "Clee Hill effect". As its name suggests, the site is on a hill and this overlooks a wide valley. Aircraft in a nearby airway give rise to signals reflected from the valley floor. When the extra path length is 435 m the time delay is 1.45 μs. Figure 7.10 (Chapter 7) shows the geometry of this type of reflection. Because in this case the reflected ray lies in the same vertical plane as the direct ray, the horizontal angle of arrival of the reflected and direct signals is the same. Consequently an SSR system that uses monopulse direction of arrival to distinguish between the pulses of overlapping replies is unable to discriminate between direct and reflected reply pulses at sites where a reflection effect of the Clee Hill type is present. The use of signal strength measurements to overcome this problem is discussed in Chapter 7, Section 7.2.13.

## 9.1.2 Small horizontal angle between multipath signals

The problems of processing multipath signals where there is a small horizontal angle between the direct and reflected rays are possibly the most difficult to resolve. The consequences of the condition can be bearing errors and split plots.

Figure 9.7 shows a diagrammatic view of the aircraft seen from the ground antenna with the image of the aircraft shown below ground level. Because of the slope of the ground the image is offset in the horizontal direction. As the antenna sweeps across the aircraft, the ground image of the aircraft and the real aircraft are detected at different times.

**Short path difference**

If the path difference between the direct and reflected signals is short so that they overlap substantially, the result of the multipath is to produce an effective antenna beam that is highly distorted. This distortion is apparent on both the interrogation and the reply paths. Figure 9.7 gives the geometry of the reflection paths for sloping ground. The angular image displacement $\alpha$ can be related to the aircraft elevation angle $\phi$ and the ground slope angle $\theta$ by (angles in degrees)

$$\alpha = 0.035\,\phi\,\theta$$

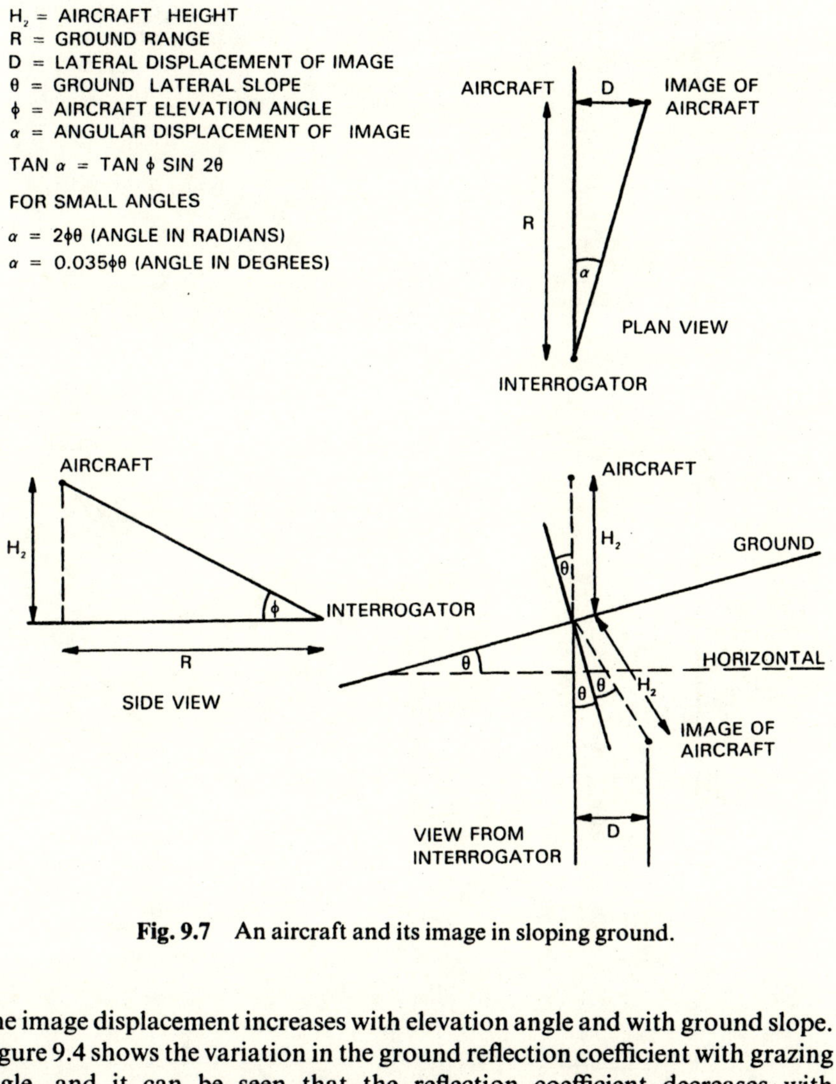


**Fig. 9.7** An aircraft and its image in sloping ground.

The image displacement increases with elevation angle and with ground slope. Figure 9.4 shows the variation in the ground reflection coefficient with grazing angle, and it can be seen that the reflection coefficient decreases with increasing angle up to the Brewster angle where it is nearly zero.

Most sites are surrounded by ground that could be classified as "fertile land", and therefore the curve labelled "fertile land" in Fig. 9.4 is most likely to apply. However, under wet conditions the reflection coefficients will move to the "moist soil" curve. Dry snow approximates to "very dry soil", reverting to moist soil when it melts. Beam shape distortion effects can vary with the seasons as well as on an hourly basis. Since there is always some reflection loss, the magnitude of any beam shape distortion can be expected to be small. It will be at its maximum when the difference in path lengths is such that the reflected

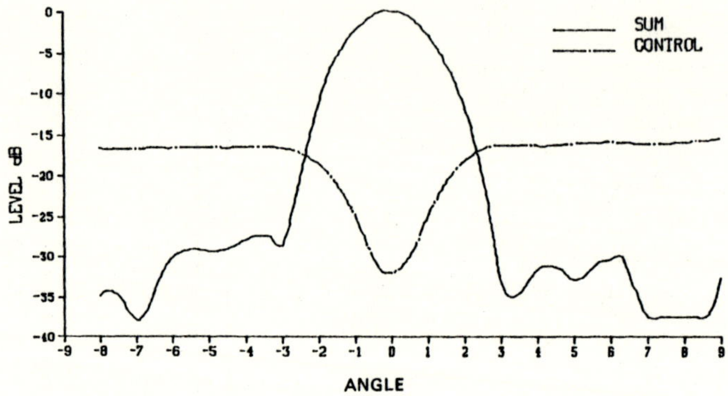
SUM
CONTROL
LEVEL dB
ANGLE

(a)

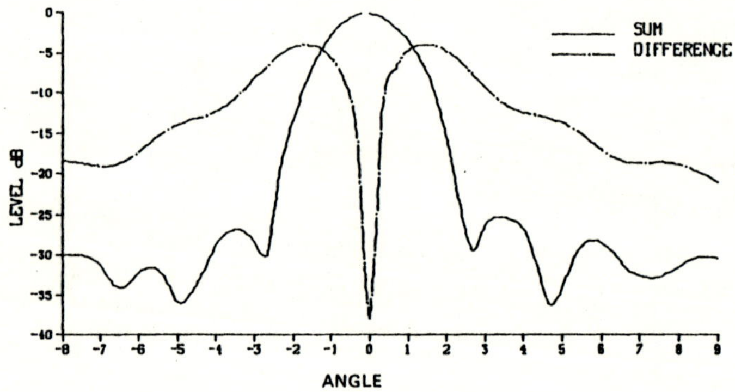
SUM
DIFFERENCE
LEVEL dB
ANGLE

(b)

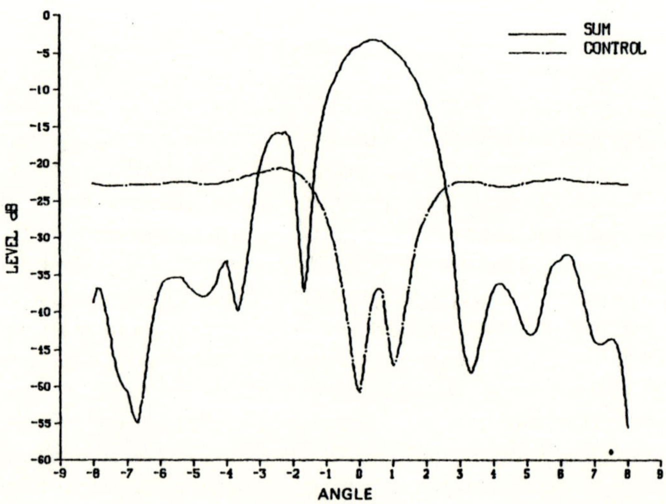
SUM
CONTROL
LEVEL dB
ANGLE

(c)

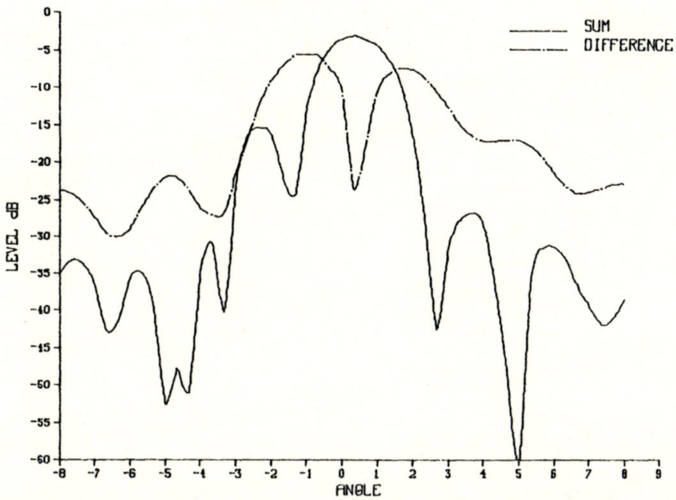


(d)

**Fig. 9.8** Free-space and reflection-distorted antenna patterns: (a) free-space sum and control patterns; (b) free-space sum and difference patterns; (c) distorted sum and control patterns; (d) distorted sum and difference patterns.

signal is in antiphase with the direct signal, i.e. when the aircraft is in a minimum in the antenna's vertical lobing pattern.

Figures 9.8(a) and 9.8(b) show the free-space sum and control patterns and the sum and difference patterns respectively for a typical SSR linear array. Such an antenna has a broad vertical beam and therefore is particularly sensitive to the effects of ground reflections. Figures 9.8(c) and 9.8(d) show the distorted beam patterns resulting from the effects of a ground image displaced 1° in azimuth and attenuated by 6 dB. These conditions would be obtained, for example, in the case of an aircraft at 6° elevation above fertile land which has a lateral slope of 5°. It can be seen that the effect of the displaced ground image is to produce a sidelobe punchthrough (Fig. 9.8(c)) and a disturbed monopulse sum–difference characteristic (Fig. 9.8(d)).

Figure 9.9 shows a sequence of replies received from an aircraft as the antenna scanned across it. The dotted line attached to each reply is a graphical way of illustrating the monopulse bearing as measured on those individual replies. The monopulse performance in this example is very good with little bearing measurement difference between any of the replies. The open squares trailing to the upper right of the replies show the aircraft's previous positions, and these data are used to predict the aircraft's present position which is shown by a vertically oriented rectangle. The measured position agrees well with the predicted position.

Figure 9.10 shows replies from another aircraft which present some evidence of antenna beam distortion. The sequence of replies lasts for much

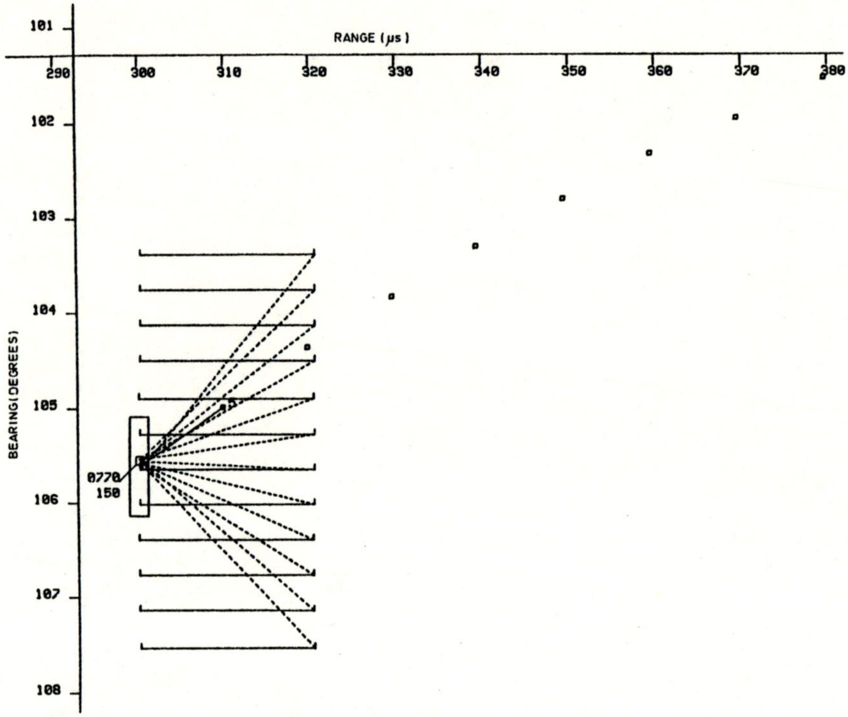


**Fig. 9.9** Sequence of replies from an aircraft (courtesy of the CAA).

longer than the 4.5° antenna beam width and the lower group appears to result from an apparent "sidelobe" 13 dB weaker than the main beam. The site was substantially flat but had a small stream in one direction with sloping ground along its banks. The top sequence of replies shows reasonable monopulse performance with most of the replies lying within the predicted zone. It is interesting to note that the "sidelobe" replies have monopulse measurements with the same sign, indicating that the antenna boresight has not been crossed. This can be anticipated from Fig. 9.8(d) since no null exists in the difference pattern in the "sidelobe" direction.

Figures 9.11(a), 9.11(b) and 9.11(c) show replies from one aircraft on successive scans of the antenna. The replies in Fig. 9.11(a) are reasonable. However, those in Fig. 9.11(b) are greatly disturbed. Reply correlation rules cause the target bearing to be taken from the average of the two replies on either side of the sign change and this places the measured target bearing outside the prediction zone. Most of the replies show evidence of overcorrecting the angle of the reply from beam center, and this can be expected from Fig. 9.8(d) since the sum–difference ratio is reduced in this multipath condition.

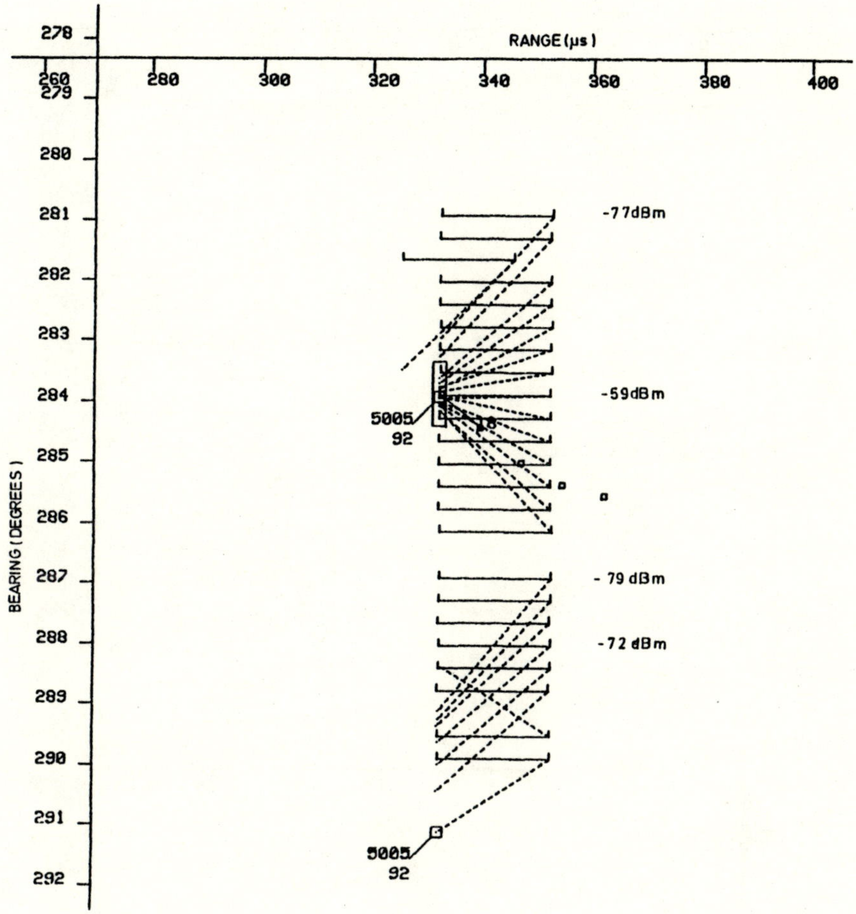


**Fig. 9.10** Sequence of replies showing antenna beam distortion (courtesy of the CAA).

Inspection of the reply signal strength shows that signal amplitude at beam center has dropped by 12 dB from the previous scan. The "sidelobe" (the top four replies) show a further loss of 6 dB. In the next scan (Fig. 9.11(c)) the situation has recovered and the signal amplitude has risen by 15 dB. Clearly the aircraft flew through a sharp minimum. Unfortunately, because of the previous bad plot the predicted position is considerably in error.

The geometry of Fig. 9.7 indicates that the error is small for low aircraft elevation angles. The reflectivity curves of Fig. 9.4 show that little signal is reflected from the ground for aircraft elevations of about 10° or more. At these elevations ground surface roughness would be an additional loss factor.

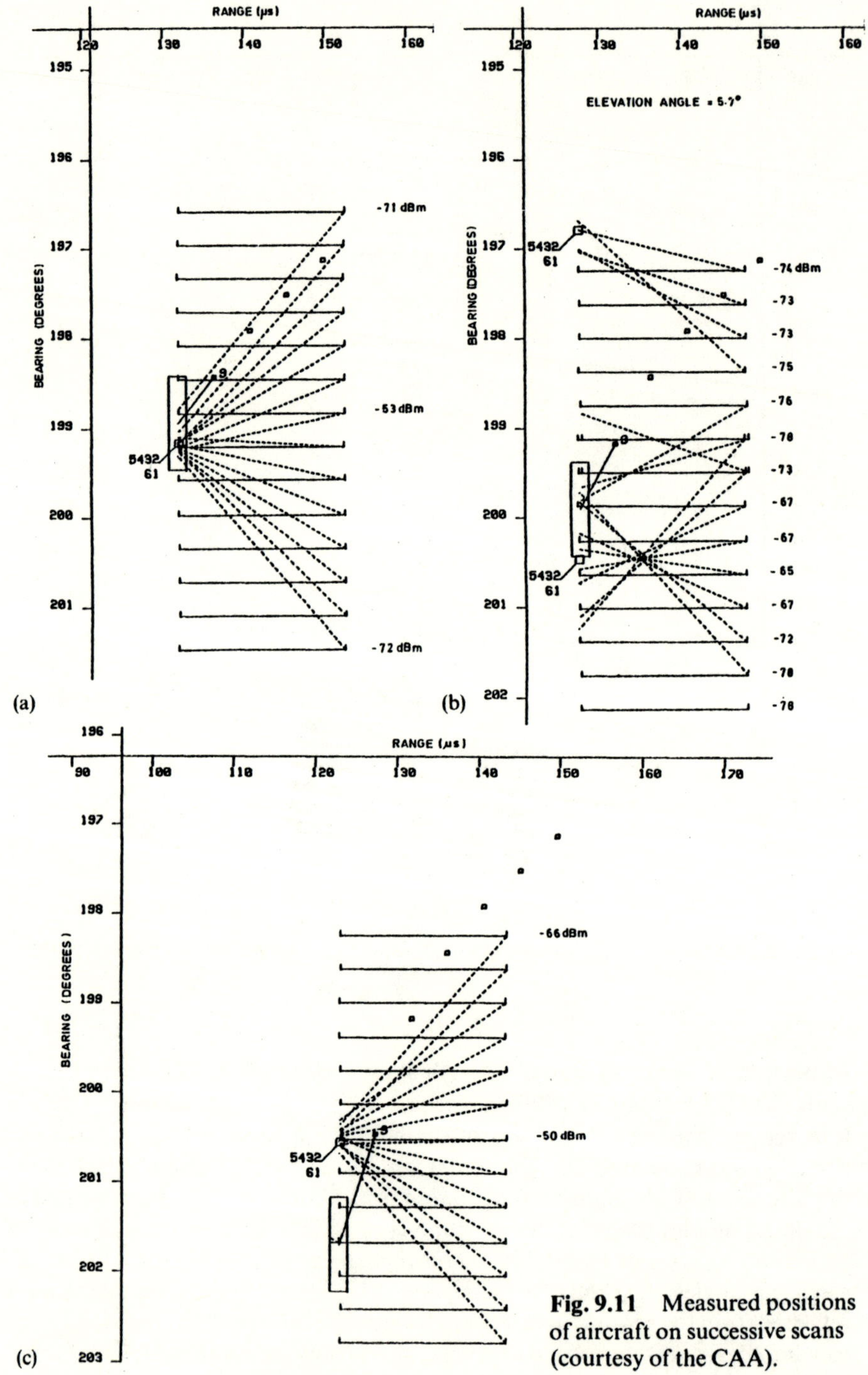


**Fig. 9.11** Measured positions of aircraft on successive scans (courtesy of the CAA).

Consequently beam distortion errors are only of significance for aircraft elevations between 3° and 10°.

The above measured results were obtained from a small vertical aperture linear array antenna. A large vertical aperture antenna, which offers better discrimination against ground reflections, largely overcomes the problems described. The results obtained using an antenna exhibiting similar ground discrimination are discussed further in Section 9.1.4. Reflections from high ground or buildings remain a source of beam distortion.

Although this discussion has concentrated on the effects of multipath on the reply signals, similar phenomena occur on the transmission path from ground to air.

An aircraft will reply to a valid interrogation if it is not sidelobe suppressed by a strong P2 pulse. Transponder sidelobe suppression performance is discussed in Chapter 8, Section 8.3.5.

Figure 9.8(c) shows the distortion of the antenna sum and control patterns under the multipath conditions assumed earlier. A transponder may reply when the sum pattern, which transmits P1, exceeds the control pattern, which transmits P2. Figure 9.8(c) shows two positions where this occurs with the result that an aircraft will start replying and stop replying twice as the antenna sweeps across the aircraft. Such an occurrence can cause two separate aircraft to be indicated.

In the sliding window method of measuring aircraft bearings the average of the positions where the replies start and stop is used to deduce the aircraft direction. This process is discussed in Chapter 6. Inspection of the undistorted patterns shown in Figs 9.8(a) and 9.8(b) indicates that replies should start about 2.3° before beam center and stop at a similar position after beam center. For the main lobe of Fig. 9.8(c), however, the start position has moved to about 1.4° before beam center and the stop position to 2.5° after beam center. This distortion shifts the measured position about 0.6° to the right.

The sliding window method of bearing measurement has been rightly criticized for producing poor bearing measurement accuracies, and this aspect has often been related to the loss of replies at a beam edge due to transponder reply capture. While capture is a source of the problem, beam distortion effects are more likely to cause bearing errors but until recently this mechanism has received less recognition.

Monopulse direction finding is also affected by antenna beam distortion but the measurements are made near to the beam center instead of at the beam edge. Figure 9.8(d) shows the distortion of the sum and difference patterns and it can be seen that the difference null is shifted by about 0.2°.

**Long path difference**

Long path difference multipath conditions result in a sufficient path difference between the direct and reflected rays for the same pulses in the reply not to overlap significantly. These can cause multiple reply and code corruption effects similar to those that occur when the signals are in the same vertical plane (see Section 9.1.1). In this case the horizontal separation between the aircraft and its image gives rise to different monopulse directions of arrival so that direction measurements can be helpful in separating the two replies.

Different effects can occur, however, on the interrogation path. Figure 9.12(a) shows an interrogation, transmitted from the ground, as received by an airborne transponder. The two large pulses are the P1 and P3 pulses of the interrogation. It can be seen that immediately following P3 there is a smaller pulse which is actually an echo of P3 received by way of reflection from the ground. A similar echo of P1 can be seen following P1, although in this case it has merged with the sidelobe suppression pulse P2.

Figure 9.12(b) shows a further example. In this case, however, three echoes can be seen following pulse P3 and the second and third pulses have greater amplitudes than P3 itself. Superficially this might suggest that the reflection surfaces responsible for the second and third pulses are imparting some gain to the signal. The explanation is that the ground antenna is pointing in the direction of the reflecting surfaces and slightly away from the aircraft. Consequently the direct P3 pulse is received from a weak part of the antenna beam. The second and third echo pulses are received near the peak of the beam so that the amplitude of these pulses is strong even though some reflection loss has occurred. Similar echoes follow P1, but the first two are partly cancelled by the P2 pulse which is also present. Indeed it is possible that the transmitted P2 pulse was stronger than P1 and transponder suppression was intended. In this instance the transponder replies persist for longer than a normal beam dwell-time and would cause a bearing measurement error in a sliding window processor.

Figure 9.12(c) is a further example obtained during the early Mode S trials using a Mode S transponder. The echo of P1 is delayed on the true P1

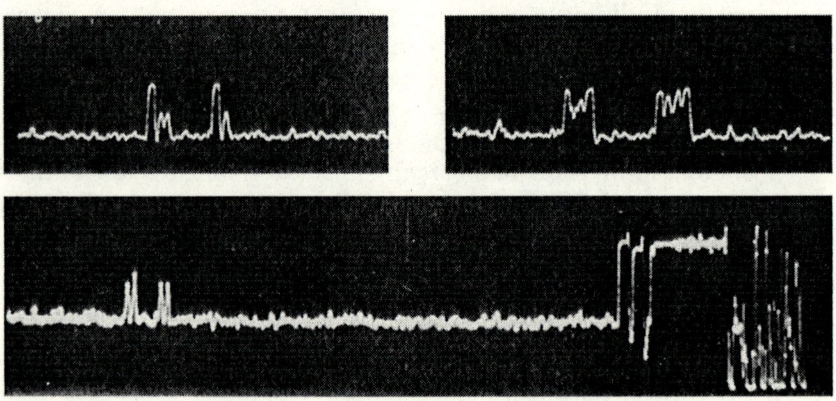

(a) (b) (c)

**Fig. 9.12** Examples of delayed uplink multipath.

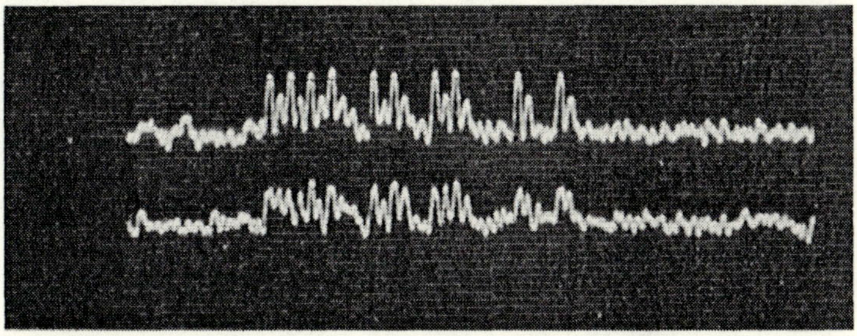

**Fig. 9.13** Sum and difference reply signals with an off-angle echo.

pulse by 1.5 μs. The leading edge of the suppression pulse P2 is masked so that it is not recognized by the transponder and suppression does not occur. Similarly the echo of P3 is also delayed by 1.5 μs at what was the original location of the Mode S P4 All-call pulse. In the Mode S system described in Chapter 12 an All-call interrogation is a modified SSR interrogation to which a Mode S transponder responds with a Mode S reply and not an SSR reply.

The signal at the extreme right of Fig. 9.12(c) is the Mode S All-call reply from the transponder. The SSR station originating the interrogation misinterpreted the Mode S reply to form a false target at an erroneous range. The specification of the Mode S P4 pulse position has now been changed to 2 μs after P3 and doubled in width so that this phenomenon should not occur for the operational Mode S system. Multipath signals which could trigger an All-call reply with this new specification should also generate a sidelobe suppression pulse pair.

Figure 9.13 shows a reply signal with a delayed multipath signal. The monopulse bearing is determined from the ratio of the sum- and difference-channel signals. In this example the direct and delayed signals are arriving from different directions.

### 9.1.3 Large horizontal angle between multipath signals

Objects such as buildings and chain-link fences can produce "mirror" surfaces for radar signals in which images of aircraft are formed. If an aircraft is interrogated via a reflecting object, a ghost aircraft can appear at the bearing of the reflecting surface. This effect is illustrated in Fig. 9.14.

When the ground antenna is pointing at the reflector, the interrogation is reflected to a different bearing and an aircraft at this new direction may reply. Although there may be some reflection loss which will attenuate the signal, there is usually sufficient power to trigger a transponder into replying. The reply will be at full power and usually sufficient energy reaches the ground antenna for it to be detected and converted into an aircraft plot.

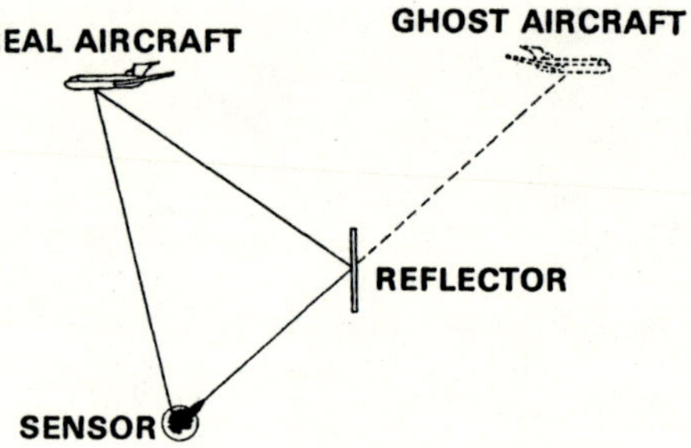


**Fig. 9.14** Reflection geometry.

The distance between the ground interrogator station and the reflecting object rarely exceeds 5 miles and is usually very much less, although the aircraft involved can have a range of over 100 miles. The actual geometry of each particular instance will determine whether any appreciable extra distance is travelled on the reflection path.

Adjustment of transmitter power to the minimum necessary for good target detection at the required maximum range will reduce the number of false targets produced.

**Short path difference**

The use of the antenna control pattern to radiate the sidelobe suppression pulse P2 is discussed in Chapter 4, Section 4.2. However, if this antenna port is also connected to a receiver channel, it is possible to detect the transponder reply as received by both the control pattern and the sum pattern. As long as there is sufficient reflection loss on the return path, the signal received by the control pattern will be stronger than the signal received by the sum pattern. By rejecting all replies for which the control pattern signal is stronger than the sum pattern signal, some of the ghost aircraft replies can be removed at an early stage in the processing.

The position of real aircraft subject to possible interrogation via the reflection can often be predicted, and this permits an alternative approach to the suppression of ghost aircraft [9.5, 9.6]. An additional directional antenna is placed to radiate P1 and P2 suppression pulse pairs in the direction of the real aircraft before the interrogation proper is transmitted by the main antenna. The transponder is thereby suppressed and unable to reply to the interrogation via the reflection path. This solution is very effective but, unfortunately, is only practical where there are few reflecting objects.

Computer techniques can also be applied by searching for multiple examples of aircraft with the same identity code and flight level. This aspect is discussed in Chapter 7, Section 7.5.

**Long path difference**

The consequence of a long path difference is that the direct and reflected replies arrive at separate times at the ground antenna and are detected separately by subsequent processing. The direct path signal will be detected first as it is shorter. A receiver connected to the control pattern of the antenna will indicate a sidelobe response (control signal stronger than sum signal) before the main beam reply (control signal weaker than sum signal) is received via the reflector. This method is not a complete solution for the identification of false targets caused by this class of multipath. However, it can be used as an indicator of a possible false target (see Chapter 7, Section 7.5).

When the extra path length of the reflected path contributes a delay of 2–25 μs, a further technique can be applied. The direct path to the aircraft will lie outside the main beam of the antenna so that the P1 pulse will be weaker than the P2 pulse and the transponder should be suppressed. When the interrogation from the main beam, with pulse P1 stronger than pulse P2, arrives via the reflecting surface the transponder is already in a suppressed state and unable to reply. The technique is not fully effective since the transponder can fail to be suppressed by the direct path signal if the pulse P1 from the sidelobe of the sum channel is too weak to be detected by the transponder. Without pulse P1 the transponder will not be able to recognize pulse P2 as the second of a suppression pair and take the correct action.

A technique known as "improved interrogator sidelobe suppression" (IISLS) has been developed to overcome the problem of possibly not detecting the sidelobe P1 pulse. In this system pulse P1 is radiated from the control pattern as well as from the conventional sum pattern. As a result, the aircraft in question can detect a pulse P1 of sufficient strength to enable the P1–P2 pair to be recognized and suppression action to be taken. One disadvantage of the IISLS technique is that it is only effective if the extra path length of the reflection is sufficient for the suppression signal to be detected separately by the transponder. A second disadvantage is that the P1–P2 suppression pairs are radiated over a wide area, causing the unnecessary suppression of many other aircraft.

Long path difference multipath can also corrupt valid interrogations. Figure 9.15 shows an example where a reflected control pulse P2, delayed by 11 μs, combines with main beam pulses P1 and P3 of a Mode C interrogation. This delay value results in the appearance of the reflected pulse 8 μs before pulse P3, thereby forming a Mode A interrogation. The transponder decoder is presented with what appear to be simultaneous Mode A and Mode C interrogations. Early editions of ICAO *Annex 10* left the transponder response to this possible signal combination unspecified and untested. Transponders of various design and manufacture perform differently and the following responses to a combined Mode A and Mode C interrogation have been measured: Mode A reply, Mode C reply, reply codes ANDed, reply codes ORed and code 7777.

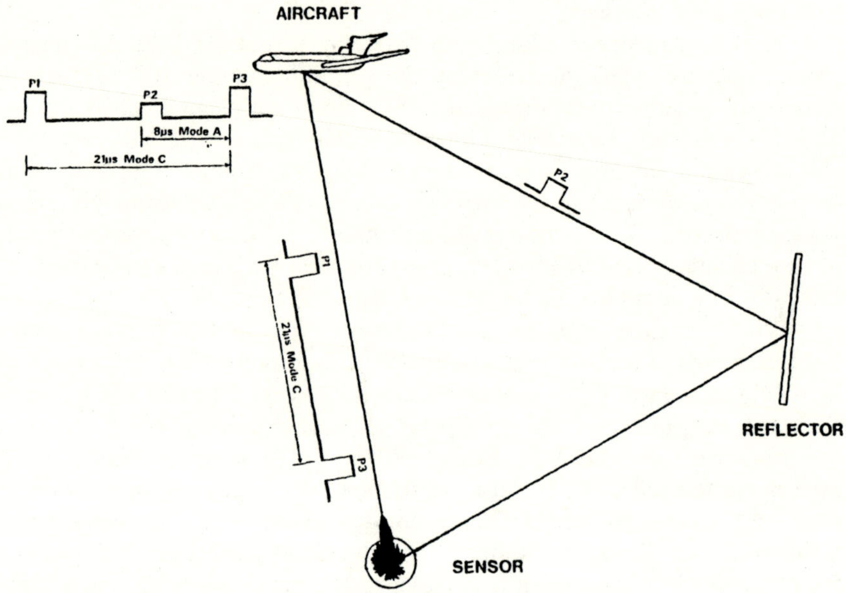


**Fig. 9.15** Multipath causing simultaneous Mode A and Mode C interrogations.

The multipath condition will persist during the antenna beam dwell-time and frequently over several scans. The interrogator can receive erroneous height data, and one consequence can be that an incorrect slant range correction is applied. This, in turn, can place the aircraft at a range where it does not correlate with the previously established track and a target is generated at a false position.

Similar effects can occur if the reflected P2 pulse is delayed by 6 μs or the reflected P1 pulse is delayed by 8 μs. The reflected pulse forms a false Mode A interrogation with the direct P1 pulse and triggers a Mode A transponder reply which is 13 μs early.

### 9.1.4 Example using an improved antenna

Most of the discussion of multipath effects above is concerned with the problems caused to the efficient detection and decoding of replies and the elimination of false responses. The impact on bearing measurement accuracy is considered in Section 9.1.2. A practical example of such a cause of inaccuracy is presented in this section together with a demonstration of its solution by the use of an improved antenna.

One of the operational SSR sites in the UK is located at Clee Hill in the West Midlands. The site had experienced multipath problems for many years,

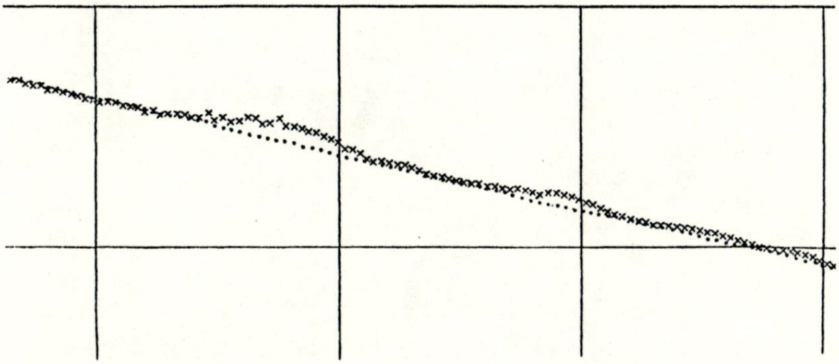

**Fig. 9.16** Ground and airborne recordings of an aircraft track (courtesy of the CAA).

notably that of code corruption due to delayed reflections from the ground. During the development of monopulse SSR in the UK improvement in performance at sites such as Clee Hill was a particular requirement and the technique of using signal amplitude as an additional pulse discriminant was incorporated as described in Chapter 7, Section 7.2.13. Naturally Clee Hill was one of the first sites selected for the trials of the new equipment and, indeed, the pulse corruption problems had been resolved.

Clee Hill was also known for its poor azimuth performance, but this had initially been attributed to the use of a sliding window plot extractor. Further investigations revealed that multipath from uneven ground close to the site was causing severe beam distortion. It was somewhat disappointing to discover that the monopulse bearing accuracy achieved at Clee Hill was little improved over that obtained with the previous equipment.

Figure 9.16 shows the recorded track of a calibration aircraft flying radially from Clee Hill. The crosses indicate the positions as measured by the SSR installed at the site. The dots show where the onboard equipment determined the position of the aircraft to be using multiple DME measurements. Clearly the departures from the straight line cannot be explained by temporary off-course wandering of the aircraft.

The rotating ground antenna obtains a number of replies from an aircraft on each scan. In the monopulse system a measurement of each reply should indicate the same aircraft bearing. Inspection of the recorded replies showed that the deviations from the straight line were accompanied by dispersion in the measurement of the individual replies during the scan.

The antenna being used at the time was a linear array (Chapter 4, Fig. 4.6) which had a small vertical aperture and consequently a large vertical beam width. Such an antenna is almost as sensitive to signals from the ground as to signals from angles above the horizon. A conical screen [9.7] was constructed around the antenna to mask it from the ground (Fig. 9.17). The flight trial was repeated and the results are shown in Fig. 9.18. The agreement between the

Fig. 9.17 Clee Hill screen (courtesy of the CAA).

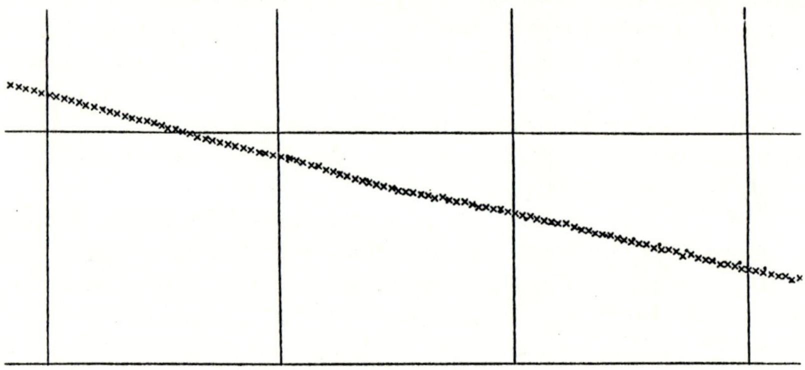

Fig. 9.18 Repeat of flight trial (courtesy of the CAA).

radar measurement of the aircraft position and the aircraft's own measurement using multiple DME is very close. The analysis of track smoothness showed that Clee Hill had been transformed from the worst site for monopulse performance to the best at that time.

It is not usually practical or economical to construct a screen around such antennas. For example, it cannot be done if the SSR antenna is mounted on top of a primary radar antenna. The alternative is to use an antenna with an equivalent performance such as the large vertical aperture antenna shown in Fig. 4.1 (Chapter 4). This antenna's vertical radiation pattern is shown in Fig. 9.19 superimposed on the vertical pattern of the linear array as modified by the screen. It can be seen that the pattern of the large vertical aperture antenna is as good as or better than the screen. Experience has shown that the large vertical aperture antenna is equally effective at minimizing the effects of multipath due to sloping ground.

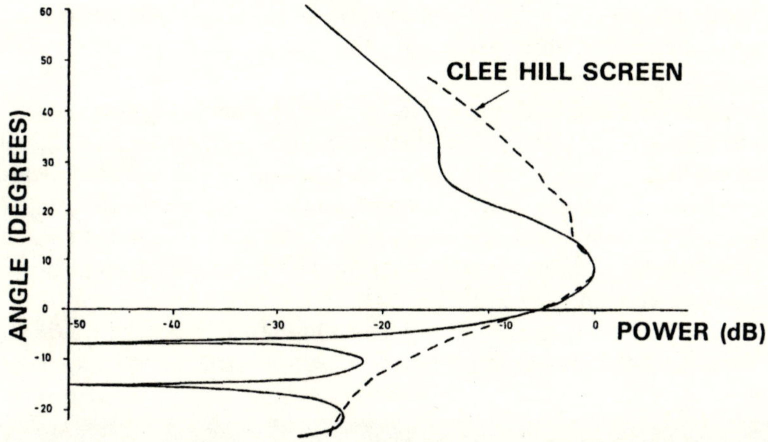


**Fig. 9.19** Vertical radiation patterns of the large vertical aperture antenna and the screen.

# 9.2 Types of interference

For the SSR system, interference can be divided into two classes—unwanted signals and the absence of wanted signals. Unwanted signals include fruit SSR replies, which are transponder responses to other interrogators, IFF Mk XII replies and DME signals. Other future unwanted signals may arise from the military JTIDS and IFF Mk XV systems. The absence of wanted signals includes the lack of aircraft replies owing to transponder occupancy with other interrogations and the suppression of the transponder by other onboard equipment. Multipath can be regarded as a form of interference, but this subject is treated separately in Section 9.1.

## 9.2.1 Unwanted signals

**Fruit**

An aircraft's antenna is sensitive to interrogation signals received over a 360° arc in azimuth and from sources with a range out to the radar horizon. If the received signal exceeds the transponder minimum trigger level, the transponder emits a reply transmission which is radiated omnidirectionally. Consequently a reply initiated by an interrogator may be detected by many other interrogators over a radius of up to 250 n miles from the aircraft and 500 n miles from the originating interrogator. These extraneous replies are known as "fruit".

By their nature, fruit replies are unwanted by all interrogators except that which triggered the reply. Fruit replies can combine to form a target report for an aircraft which does not exist at that range or bearing. Fruit can also corrupt synchronous replies, thus preventing the correct decoding of an aircraft's identity code and flight level.

Figures 9.20 and 9.21 show a radar display presenting raw video replies including fruit. The replies are characterized by the F1 and F2 framing pulses with some intervening code pulses. The location of aircraft is indicated by synchronous replies, e.g. just north and south of due west in Fig. 9.20. Due east synchronous replies can be seen which persist for an extended arc of about 90°. The latter is an example of ring-around and is caused by a failure of the sidelobe suppression system.

Figure 9.21 shows an example where a high fruit rate was experienced. In this case the interrogator transmitter was inhibited so that there are no

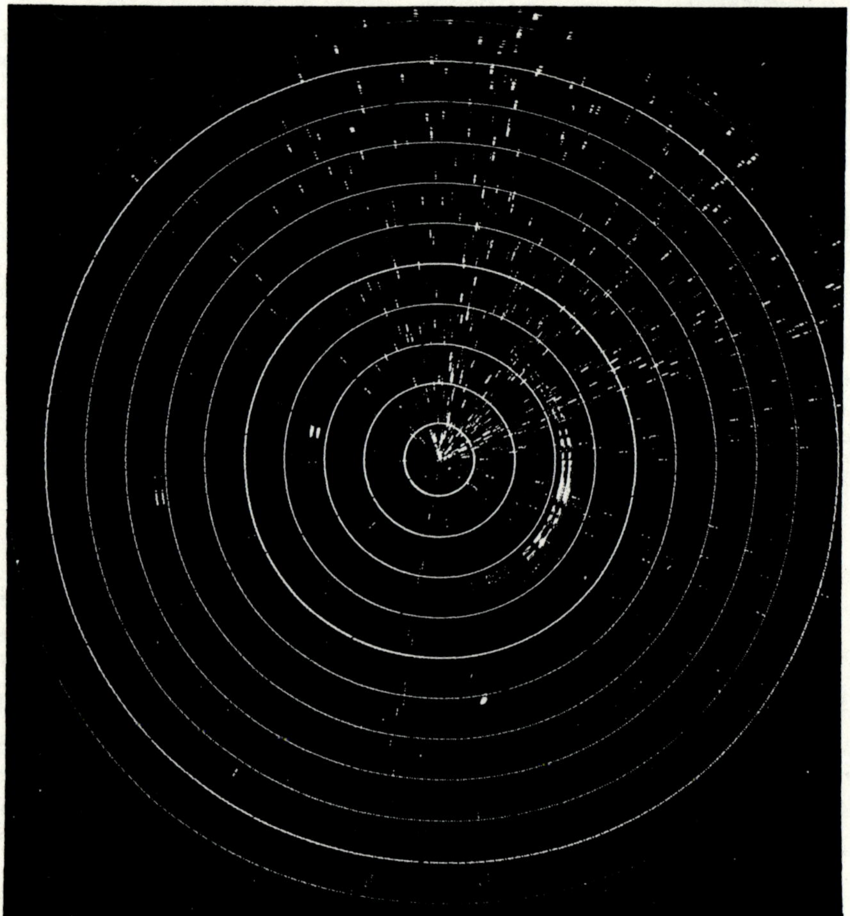

**Fig. 9.20** Raw video display showing fruit and synchronous replies.

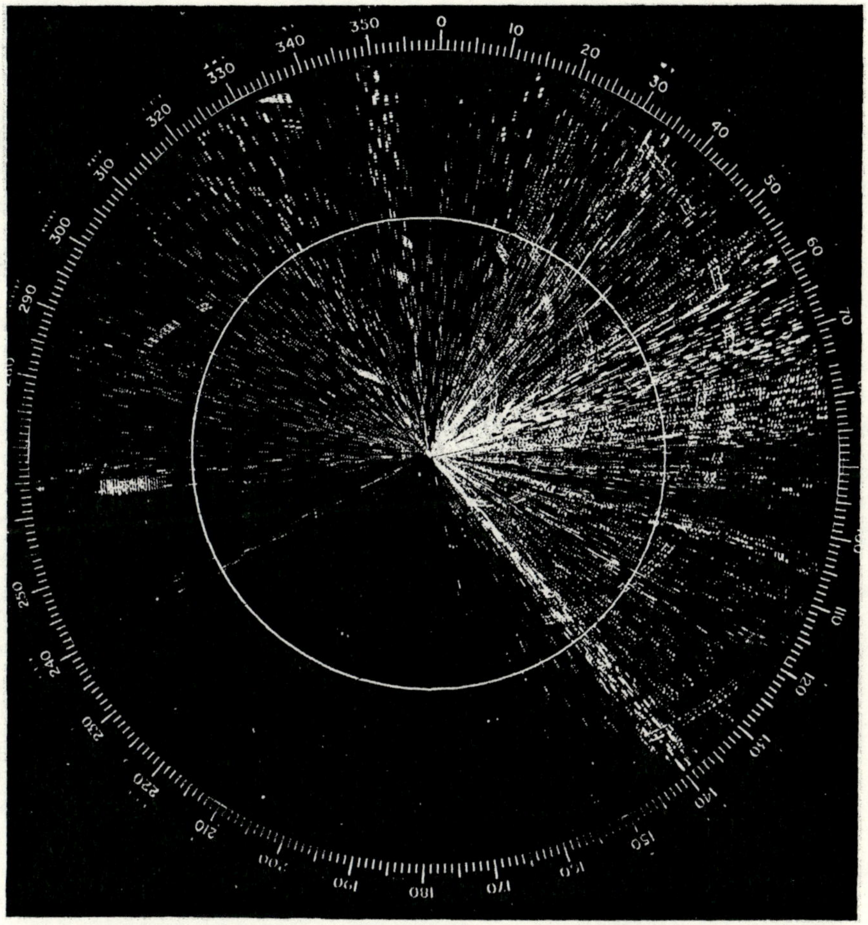


**Fig. 9.21** Raw video display showing near-synchronous fruit.

synchronous replies. However, to the east there are examples of replies which are producing a spiral pattern on the display and are caused by a nearby interrogator operating on a similar interrogation pulse repetition frequency. A closer degree of synchronism with this other interrogator would produce numerous false aircraft, particularly if a plot extractor were employed.

About two-thirds of the fruit is received in sidelobes and can be removed if a receiver sidelobe suppression (RSLS) facility is employed. Some of the remaining fruit is weak in signal strength and, for short ranges, will fall below the threshold set by the receiver gain–time control (GTC).

Fruit is caused by real aircraft replying to real interrogators and therefore the fruit rate is deterministic. However, it will vary with antenna direction and with time of day as the disposition of air traffic changes, and furthermore the necessary characteristics of the other interrogators are usually not known.

As an alternative to calculating the fruit rates that can be experienced by an SSR site, it is more usual to assume a value for the purposes of equipment design. Measurements at typical sites suggest a practical specification of 4000 fruit per second mean and 8000 fruit per second peak. For military applications *STANAG 4193* [9.8] suggests between 5000 and 20 000 fruit per second depending upon the range of the system.

The effect of fruit is usually small as the probability of sufficient fruit appearing at the same range to form a target report is itself very small.

**Distance-measuring equipment**

DME can be likened to inverted SSR where the interrogator is airborne and the transponder is ground based. The system is used by aircraft for purposes of navigation and provides the distance to the charted location of the ground equipment. Unlike SSR, the bearing is not measured but the position can be determined by measuring the ranges of a number of ground installations. Also unlike SSR, DME uses multiple frequency channels so that the distance to a required location necessitates operation on the frequency of that ground facility.

The operating frequency of DME extends from 960 to 1215 MHz and includes the SSR frequencies of 1030 and 1090 MHz. Interference with SSR is reduced by not normally using the DME channels within the bands 1021–1040 MHz and 1085–1093 MHz. The 20 MHz guardband around 1030 MHz reflects the fact that a DME interrogator and SSR transponder can exist in close proximity on an aircraft. DME channel 70X uses an airborne interrogation frequency of 1094 MHz which is within the passband of most ground SSR receivers. Inspection of an SSR receiver video signal using an oscilloscope will often show evidence of DME pulses which may be due to mistuned airborne DME equipment.

A DME pulse is approximately Gaussian in shape with a rise and fall time of about 2.5 μs and a duration of nominally 3.5 μs. The rise time for the precision mode of DME (DME/P) is approximately 1.5 μs. Two pulses are transmitted with various pulse spacings including 21 μs for Mode Z. The pulse repetition frequency can be 150 Hz during search mode, falling to 30 Hz when the DME transponder is acquired.

DME pulses could be mistaken for multiple overlapping SSR reply pulses. Some SSR equipment contains long pulse rejection circuits, but these also reject multiple overlapping SSR pulses or pulses elongated by multipath and so degrade the SSR performance. Although DME interrogation signals could be misinterpreted as SSR fruit replies there is little evidence that any system degradation results. This conclusion could change when DME/P Mode Z receives more common usage.

During a DME interrogation it is common to suppress the onboard SSR transponder, and this is a cause of the failure of the transponder to reply to an SSR interrogation from the ground.

### 9.2.2 Loss of wanted signals

**Capture**

When a transponder receives an interrogation and transmits a reply, a short time is required to allow the power supply to recover and the transmitter output stage to cool. ICAO *Annex 10* permits this dead time to last for a maximum of 125 μs after the transmission of the last reply pulse. A typical value is 25 μs. The time for the interrogation and the time to transmit the reply, which is typically 40 μs, must be added to the above dead time. This gives a total capture period of 65 μs.

Capture reduces the opportunity to reply to another interrogator. If an interrogation rate of 200 Hz and an effective antenna beam width of 5° are assumed, the transponder reply probability is

$$1-\frac{65\times 10^{-6}\times 200\times 5}{360}=1-181\times 10^{-6}=0.99982$$

Capture can also result from a sidelobe suppression signal, and the suppression period can last from 25 to 45 μs with a typical value of 35 μs. Adding to the 2 μs duration of the interrogator P1, P2 suppression pair gives a capture period of 37 μs during which the transponder cannot reply to another interrogation. If an interrogation rate of 200 Hz and an angular extent of the antenna suppression signals of 90° are assumed, the transponder reply probability is

$$1-\frac{37\times 10^{-6}\times 200\times 90}{360}=1-1850\times 10^{-6}=0.99815$$

Adding the transmitter recovery and the sidelobe suppression periods together gives

$$1-2031\times 10^{-6}=0.99797$$

If the aircraft were within range of 10 such interrogators and their interrogations were randomly related, the reply probability would be

$$1-10\times 2031\times 10^{-6}=0.9797$$

Aircraft are often within cover of a large number of interrogators, sometimes in excess of 100. However, most would be beyond the range of the antenna sidelobes and so the loss of replies due to suppression would not occur. For example, an aircraft within range of only the main beams of 100 interrogators would present a reply probability of

$$1 - 100 \times 181 \times 10^{-6} = 0.982$$

In the above discussion it is assumed that all interrogations are randomly related. This clearly does not apply in the case illustrated in Fig. 9.21 where another interrogator is generating near-synchronous fruit. Capture by this other interrogator could extend over several interrogation periods. Given one lost reply there is a higher probability that the subsequent reply will also be lost. This effect is known as group capture [9.9]. The likelihood of group capture can be avoided if interrogation timing is sufficiently randomized, or jittered, to escape synchronism with other interrogations.

Capture is not the sole mechanism which can deny replies to an interrogation.

**Reply rate limiting**

The transponder transmitter and power supply must be protected from overload should the transponder be subjected to excessive interrogations. Protection is effected by progressively desensitizing the transponder receiver so that replies are denied to the weakest interrogations first. The onset of reply limiting can be set in the transponder to between 500 and 2000 replies per second. If the interrogation rate is in excess of 150% of the selected reply limit, the receiver desensitization is greater than 30 dB.

**Onboard suppression**

Other systems on aircraft operate on similar frequencies to SSR and there is a possibility of coupling between the antennas such that the transmission of one system could be detected by the receiver of another system and cause it to generate interference.

One such system to which this comment applies is DME. By interconnection between the suppression terminals on the DME and on the transponder, the SSR transponder can be inhibited from producing a reply during a DME transmission. The period of the suppression can vary between about 7 and 60 μs (average value, 45 μs), depending upon the DME mode, and at rates between 30 and 150 Hz for the track and search phase respectively. If operation on DME mode Y during the search phase is assumed, the SSR transponder reply probability is

$$1 - 150 \times 45 \times 10^{-6} = 0.9325$$

TCAS equipment (see Chapter 13) includes an airborne SSR interrogator in addition to a transponder. During an interrogation the transponder is suppressed so that it cannot itself reply. The TCAS interrogation rate is nominally 1 per second and the effect on the transponder reply probability is small.

Many military aircraft use radio communication equipment operating in the u.h.f. band, and the SSR transponder can be suppressed during radio transmission. The operation is sporadic and its impact on the availability of the transponder to reply to interrogations is difficult to quantify but is normally neglected.

**Antenna switching**

Some military aircraft employ an upper and a lower SSR antenna so that at least one antenna is in view of a ground interrogator when the aircraft is maneuvering. An example of the effect on the replies received is shown in Fig. 8.1 (Chapter 8). Aircraft equipped in this way can present reply probabilities of less than 50%.

# References

9.1 Stevens, M. C. Multipath and interference effects in secondary surveillance radar systems, *Proc. Inst. Electr. Eng., Part F*, **128** (1), 43–53, 1981.

9.2 Skolnik, M. I. (ed.) *Radar Handbook*, pp. 2–38, McGraw-Hill, New York, 1970.

9.3 Zatkalik, J., Sengupta, D. L. and Tai, C. T. Sidelobe suppression mode performance of ATCRBS with various antennas, *Rep. FAA-RD-75-31*, FAA, Washington, DC, 1975.

9.4 Sengupta, D. L. and Zatkalik, J. On the performance of air traffic control radar beacon system, *IEEE Trans. Aerosp. Electron. Syst.*, **12** (4), 494–502, 1976.

9.5 Freedman, J. E. Dealing with false targets in the air traffic control radar beacon system, *Eastcon 77*, pp. 10.3A–10.3J, IEEE, New York, 1977.

9.6 Brook-Footit, E. Equipment for suppressing reflected secondary radar signals, *IEE Conf. Publ.*, **105**, 38–42, 1973.

9.7 Hocking, T. T., Plant, R. J. and Wyndham, B. A. Evaluation of the Clee Hill near-in ground screen, *CAA Pap. 82012*, CAA, London, August 1982.

9.8 Standardization Agreement "Technical characteristics of IFF Mk XA and Mk XII interrogators and transponders", *STANAG 4193, Edition 3*, NATO, Brussels, September 1986.

9.9 Stevens, M. C. and Jenkins, V. Model of the secondary surveillance radar environment of the United Kingdom, *Proc. Inst. Electr. Eng.*, **121** (5), 340–344, 1974.

# Chapter 10

# System Design

## 10.1 The radar equation as applied to secondary surveillance radar

**Path attenuation**

Unlike primary radar, SSR is involved in only a one-way path loss since sufficient signal, received at the aircraft, will trigger the transponder into replying at full power. The following equation gives the radar range:

$$R = \frac{\lambda}{4\pi}\left(\frac{P_t G_t G_r}{L P_r}\right)^{1/2}$$

where $R$ is the radar range (the same units as $\lambda$), $\lambda$ is the wavelength, $P_t$ is the transmitter power at the antenna terminal, $G_t$ is the transmitter antenna gain, $G_r$ is the receiver antenna gain, $P_r$ is the receiver sensitivity at the antenna terminal and $L$ is the additional system losses such as atmospheric, beam shape effects etc. This can be rearranged to express the power at the receiving antenna terminals in terms of the other system variables:

$$P_r = \frac{P_t G_t G_r}{L}\left(\frac{\lambda}{4\pi R}\right)^2$$

The factor $(\lambda/4\pi R)^2$ can be expressed as a loss:

$$a = -37.801 - 20 \log f - 20 \log d$$

where $f$ is the frequency in megahertz ($c = 299\,792$ km s$^{-1}$; this ignores the ground level air dielectric constant of 1.0003), $d$ is the range in nautical miles and $a$ is the loss in decibels due to the dilution of the signal with range. The loss $a$ is reduced by the transmitting and receiving antenna gains and is increased by any r.f. cable losses, the reduction in antenna gain due to the necessity to operate over a given beam width and any additional losses due to atmospheric absorption.

Since only two frequencies are used in SSR the equation can be written as follows:

$$\text{uplink (1030 MHz)} \quad a = -98.058 - 20 \log d$$
$$\text{downlink (1090 MHz)} \quad a = -98.550 - 20 \log d$$

The equation for the uplink loss is plotted in Fig. 10.1, and the value of the downlink loss can be obtained from this by adding $-0.5$ dB.

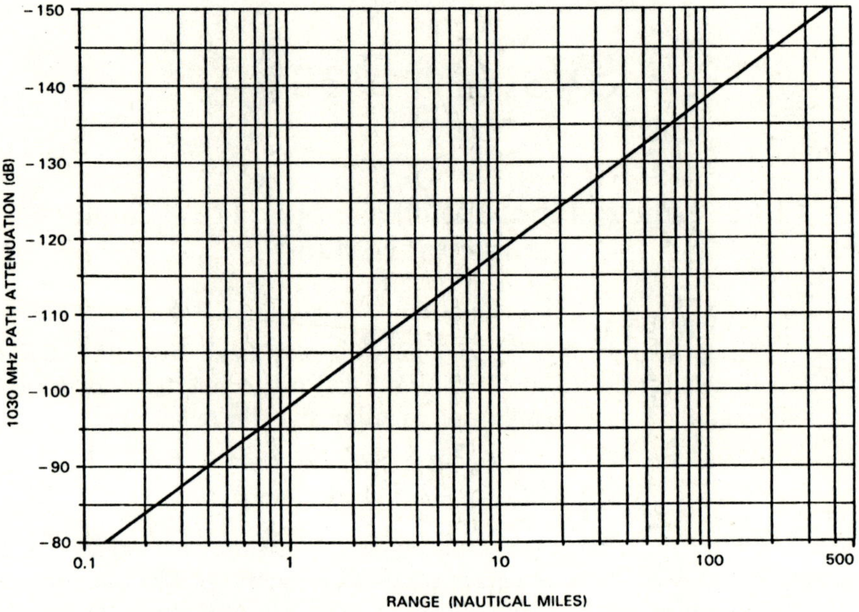


**Fig. 10.1** Uplink loss versus range.

**Atmospheric absorption**

The atmosphere introduces a small amount of loss to the SSR signals which can be neglected for short distances but not for long distances. Absorption is due to the presence of oxygen and is proportional to the atmospheric density; it is greater for long ranges and low elevation angles. The lowest practical elevation angle for SSR is about 0.25°. Figure 10.2 shows a plot of atmospheric absorption against distance for various elevation angles and is valid for both SSR frequencies.

**Refraction and lens loss**

SSR signals experience a small amount of refraction as they pass through the atmosphere as a result of the air pressure differential with height. The refraction bends the SSR signals towards the ground so that their range is beyond the physical horizon. An assumption of a linear decrease in refractive index with height of $3.93 \times 10^{-5}$ per kilometer results in a radar range performance which would apply if the Earth's radius was four-thirds greater than its true value. The assumption is faulty since it results in the refractive index's decreasing below unity at a height of about 8 km, but the simplicity of the four-thirds model sustains its widespread use. A more accurate law for the change in refractive index with height is an exponential function. Figure 10.3 is based on the four-thirds Earth model and shows the height that an aircraft must be above the ground to subtend a given elevation angle.

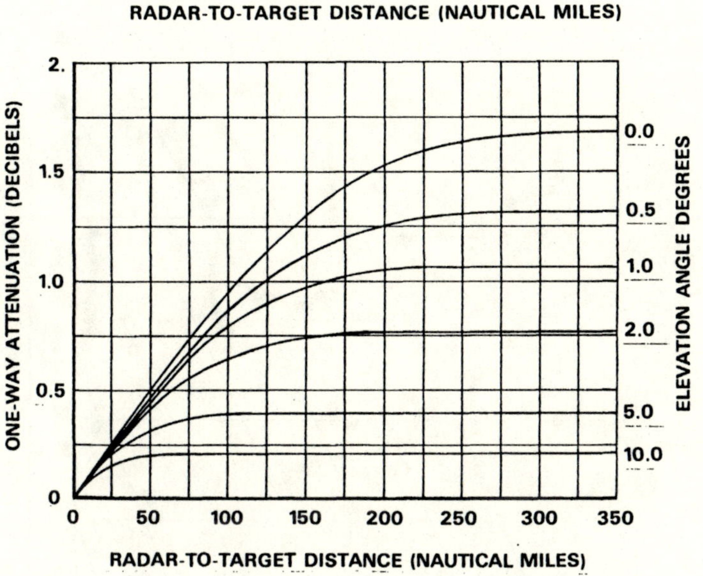


**Fig. 10.2** Atmospheric absorption versus range for various signal elevation angles (interpolated from Figs 5.15 and 5.16 of ref. 10.1).

Weil [10.2] has shown that refraction can also defocus the radar beam to produce an effective signal strength loss. In free space a wavefront expands proportionally to the square of the distance to cause the familiar range-squared attenuation effect usually assumed. In air the change in refractive index with height causes an additional wavefront expansion and the consequential additional loss is plotted in Fig. 10.4.

**Anomalous propagation**

Anomalous propagation effects which temporarily increase the range of the radar can sometimes occur. Atmospheric temperature inversion or the occurrence of moisture-laden air close to a water surface can increase the refraction of radio waves so that they closely follow the Earth's surface. During instances of temperature inversion, aircraft have been detected leaving airports 230 miles distant, and well below the horizon, although the targets are usually lost again as they climb. False targets due to reflections from objects normally too far away to present a problem can also occur.

Over the sea, an evaporation duct which alters the vertical refractivity gradient and traps low angle radiation can be produced. Experiments using SSR over the North Atlantic [10.3] show that target ranges can double owing to this effect as shown in Fig. 10.5. Since the radar signal is confined within the upper and lower boundaries of the ducts it can diverge only in the horizontal plane. As a result the reduction in signal strength with range more closely

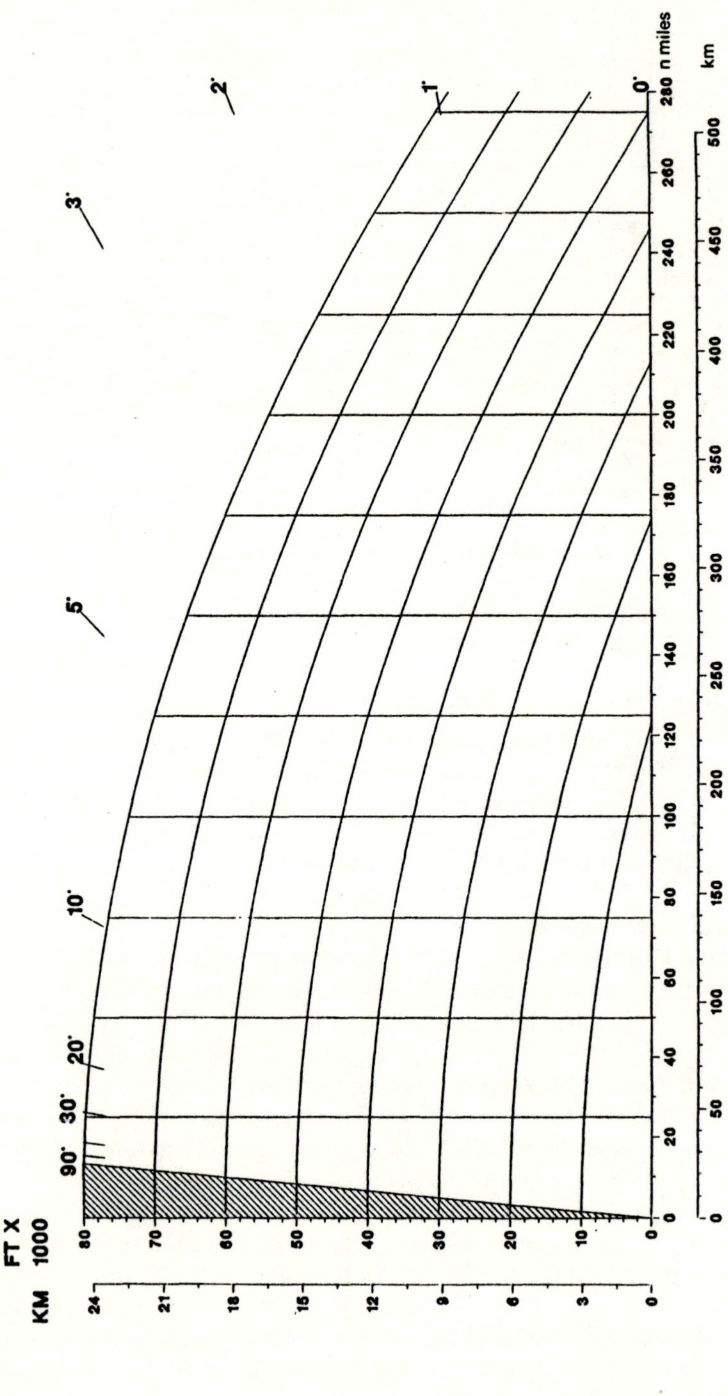


Fig. 10.3 Range–height diagram.

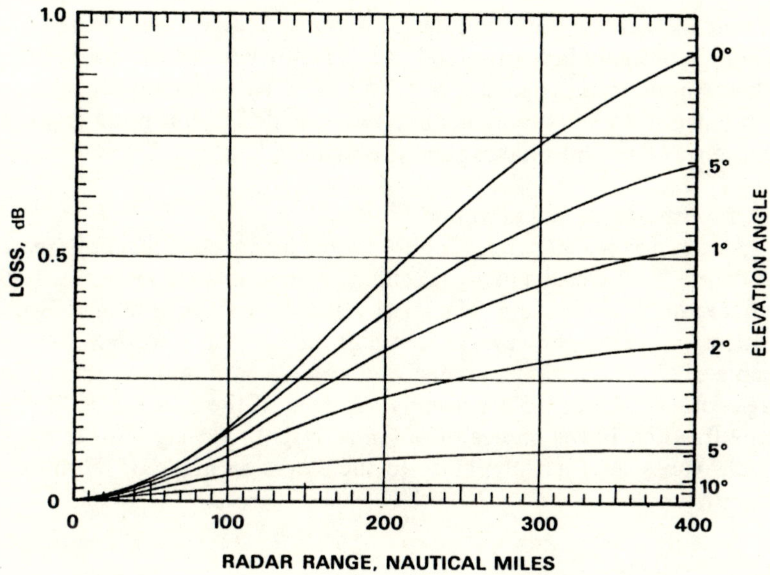


**Fig. 10.4** Lens loss for various signal elevation angles (from ref. 10.1).

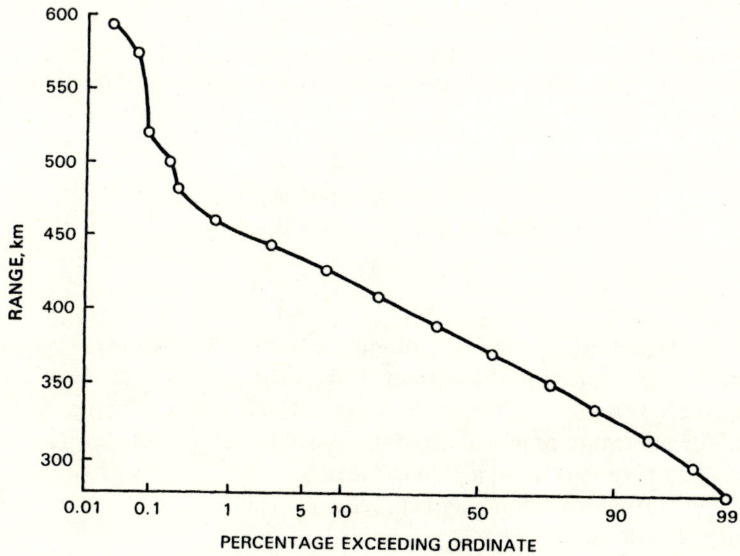


**Fig. 10.5** Distribution of maximum ranges over the North Atlantic (from ref. 10.3, courtesy of the CAA).

follows a linear law instead of the more normal square law. As a consequence the range of the radar can extend considerably beyond the normal free-space value. Anomalous propagation is influenced by too many variables for this phenomenon to be reliably included in SSR design but point is given to the avoidance of second-time-around problems (Chapter 5, Section 5.1.3).

**Loss due to ground reflection**

Antenna vertical pattern lobing caused by ground reflection (see Chapter 9, Section 9.1.1) results in signal enhancement at some elevation angles and signal reduction at others. The large vertical aperture antenna is valuable in reducing the effect; however, the null at the horizon (zeroth null) is still very deep and the radar range is much reduced. A minimum practical elevation angle, often 0.25° or 0.5°, is usually assumed. If the SSR system designer has some freedom in the choice of antenna height, he may wish to choose one which places the first lobe close to the elevation angle of the most distant aircraft.

## 10.2 Uplink and downlink power budgets

The power budget relates the effective radiated power (e.r.p.) of the transmitter, the propagation losses and the power at the receiving antenna to give a required signal at the input to the receiver.

**Ground antenna gain and feeder loss**

For the purpose of this discussion the ground system is assumed to have the following characteristics (losses are shown as negative and gains are shown as positive):

| | | |
|---|---|---|
| feeder losses, interrogator to antenna (including 0.8 dB for the rotating joint) | = | −3.5 dB |
| antenna gain at 0.25° elevation | = | 24.0 dBi |
| total | = | 20.5 dB |

(1) A large vertical aperture antenna is assumed with a rapid rate of gain reduction at the horizon. The antenna has a peak gain of 27.5 dB and is tilted so that there is a gain reduction of 4 dB at the horizon (Chapter 9, Section 9.1.1). The minimum practical elevation angle for target detection is assumed to be 0.25° where the antenna gain is 24.0 dB.

(2) A further loss, typically of 0.2 dB, should be included if the antenna is housed in a radome.

**Airborne antenna gain and feeder loss**

The airborne antenna is usually the equivalent of a quarter-wave monopole mounted above a reflecting surface (the fuselage). The gain of such an antenna is influenced by the fact that the fuselage is not an ideal ground plane and by obstruction by engines and wings. The result is an irregular antenna radiation pattern (see Chapter 8, Fig. 8.2). An effective gain of 0 dB is normally assumed. Airborne cable losses of 3 dB are conventionally assumed, but these are usually included in the specification of parameters of the transponder so that they can be ignored.

**Transmitter power and range calculations**

The available transmitter power may be limited by equipment or ICAO e.r.p. restrictions. The system designer's task is to determine the antenna beam width over which replies will be obtained and then the necessary number of interrogations to be made in that beam width to realize the required probability of detection. At short ranges the beam width will be determined by the sidelobe suppression function, but at longer ranges it will be limited by path attenuation and only signals near the peak of the antenna beam will reach the aircraft. In most cases the range of the SSR is dictated by the uplink power budget.

## 10.2.1 Uplink

For many years ICAO *Annex 10* restricted the e.r.p. at a ground interrogator antenna to 52.5 dBW in order to limit the interrogations received by aircraft at extreme range. This limitation was appropriate when the common form of ground antenna was a linear array with the peak gain of the vertical beam close to the horizon as shown in Fig. 4.8 (Chapter 4).

The recent adoption of antennas with a large vertical aperture and with a peak gain several degrees above the horizon (see Chapter 4, Fig. 4.8) have caused the *Annex 10* restriction to be removed. These antennas do not radiate their maximum power at the low elevation angles presented by distant aircraft. However, in most circumstances the limit of 52.5 dBW can still be used. If an antenna peak gain of 27.5 dB and an e.r.p. limit of 52.5 dBW are assumed, the maximum transmitter power at the antenna is 25 dBW. Cable losses, which have been assumed to be 3.5 dB in the present example, must be accommodated, and so the transmitter power at the output of the interrogator cabinet is 28.5 dBW.

The interrogator signal strength, as detected at the airborne transponder receiver, can be calculated for various aircraft ranges assuming an elevation angle of 0.25° (Table 10.1).

The nominal minimum trigger level (m.t.l.) of the transponder receiver is −71 dBm and the excess received signal over the m.t.l. can be found. This excess signal level only applies at the peak of the ground antenna horizontal interrogate beam (Chapter 4, Fig. 4.2). The ground antenna gain is less on

**Table 10.1**

| | Range | | | |
|---|---|---|---|---|
| | *0.5 n miles* | *100 n miles* | *200 n miles* | *250 n miles* |
| Effective radiated power | +52.5 dBW | +52.5 dBW | +52.5 dBW | +52.5 dBW |
| Antenna gain decrease at 0.25° | −3.5 dB | −3.5 dB | −3.5 dB | −3.5 dB |
| Path loss at 1030 MHz | −92.0 dB | −138.1 dB | −144.1 dB | −146.0 dB |
| Atmospheric attenuation | 0.0 dB | −0.9 dB | −1.4 dB | −1.6 dB |
| Lens loss | 0.0 dB | −0.1 dB | −0.2 dB | −0.3 dB |
| Transponder received signal | −43.0 dBW or −13.0 dBm | −90.1 dBW or −60.1 dBm | −96.7 dBW or −66.7 dBm | −98.9 dBW or −68.9 dBm |

each side of the peak but the transponder will reply when the m.t.l. is exceeded. The beam width over which replies can be triggered can be calculated (Table 10.2).

**Notes**

(1) The gain reduction at beam edges refers to the horizontal beam width, the plane in which the antenna rotates. An aircraft at a higher elevation angle than the assumed 0.25° will receive a stronger signal and reply over a wider beam width.

(2) If the transponder minimum m.t.l. of −69 dBm is used instead of the nominal value of −71 dBm, the range of 250 n miles is barely achievable with the assumed e.r.p. of 52.5 dBW unless benefit can be obtained from ground reflection.

(3) The working beam width at a range of 0.5 n miles is determined by the ISLS function and not by antenna gain reduction at the beam edge.

## 10.2.2 Downlink

ICAO *Annex 10* defines the typical transponder transmitter output power, as measured at the antenna end of the r.f. cable, as 24 dBW. With an assumed antenna gain of 0 dB, the transponder effective radiated power for a reply is 24 dBW. The transponder reply signal strength at the interrogator receiver can be calculated for an elevation angle of 0.25° (Table 10.3).

It is interesting to note that, with the exception of the 0.5 n miles example where the reply is controlled by the sidelobe suppression function, the beam

**Table 10.2**

| | Range | | | |
|---|---|---|---|---|
| | *0.5 n miles* | *100 n miles* | *200 n miles* | *250 n miles* |
| Transponder m.t.l. | −71.0 dBm | −71.0 dBm | −71.0 dBm | −71.0 dBm |
| Transponder received signal | −13.0 dBm | −60.1 dBm | −66.7 dBm | −68.9 dBm |
| Excess signal | 58.0 dB | 10.9 dB | 4.3 dB | 2.1 dB |
| Working beam width | 5.5° | 4.3° | 2.8° | 2.0° |

**Table 10.3**

| | Range | | | |
|---|---|---|---|---|
| | *0.5 n miles* | *100 n miles* | *200 n miles* | *250 n miles* |
| Effective radiated power | +24.0 dBW | +24.0 dBW | +24.0 dBW | +24.0 dBW |
| Path loss at 1090 MHz | −92.5 dB | −138.6 dB | −144.6 dB | −146.5 dB |
| Atmospheric attenuation | 0.0 dB | −0.9 dB | −1.4 dB | −1.6 dB |
| Lens loss | 0.0 dB | −0.1 dB | −0.2 dB | −0.3 dB |
| Ground antenna gain | +27.5 dB | +27.5 dB | +27.5 dB | +27.5 dB |
| Antenna gain decrease at 0.25° | −3.5 dB | −3.5 dB | −3.5 dB | −3.5 dB |
| Ground feeder losses | −3.5 dB | −3.5 dB | −3.5 dB | −3.5 dB |
| Ground received signal | −48.0 dBW or −18.0 dBm | −95.1 dBW or −65.1 dBm | −101.7 dBW or −71.7 dBm | −103.9 dBW or −73.9 dBm |
| Gain reduction at beam edge | −22.0 dB | −10.9 dB | −4.3 dB | −2.1 dB |
| Signal at beam edge | −40.0 dBm | −76.0 dBm | −76.0 dBm | −76.0 dBm |

edge reply signal is constant regardless of the range of the aircraft. This somewhat surprising result is explained by the change in the working beam width with range caused by the necessity for the interrogation to exceed the m.t.l. of the transponder.

The minimum signal level of the interrogator receiver is typically −80 dBm, and the calculation in Table 10.3 shows that this level is exceeded by 4 dB. This example shows that the downlink has a greater margin than the uplink, and this is frequently the case. Replies could be received over a wider antenna beam width if they were transmitted by the transponder.

# 10.3 Reply and detection probabilities

## 10.3.1 Transponder reply probability

A transponder may be prevented from replying to an interrogation for the following reasons.

(a) It is already replying or recovering from a reply to another interrogation.

(b) It is suppressed from replying by a sidelobe suppression interrogation.

(c) It is suppressed from replying by other onboard equipment, e.g. DME.

All these causes are external to the ground interrogator in question. The effects can be estimated if the number of other interrogators within range of a given aircraft are known together with their interrogation rates and the antenna main beam and sidelobe e.r.p.s. As long as the ground interrogator pulse repetition frequency is unlike the pulse repetition frequency of any other interrogator, the effects can be assumed to be random [10.4].

However, the data needed for such an analysis are often difficult to obtain with sufficient accuracy. Usually the equipment designer is required to work with a transponder reply probability stated in a requirement specification. A typical stated reply probability is 0.90 although measured values often exceed 0.95. Lower values of reply probability were appropriate for sites where a large number of other SSR interrogators were also operating locally with high interrogation rates and without sidelobe suppression, but such conditions rarely apply today.

With the developing widespread use of monopulse SSR and its lower interrogation rate the transponder reply probability is likely to improve still further.

## 10.3.2 Detection probability

The target detection probability depends upon the transponder reply probability and the minimum number of replies necessary to determine that a target exists. These parameters are related by the following expression:

$$P_d = \sum_{j=t}^{n} \frac{n!}{j!(n-j)!} p_j (1-p)^{n-j}$$

where $P_d$ is the probability of target detection, $n$ is the sliding window length, $t$ is the minimum reply threshold and $p$ is the probability of a single reply (note that the assumption is made that this is solely dependent on the transponder reply probability and not a function of the signal-to-noise ratio).

Good target detection probability has to be balanced against the probability of false alarms. In the case of SSR, false alarms are not due to receiver noise or clutter but can be caused by fruit replies or ghost aircraft due to reflections. Fruit will only be a significant cause if the fruit rate is very high and the range bins are very wide.

If a sliding window plot extractor is employed (Chapter 6), a relatively high number of replies are required in order to obtain reasonable target bearing measurement accuracy and to discriminate against ghost aircraft. Ghost aircraft tend to have fewer replies because of the signal attenuation caused by the reflector. A typical requirement might be a minimum of six replies as a compromise between good true target detection and good ghost aircraft rejection. Such a number would also produce sufficient Mode A and Mode C replies for repetition to confirm the accuracy of the reply data. Table 10.4 shows the probability of obtaining six replies, within a given window length, for three different transponder reply probabilities.

In the case of monopulse SSR (Chapter 7) detection probability is higher using fewer interrogations because more information is available to assist video processing. In particular, the use of monopulse direction finding, tracking and improved rejection of ghost aircraft removes the need for a high interrogation rate.

A typical monopulse SSR requires only two replies to detect a target and reject single fruit replies. Other processing reduces false alarms to a very low value. Table 10.5 shows the probability of obtaining two replies, within a given window length, for three different transponder reply probabilities.

## 10.4 Transmitter pulse repetition frequency

The interrogator pulse repetition frequency is dependent upon the required detection probability, the antenna working beam width at the maximum target range and the antenna turning rate. The antenna turning rate controls the target position update rate on the ATC display and can vary from 5 rpm for an en route radar to 15 rpm for a terminal area radar. These turning rates give a target position update rate of 12 s and 4 s respectively.

For the purposes of discussion it will be assumed that an antenna turning rate of 12 rpm applies and a target detection probability of 99% is required at a range of 200 n miles with a transponder reply probability of 0.9. The power budgets and antenna characteristics of Section 10.2 and the detection probabilities of Tables 10.4 and 10.5 apply. At 200 n miles the antenna working beam width is 2.8°.

The required interrogation rate is found using the following simple equation:

$$\text{interrogations per second} = \frac{6 \times \text{antenna rpm} \times \text{window length}}{\text{working beam width}}$$

For a sliding window process, nine interrogations in the working beam are required. Therefore the interrogation rate is 231 per second. In contrast, for a monopulse process, four interrogations in the working beam are required with an interrogation rate of 103 per second. In the case of the sliding window detector it is usual to employ a higher interrogation rate than that given above in order to improve the bearing measurement accuracy.

**Table 10.4** Sliding window SSR target detection probabilities

| *Window length* | *Probability of at least six replies* | | |
|---|---|---|---|
| | *Reply probability* | | |
| | *0.80* | *0.90* | *0.95* |
| 6 | 0.262 | 0.531 | 0.735 |
| 7 | 0.577 | 0.850 | 0.956 |
| 8 | 0.797 | 0.962 | 0.9942 |
| 9 | 0.914 | 0.9917 | 0.99936 |
| 10 | 0.967 | 0.9984 | 0.999936 |
| 11 | 0.988 | 0.99970 | 0.9999942 |
| 12 | 0.9961 | 0.999950 | 0.9999995 |

**Table 10.5** Monopulse SSR target detection probabilities

| *Window length* | *Probability of at least two replies* | | |
|---|---|---|---|
| | *Reply probability* | | |
| | *0.80* | *0.90* | *0.95* |
| 2 | 0.64 | 0.81 | 0.903 |
| 3 | 0.896 | 0.972 | 0.9928 |
| 4 | 0.973 | 0.9963 | 0.99952 |
| 5 | 0.9933 | 0.99954 | 0.999970 |
| 6 | 0.99840 | 0.999945 | 0.9999982 |

# 10.5 Bearing accuracy

## 10.5.1 Sliding window

In a sliding window plot extractor the aircraft bearing is determined by averaging the antenna direction when the aircraft is first detected with the antenna direction when it ceases to be detected. Aircraft detection is declared when a minimum number of replies exist in the window. Once target detection is achieved the threshold is reduced to maintain target acquisition. When the number of replies in the window falls below this reduced threshold the target is assumed to have passed out of the beam. Typically the difference in detected replies for the trailing edge threshold is one less than that for the leading edge threshold. However, if the aircraft reply probability is low, the difference required between the two thresholds is greater if erroneous trailing edges and split plots are to be avoided.

For discussion purposes, consider the example used in Section 10.4 where the requirement is for at least six replies in a window length of nine to give a target detection probability of 99% for an individual reply probability of 0.9. The required six replies could be obtained after six, seven, eight or nine interrogations. Therefore the delay in recognizing the target can vary between six and nine interrogations depending upon the reply performance of the aircraft. There is one possibility that the leading edge criteria is reached in the first six interrogations if chance dictates that all six replies are received. In addition to that one possibility there are seven ways that six replies could be obtained from seven interrogations, 56 ways from eight interrogations and 504 ways from nine interrogations.

Superficially it could be argued that since 89% of detection will require nine interrogations the uncertainty of the angle of the target leading edge is reduced. In practice the reply probability varies over a wide range over a short period of time as it is dependent on the instantaneous number of interrogations and suppressions being received from other interrogators. The result is that the leading edge detection should be regarded as requiring between six and nine interrogations with equal weight. Using an interrogation rate of 231 per second, as determined above, gives an interval of 0.31° between interrogations. In turn this gives an angular uncertainty of $3 \times 0.31 = \pm 0.47°$ in the leading edge detection, with a similar uncertainty in the trailing edge position.

Note that if the individual reply probability is 0.8, instead of the 0.9 assumed, a window length of 12 is required for the same target detection probability. In this case the angular uncertainty is doubled.

It is possible that a fruit reply from another aircraft falls at the same range as the target and distorts the detected position of the leading or trailing edge. The angular error that results is equivalent to the angular interval between successive interrogations. Unless another fruit reply is received near the other edge, the error will be halved by averaging the leading and trailing edge angles.

The interrogation rate determined in Section 10.4 and used in this example relates to a required target detection probability. Accurate bearing measurement would dictate a higher interrogation rate, and a value closer to 400 per second is typically used depending on the maximum range required and the antenna turning rate. A reduced angle between interrogations would improve the target bearing accuracy but at the expense of reducing the reply probability and increasing the fruit rate for other users of the system.

The sliding window process of measuring aircraft bearing is very dependent on the system performance at beam edges where the interrogation and reply signals are weakest and close to the transponder minimum sensitivity. Weak interrogations are most likely to be affected by transponder desensitization in regions of high interrogation rates. The edge is also the part of the antenna beam most likely to be disturbed by multipath interference (see for example Chapter 9, Fig. 9.8). Consequently the measured bearing performance of the sliding window process can be much worse than the calculated performance. Some measured data are presented in Chapter 11, Table 11.1.

## 10.5.2 Monopulse

Jacovitti [10.5] has performed a mathematical analysis of the L1 and P1 receivers described in Chapter 5, Section 5.4. In this section an error budget for the L1 receiver is discussed in terms of a typical SSR antenna with a 3 dB beam width of 2.35°.

The sources of error in the monopulse bearing measurement system are as follows:

(a) antenna pattern errors
(b) effects of receiver noise
(c) receiver errors dependent upon reply signal strength
(d) receiver errors dependent upon reply frequency
(e) errors in the conversion of video to angle units
(f) mechanical errors in the turning gear and azimuth encoder

Most of the errors are random in nature and a repeat of the measurement is likely to produce a different degree of error. Other errors are fixed, at least for a given aircraft, so that repeated measurements produce the same error.

Many of the errors can be more easily stated in terms of maximum limits. However, when determining the aggregate of all the possible errors it is recognized that it is very improbable that all errors will be at the same extreme simultaneously. To obtain a meaningful summation the errors are converted to a root mean square (r.m.s.) value and then added by finding the root sum of the squares (r.s.s.). Errors due to noise are derived from the noise power and are inherently in r.m.s. form. Some errors are approximately sinusoidal in distribution and the r.m.s. value is found by dividing the peak error by $\sqrt{2}$. Other errors have a linear distribution, with the r.m.s. value being the peak value divided by $\sqrt{3}$.

The receiver processes the monopulse angle measurement in voltage form, and this must subsequently be converted to angle form using a calibration table. The sensitivity to measurement error is dependent upon the position of the target in the antenna beam as indicated in Fig. 10.6 and is derived from the slope of the ratio of the sum and difference antenna patterns.

The deep null in the antenna difference pattern at boresight results in this location's being most immune to voltage measurement errors and provides the region of highest accuracy. The slope of the difference pattern reduces with distance from boresight, and a given voltage measurement error translates to an increasing bearing error up to approximately the crossover point of the sum and difference beams. Beyond this position in the antenna beam the slope of the sum beam increases resulting in a partial reduction of sensitivity to measurement error.

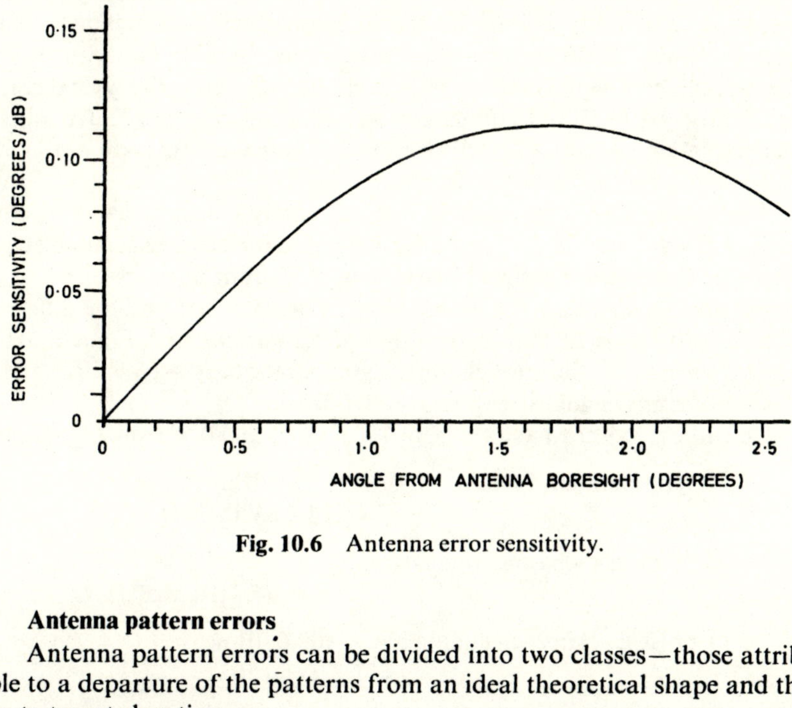


**Fig. 10.6** Antenna error sensitivity.

**Antenna pattern errors**

Antenna pattern errors can be divided into two classes—those attributable to a departure of the patterns from an ideal theoretical shape and those due to target elevation.

Experience has shown that, in the boresight region of the antenna, the patterns are very reproducible from antenna to antenna. In the region between 0.8° from boresight to beam edge at approximately 2.5° from boresight, antennas tend to exhibit some asymmetry between each side of boresight which differs from antenna to antenna.

Ideally, the use of a calibration table removes any errors due to the antenna pattern. Consequently no error contribution is specifically allocated to this source. However, allowance is made for an error in the calibration and this aspect is discussed below.

The second type of antenna pattern error that can arise occurs at high elevation angles. It is caused by geometrical distortion that results because the bearing measurement is not made in the horizontal plane. This effect is described in Chapter 4, Section 4.2.3. The error due to this effect is less than 10% up to an elevation angle of 25°. The effect can be reduced to a low level by averaging measurements made equidistantly on each side of boresight or by correcting the off-boresight angle according to the elevation angle (deduced from the height and range measurement).

**Effects of receiver noise**

The data given in Section 10.2.2 show that the typical sum-channel signal strength at beam center for an aircraft at a range of 100 n miles, ignoring any antenna pattern lobing, is −65.1 dBm. At beam center the difference-channel signal is about 30 dB below the sum-channel signal; consequently the difference-channel signal is at −95 dBm, about 5 dB above the typical receiver noise level of −100 dBm. In the case of an aircraft at a range of 250 n miles the sum-channel signal is at −73.9 dBm and the difference signal is at −104 dBm, or 4 dB below the receiver noise level.

The effect of receiver noise is approximately constant over the center region of the antenna beam, and a typical value for bearing accuracy can be derived by considering a signal arriving at 0.5° from boresight. At 0.5° the sum-channel sensitivity is 0.8 dB less than at beam center and the difference-channel sensitivity is 7.8 dB less than that of the sum channel. For a target at a range of 100 n miles the sum-channel signal strength is −65.9 dBm and the difference-channel signal strength is −73.7 dBm.

Consider an aircraft at a range of 100 n miles and 0.5° from boresight:

$$\text{sum-channel signal strength} = -65.9\ \text{dBm} = 113.4\ \mu\text{V in } 50\ \Omega$$

$$\text{difference-channel signal strength} = -73.7\ \text{dBm} = 46.2\ \mu\text{V in } 50\ \Omega$$

$$\text{noise level in each channel} = -100.0\ \text{dBm} = 2.2\ \mu\text{V in } 50\ \Omega$$

The noise voltage in each channel has a random phase relationship with the signal in that channel. The detected video amplitude is affected by noise if the noise is in phase or in antiphase with the signal. For signal-to-noise voltage ratios of better than about 2:1, or 12 dB, quadrature phase noise has only a small effect on the measured signal amplitude and can be neglected. This restriction on the phase of the noise reduces the effective noise level by 3 dB.

A further restriction on the phase of the noise could be applied since noise of the same phase in the two channels would tend to compensate and thus reduce the error. By using this restriction the effective noise could be further reduced, depending on the relative signal-to-noise ratios in the two channels. For simplicity this aspect is not considered further here but the worst-case phase relationship (opposing phases in the two channels) is assumed:

$$\text{effective noise level in each channel} = -103.0\,\text{dBm} = 1.6\,\mu\text{V in } 50\,\Omega$$

therefore

$$\text{sum-channel signal minus noise} = 113.4\,\mu\text{V} - 1.6\,\mu\text{V} = -66.0\,\text{dBm}$$

$$\text{difference-channel signal plus noise} = 46.2\,\mu\text{V} + 1.6\,\mu\text{V} = -73.4\,\text{dBm}$$

The ratio of sum to difference signal is therefore 7.4 dB instead of the noise-free ratio of 7.8 dB. Figure 10.6 gives the bearing error sensitivity to a signal amplitude error for the antenna shown in Fig. 4.1 (Chapter 4). At 0.5° from boresight the error sensitivity is 0.053° bearing error per 1 dB of signal error. An error of 0.4 dB produces a bearing error of 0.021° r.m.s.

For the case of a signal arriving at the antenna boresight the coherence between the difference signal and the noise cannot be considered since no signal is present in the difference channel. In amplitude-processing receivers a phase detector is employed to determine the sign of the angle measurement from boresight. This phase detector can also be used for boresight signals to determine the phase of the noise compared with the sum signal (see Chapter 5, Section 5.4.2). The indicated sum–difference ratio can be modified according to the phase measured. This process restores the restriction on noise phase which applies in the off-boresight case, so that a 3 dB reduction in receiver noise can be assumed.

The bearing measurement errors due to receiver noise are plotted across the antenna beam and for targets of various ranges in Fig. 10.7.

**Receiver errors dependent upon reply signal strength**

The monopulse receiver contains a sum-channel i.f. amplifier and a difference-channel i.f. amplifier, both with a logarithmic characteristic. This feature enables the ratio of the sum- and difference-channel signals to be determined by subtracting the video outputs from the two channels. Any departure from the ideal logarithmic response will produce an error in the measured sum to difference signal ratio and produce a corresponding bearing measurement error.

Integrated circuit amplifier stages are available which allow the construction of logarithmic i.f. amplifiers with nonlinearity less than $\pm 0.5$ dB. However, when the output video signal is to be converted to a digital number, the highest accuracy can be achieved if the output characteristic is calibrated by a programmable read-only memory after the digital conversion. Practical realization of this process shows that amplifiers can be matched to within $\pm 1$ least significant bits (l.s.b.) or $\pm 0.314$ dB. (This value corresponds to an 8-bit

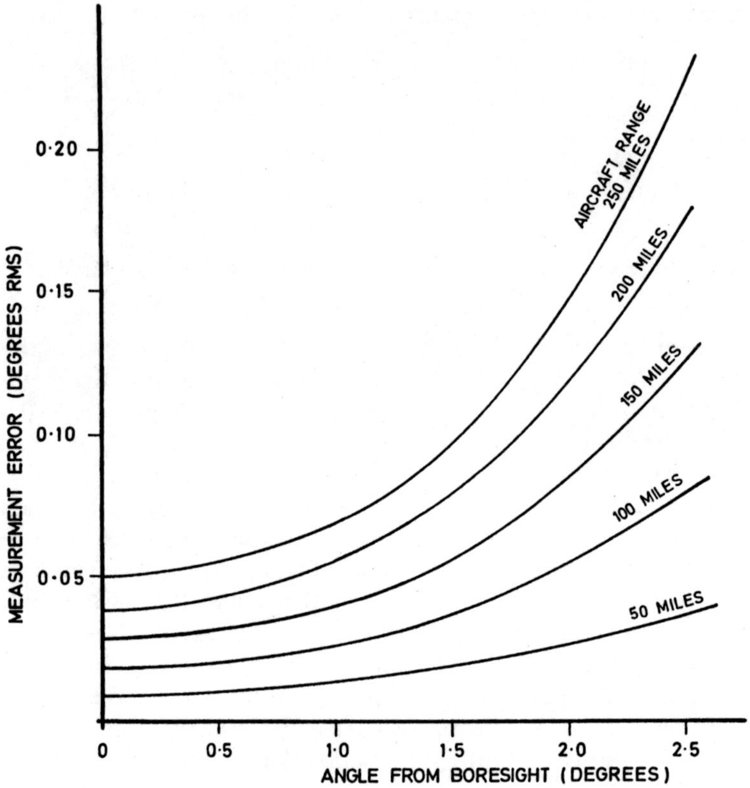


**Fig. 10.7** Bearing errors due to receiver noise (single pulse).

analog-to-digital converter giving 255 different intervals between −16 dBm and −96 dBm.)

An allowance must be made for the quantization step of the analog-to-digital converter which is ±0.5 l.s.b. or ±0.167 dB. A further allowance of ±1 l.s.b. (±0.314 dB) is made for variation in the timing of the measurement of pulse amplitude which might be subject to some overshoot or ringing.

The above three sources of error in the measurement of the signal sum–difference ratio are uncorrelated and can be separately converted to r.m.s. form and added by finding the r.s.s. to give 0.274 dB r.m.s. This error is converted to angle units and plotted in Fig. 10.8.

### Receiver errors dependent upon reply frequency

The frequency response of the sum and difference channels is determined by the bandpass filters in the two channels. A mismatch between the two bandpass filters will produce an amplitude error in the two video output signals that will translate into a bearing measurement error. The bandpass

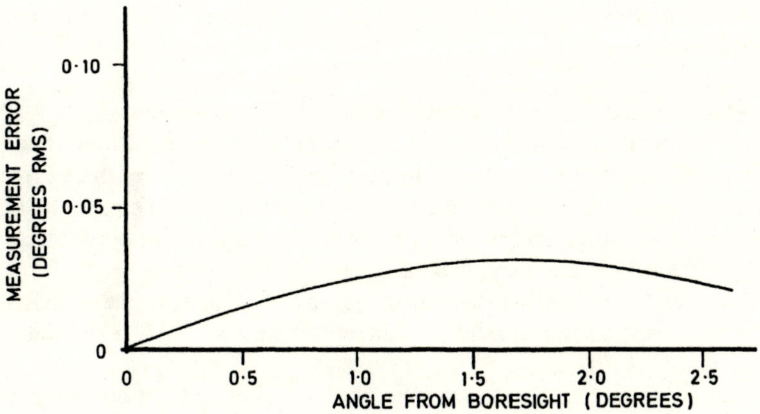


**Fig. 10.8** Bearing errors due to receiver amplitude errors.

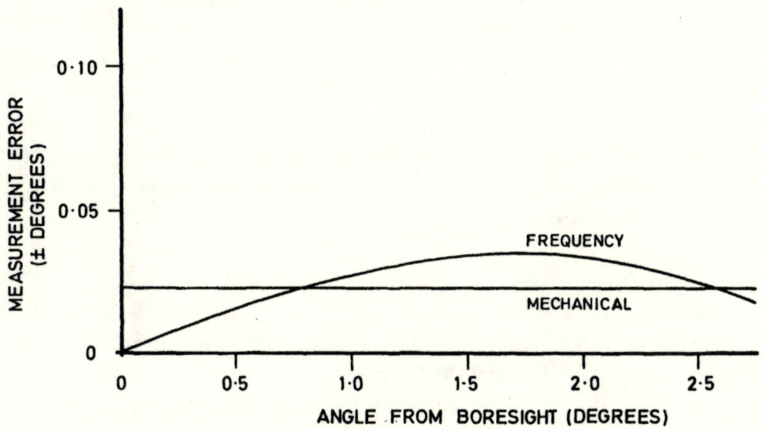


**Fig. 10.9** Bias errors.

filters are typically matched to within $\pm 0.3$ dB over the $1090 \pm 3$ MHz frequency band for SSR replies. A bearing error due to frequency offset will be maintained for a given aircraft over a large number of antenna scans and will be manifest as a bias error. This error is plotted, as a maximum limit, across the antenna beam in Fig. 10.9.

**Errors in the conversion of video to angle units**

The sum- and difference-channel signal ratio is converted to angle units using a calibration table. Typically the resolution of the table will be 0.022°. The finite resolution of the table will introduce an error of $\pm 0.011°$, and a similar contribution can be allocated to an error in the calibration table itself.

Both these errors are subject to a linear distribution and are independent, giving a total error of 0.009° r.m.s.

**Mechanical errors in the turning gear and azimuth encoder**

Mechanical errors present in the turning gear and the antenna azimuth encoder will also contribute to the total system error. Some of these errors will be fixed, at least at a given azimuth, and will contribute to a bias in the bearing determination. Other errors are random and uncorrelated between antenna scans, and contribute to target track jitter.

It is usual to use a 14-bit azimuth encoder with a monopulse SSR system so that it supports the maximum bearing accuracy. Such a digitizer has a resolution of 0.022°.

The antenna direction cannot be known to better than one angle unit or $\pm 0.011°$. This error is linearly distributed and is equivalent to 0.006° r.m.s.

The encoder zero cannot be aligned to north by better than one angle unit or $\pm 0.011°$. This error is fixed and contributes a bias to all bearing measurements. Manufacturers of digitizers often quote an accuracy of one unit to allow for some eccentricity in the encoding disk. This error is also $\pm 0.011°$ and will be slowly varying with direction and therefore manifest as a bias common to all aircraft in a given direction. These two errors together contribute a maximum bias error of $\pm 0.022°$.

If gearing is used between the antenna shaft and the angle digitizer, that gearing can be assumed to have some eccentricity which will give rise to a bias error at some azimuths.

**Total errors**

The bias errors are shown in Fig. 10.9. Figure 10.10 totals the system random errors described above for single pulse measurement. Replies are unlikely to be received in the upper right-hand area of Fig. 10.10 because the path losses are likely to be too great for the interrogation to trigger the transponder. For example, Table 10.2 shows that for a target at a range of 250 n miles replies will only be obtained within $\pm 1°$ from boresight.

The random errors shown in Fig. 10.10 will tend to a normal distribution through the central limit theorem: 68.3% of errors will be less than the value shown ($1\sigma$), 95.5% less than twice the value ($2\sigma$) and 99.7% less than three times the value ($3\sigma$). The $3\sigma$ value can be sensibly taken to be a maximum error when this quantity is required.

The error contribution due to receiver noise can be reduced by averaging the measurements on a number of pulses, and the reduction in equivalent noise will be equal to the square root of the number of measurements. Typically an SSR Mode A or C reply contains eight pulses and a Mode S reply contains 56 pulses. The multiple replies obtained with SSR Modes A and C permit measurements obtained on replies close to but on either side of boresight to be averaged. This process reduces the errors due to receiver

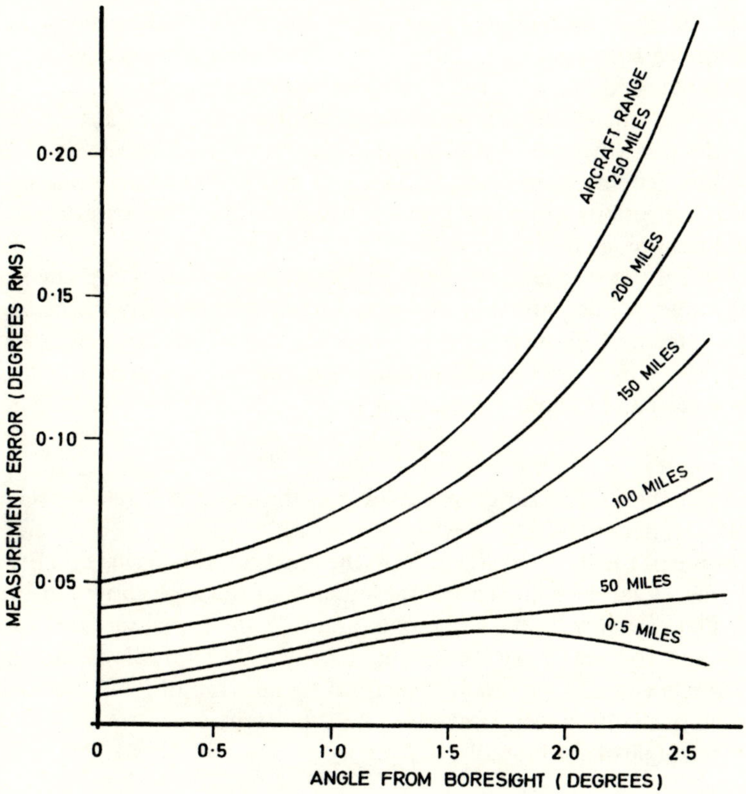


**Fig. 10.10** Total random errors.

nonlinearity and reply frequency offset. In the case of Mode S, only a single reply is normally received and this may be at any position within the antenna beam. Consequently greater bearing measurement errors can be expected. Some improvement will arise owing to the more limited tolerance of $\pm 1$ MHz on the reply frequency and the greater number of pulses available for averaging.

## 10.6 Antenna height

The choice of antenna height can be restricted for a variety of reasons. The location can be environmentally sensitive and the minimum height demanded because of the landscape. Financial constraints may limit the height of the antenna tower, and the same constraints may require the SSR antenna to be mounted on a primary radar antenna. In this case the requirements of the primary radar, to reduce clutter and possibly serve a shorter target maximum range, can dictate a different optimum height.

If the SSR system designer has a free choice of antenna height, a compromise still has to be made. One of the most obvious system requirements is the need to see over local obstructions. Adequate clearance should be given so that objects do not intrude into the near-field of the antenna and thus affect the performance of the antenna. Nearby obstacles, such as buildings, chain-link fencing, and aircraft fuselages and tailfins, can be a significant cause of reflections which can give rise to false targets and azimuth errors and should be avoided if at all possible.

Even if the transmitter power and receiver sensitivity are adequate, the radar range will be limited by the screening of the target by the horizon. The range increases with increasing antenna height and aircraft height according to the following equation which assumes a four-thirds Earth (the distance to the horizon can be found by setting $h_t = 0$):

$$R = 1.23\,(h_a^{1/2} + h_t^{1/2})$$

where $R$ is the aircraft range in n miles, $h_a$ is the antenna height in feet and $h_t$ is the aircraft height in feet.

Some of the radiated signal from the antenna will impinge on the ground even when a large vertical aperture antenna of the type shown in Fig. 4.1 is used. This signal will reflect and combine with the signal radiated directly at the aircraft to modify the total signal received. At some elevation angles the two signals will add, enhancing the total signal strength, and at other angles they will partially cancel, weakening the total signal.

The angle of the first lobe is given by

$$\theta = \sin^{-1}\left(\frac{\lambda}{2h}\right)$$

It may be advantageous to place the first lobe at the minimum angle for the most distant target to provide some enhancement to its detection. If this minimum angle is 0.25°, the antenna height $h$ is 55 ft (frequency, 1030 MHz). It should be noted that at 0.5° there will be a minimum or null in the signal at which targets may be lost. A compromise antenna height may be sought whereby the improved detection in the first lobe is balanced by the loss of target detection in the first null.

## 10.7 Cross-coupling between primary radar and secondary surveillance radar

An SSR ground interrogator is commonly colocated with a primary radar operating either in L band (1215–1365 MHz) or S band (2700–3400 MHz). Usually the SSR antenna is mounted on top of the primary radar antenna. Inevitably there will be some cross-coupling of signals between the two radar systems, and careful design is required to ensure that the performance of either is not degraded.

Primary radar transmitters can emit a variety of r.f. waveforms from narrow pulses measured in a few microseconds to long frequency-modulated pulses of 100 μs duration or more. Peak powers can be measured in tens of kilowatts to megawatts. Often the interval between the pulses is insufficient to permit a synchronized SSR to receive replies from maximum range without the primary radar transmitting during the interval.

There are two ways in which a primary radar transmitter r.f. pulse can interfere with an SSR receiver.

(a) The high power transmitter pulse saturates the front end of the SSR receiver and either blocks it or generates cross-modulation products within the SSR passband. The solution is to provide sufficient out-of-band filtering at the input to the receiver.

(b) The transmitter sidebands at 1090 MHz are received above the SSR minimum signal level. Here the solution is to limit the sideband level of the transmitter.

The primary radar system designer will be equally concerned about the equivalent effects between the SSR transmitter and the primary radar receiver.

Coupling between the two systems may occur at the antennas, within the rotating joint or between the equipment cabinets. The latter path requires careful screening of equipment and attention to good grounding, and can present almost intractable problems if the original design is defective. Coupling between the antennas will occur as a result of the direct path between the SSR antenna and the primary radar feed horn. The horn will illuminate the antenna with its beam edge or sidelobes. A typical separation between the horn and the antenna is 25 ft which would provide a path loss of about 50 dB at 1250 MHz. However, the gain of the SSR antenna at this frequency and at the declination angle presented by the horn will probably be unknown. A worst-case assumption is sometimes made that the gain is the same as at 1090 MHz. The other route for cross-coupling is leakage between the channels in the common rotating joint. A typical manufacturer's specification for isolation between channels is 60 dB and may exceed the coupling between antennas.

## References


10.1 Blake, L. V. *Radar Range-performance Analysis*, Artech House, Dedham, MA, 1986.

10.2 Weil, T. A. Atmospheric lens effect: another loss for the radar range equation, *IEEE Trans. Aerosp. Electron. Syst.*, 9 (1), 51–54, 1973.

10.3 Clarke, J. and Rider, G. Long-range propagation experiment to investigate the incidence of anomalous propagation in the North Atlantic, *Proc. Inst. Electr. Eng., Part H*, **134** (1), 65–70, 1987.

10.4 Stevens, M. C. and Jenkins, V. Model of the secondary surveillance radar environment in the United Kingdom, *Proc. Inst. Electr. Eng.*, **121** (5), 340–344, 1974.

10.5 Jacovitti, G. Performance of monopulse receivers for secondary surveillance radar, *IEEE Trans. Aerosp. Electron. Syst.*, **19** (6), 884–897, 1983; **21** (2), 264–266, 1985.

# Chapter 11

# Performance Measurement and Monitoring

## 11.1 Performance measurement

Customers for radar systems invariably have a performance requirement in mind when they procure the equipment. If the procurement authority is undertaking the system design itself, it may specify technical attributes of the equipment such as transmitter power and receiver sensitivity. These parameters are quite easy to demonstrate when proof of compliance with the specification is required. However, with the increasing complexity of radar equipment and the much greater performance demanded from the equipment, the customer is now giving almost complete responsibility for the system design to the manufacturer. Proof of compliance is then a matter of measured performance at the radar site at which the equipment will ultimately be used.

The radar site imposes other constraints upon radar performance, such as multipath reflections, over which neither the customer nor the manufacturer has much control. If a flight trial is part of the acceptance procedure, the aircraft used should carry a transponder with known receiver sensitivity, transmitter power and reply frequency. What will probably not be known is the aircraft antenna pattern, the transponder reply probability through the region of the flight and the absolute position of the aircraft at all times. It is of crucial importance that, when customer and supplier are agreeing the performance of the radar system, the means of testing that performance is also agreed.

### 11.1.1 Target detection

SSR operates with good signal-to-noise ratio conditions on both the up and down signal paths and target detection can be near to the ideal of 100%. Loss of detection can occur for a number of reasons.

(1) The aircraft is flying directly overhead in the antenna "cone of silence".

(2) The aircraft is flying through a null in the vertical antenna pattern. Nulls will occur even with the improved large vertical aperture antennas and the depth of the nulls will vary with local ground conditions.

(3) The aircraft is turning so that its antenna is hidden from the ground sensor.

(4) The aircraft transponder can be prevented from replying by being captured or desensitized by other interrogators or by being suppressed by other onboard equipment.

(5) Two aircraft reply at precisely the same range, so that the two replies exactly overlap and are indistinguishable from a single reply.

The detection probability, which is also known as the blip–scan ratio, can be defined as the number of antenna scans during which a target report for that aircraft was produced divided by the number of scans that the aircraft was within range. Misses for more than three consecutive scans should be excluded from the calculation in order to avoid including the effects of antenna nulls, turns, overhead flying, and landing and taking off.

### 11.1.2 Track jitter

One of the reasons for the development of monopulse SSR was the dissatisfaction with the azimuth performance of the earlier sliding window SSR. Consequently a natural requirement of modern equipment is a defined bearing accuracy.

Both bearing and range error can be divided into two components: a fixed bias error and a random variation, or jitter, about the fixed bias. Once measured, bias in range and bearing, if constant in all antenna directions, can be compensated within the equipment. Random range and bearing error is mainly a function of signal-to-noise ratio and can vary with aircraft and with distance.

An assessment of a radar's positional measurement error presupposes a knowledge of the true position of the aircraft being measured, and therein lies the main difficulty. Two approaches to determining the true position of the aircraft can be used. One approach averages a number of measurements made by the radar under test to reduce the uncertainty in the measurements. The other approach uses an independent source of measurement with a proven accuracy much better than that of the radar under test. This source can be a precision tracking radar, onboard precision navigation equipment (such as that based on the use of multiple DME range measurements) or simultaneous measurements made by multiple ground radars.

**Averaging by curve fitting**

An accurate and convenient method for determining the random errors in position measurement uses recorded target report data obtained from the radar under test. A curve, usually second-order, is fitted to a series of measurements made on a succession of scans. The curve position at the middle of the series of measurements is then held to indicate the true position of the aircraft [11.1].

Typically 11 scans are used in the curve fitting, five before the position of current interest and five after. The fitted curve then indicates the assumed "true" position of the aircraft at the sixth position. The difference between the measured range and azimuth and the "true" range and azimuth are determined, and are squared and stored. The 11 selected scans are then moved forward in time by one scan so that the earliest scan is dropped and a later scan is added. The measurement error for the new middle position is again calculated. The square of this error is added to the previous stored values. The process is repeated for the length of the track to be assessed. The accumulated square of the errors is divided by the number of measurements and the square root of the result is found. This process determines the r.m.s. of the range and bearing measurement errors of the radar. Any reasonable degree of curve fitting can be applied, but higher orders will require the use of longer scan lengths. A second-order fit is usually adequate if care is taken to avoid tracks which include sharp turns.

The process is very useful in that it provides a good measure of the performance of the equipment. The manufacturer can readily make the measurement, using aircraft of opportunity, and apply any fine tuning necessary before presenting the equipment to the customer for acceptance.

**Use of multiple radars**

The above process does not identify small track perturbations which may be caused by multipath effects (see Chapter 9, Section 9.1.2), since the curve fitting would conform to small track deviations. The manufacturer may argue that this attribute is desirable since the technique produces an indication of the performance of his equipment and multipath effects at the customer's site are beyond his responsibility or power to control.

The customer may have a different view since his requirement is to know where aircraft actually are and he will certainly wish to know how good the radar is at measuring true position. In order to provide this information it is necessary to use an external means of determining the true position of the aircraft and use this information to check the accuracy of the radar. The external source of position data can be a precision tracking radar, a precision navigational aid on board the aircraft or the use of multiple measurements from a number of other radars. The essential requirement is that the source of the external position measurements must have a proven accuracy very much better than that expected of the radar under test.

An interesting recent development is MURATREC (MUlti-RAdar Trajectory REConstitution) [11.2, 11.3]. This technique uses multiple SSR ground stations all of which have aircraft of interest within cover. Simultaneous recordings are made at all radar sites. Given that the locations of two of the radar sites are known precisely, then errors in the location of the other radar sites, bias errors in their range and bearing calibration of all sites, and errors in their local clocks which time stamp the recordings can all be found

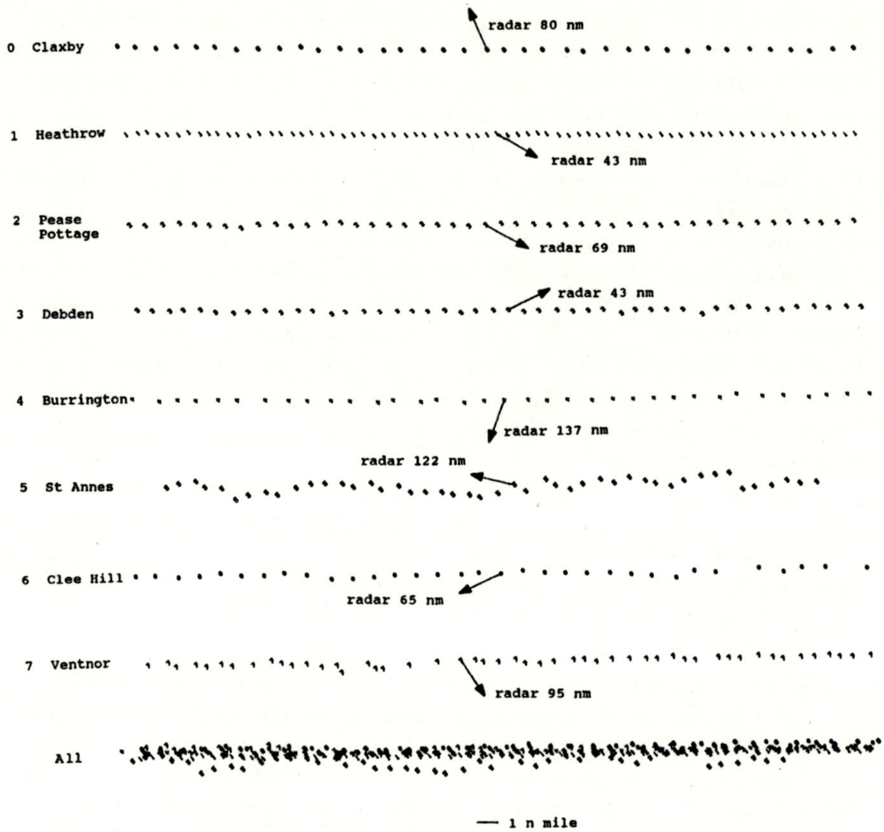


**Fig. 11.1** Aircraft track measurement from eight radars (courtesy of the CAA).

from these recordings. The precise location of the aircraft detected by the radars can be determined from these data.

Figures 11.1, 11.2 and 11.3 show details of a track of an aircraft detected by eight different SSR stations in the UK [11.4]. Of particular interest in this example is the variety of types of SSR equipment used which includes all four combinations of sliding window and monopulse SSR with linear array and large vertical aperture antennas. Table 11.1 lists the installation characteristics, the bearing jitter calculated by curve fitting as described above, and the bearing jitter referenced on the true position determined using MURATREC.

Figure 11.1 shows the plots for one aircraft track as obtained simultaneously from the eight radars. The bottom plot in the figure is the superposition of the measurements of all eight radars. The two sites which use a linear array antenna (small vertical aperture) and sliding window processing (St Annes and Ventnor) show plots with significant jitter. Clee Hill, which uses

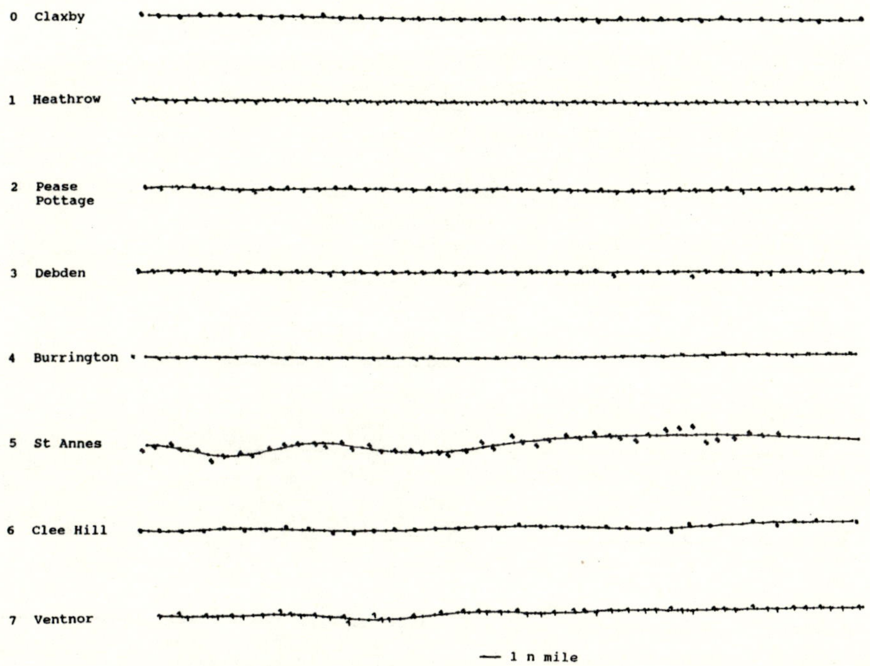


**Fig. 11.2** Aircraft track measurement from eight radars curve fitted using single-site data only (courtesy of the CAA).

sliding window processing but a large vertical aperture antenna, is significantly better though still worse than Pease Pottage, which is at a similar range and uses monopulse processing. Both Heathrow and Debden are at similar ranges and both use monopulse processing; however, Heathrow's large vertical aperture antenna shows better performance than Debden's small vertical aperture antenna.

Figure 11.2 shows the same plots to which a smoothed curve has been fitted using the technique described in Section 11.1.2. In contrast, the smoothed curve in Fig. 11.3 is derived from the multisite data from all eight radar sites using the MURATREC process. The two sets of curves in Figs 11.2 and 11.3 show a similar fit for the five monopulse sites except that MURATREC has identified some alignment errors between the radars. The three sliding window sites show a greater difference between the curve deduced from the single-site radar data and that of MURATREC, with that of St Annes being an extreme example. Table 11.1 shows the difference in track jitter as measured using the two methods.

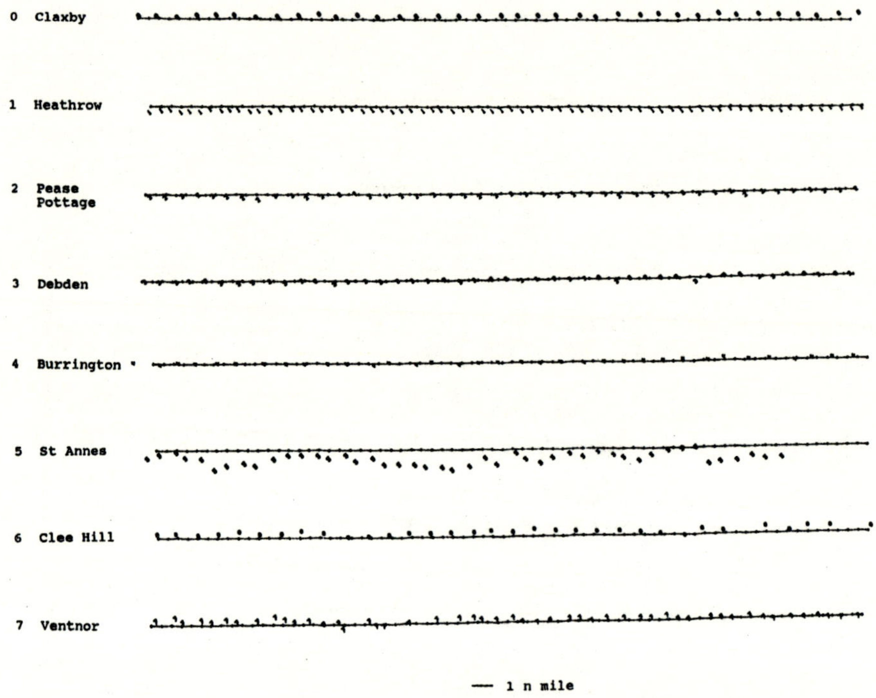


**Fig. 11.3** Aircraft track measurement from eight radars curve fitted using multisite data (courtesy of the CAA).

**Table 11.1** Characteristics of the sites and the calculated azimuth performance

| *Plot no.* | *Radar site* | *Antenna type*[a] | *Processing type* | *Range* (nm) | *Single-site track error* (deg r.m.s.) | *Multisite track error* (deg r.m.s.) |
|---|---|---|---|---|---|---|
| 0 | Claxby | LVA | Monopulse | 80 | 0.043 | 0.046 |
| 1 | Heathrow | LVA | Monopulse | 43 | 0.036 | 0.043 |
| 2 | Pease Pottage | LVA | Monopulse | 69 | 0.037 | 0.043 |
| 3 | Debden | SVA | Monopulse | 43 | 0.097 | 0.096 |
| 4 | Burrington | LVA | Monopulse | 137 | 0.040 | 0.051 |
| 5 | St Annes | SVA | Sl. window | 122 | 0.080 | 0.123 |
| 6 | Clee Hill | LVA | Sl. window | 65 | 0.070 | 0.093 |
| 7 | Ventnor | SVA | Sl. window | 95 | 0.081 | 0.122 |

[a]LVA refers to antennas, similar to that shown in Fig. 4.1 (Chapter 4), which possess a large vertical aperture with a beam shape that discriminates against signals reflected from the ground. An exception to this is Clee Hill which uses a linear array with the screen shown in Fig. 9.17 (Chapter 9) which simulates an LVA antenna. SVA refers to linear array antennas, or "hogtroughs", similar to that shown in Fig. 4.6 (Chapter 4). These antennas possess a small vertical aperture and are sensitive to signals reflected from the ground.

## 11.2 Performance monitoring

### 11.2.1 Site monitor

During operational use the continuous confirmation of the correct functioning of the SSR is clearly important. One effective and commonly employed procedure is the automatic monitoring of a fixed transponder positioned a suitable distance from the sensor antenna. This approach allows the full SSR equipment to be tested, including the antenna, in a manner which represents the operation of the system in detecting aircraft.

The fixed transponder is known as a site monitor, and colloquially as a "parrot", and is sited at a surveyed position. Correct measurement of the site monitor range and azimuth and correct decoding of its Mode A and Mode C reply data confirm the satisfactory operation of the equipment. Figure 11.4 shows a photograph of a typical site monitor which includes a dual transponder.

The site monitor antenna should be in line of sight of the sensor antenna and the path should be clear of any off-angle reflecting objects. To avoid errors due to the corruption of the site monitor replies by reflections, the terrain between the SSR station and the site monitor should be free from obstructions. No buildings likely to cause reflections should lie within an ellipse formed with the sensor and the site monitor as foci and describing a locus of excess path length corresponding to 500 ft. This restriction will ensure that at least the first pulse of the site monitor reply is free from multipath distortion.

**Fig. 11.4** Site monitor.

The range of the site monitor should place it beyond the Fresnel zone of the sensor antenna, and a common distance is 1 or 2 miles.

Since the site monitor is ground based it follows that its antenna will present a low elevation angle and the path attenuation will be significantly affected by reflections from the ground. The amount of attenuation will be greatly influenced by the physical characteristics of the ground and can be expected to change rapidly according to the current moisture content.

### 11.2.2 Fault detection and isolation

Internal performance monitoring, or built-in test equipment (BITE), is also important and is an increasing requirement [11.5]. The availability of microprocessors has inevitably permitted a degree of intelligence to be built into the equipment such that reasonably sophisticated self-testing can be included without greatly increasing the complexity of the equipment itself. The one-time fear that comprehensive built-in self-testing would significantly reduce the reliability of the equipment need no longer be true.

Equipment is also becoming too complicated for easy performance verification by external measurement. Simple routine tests, such as the checking of transmitter power and receiver sensitivity, are not sufficient to confirm correct system operation. More rigorous tests require the use of complex equipment at the site and the availability of suitably trained operators.

Regardless of whether the ATC authority is a member of the developed or less-developed part of the world, the availability and cost of appropriately skilled personnel to man the radar sites permanently and carry out repairs on the equipment is causing increasing problems. The total cost of ownership over the lifetime of the equipment, instead of just the initial cost of purchase, is becoming a selection criterion.

There is a strong trend to a requirement that a site be normally unmanned. Not only should the equipment be capable of automatically and comprehensively checking its own performance, but, if a failure should be detected, it should also be able to diagnose the cause of the failure and identify the faulty module. Means are needed to communicate the nature of the failure to a remote maintenance base so that a spare can be brought to the site to effect the repair. The BITE should then confirm that the equipment is once again operating within its specification. The faulty module is returned to the maintenance base where more skilled personnel can repair it using specifically programmed automatic test equipment.

It is unreasonable to expect that all possible faults can be detected and located automatically. The following is a typical requirement:

(i) at least 90% of all faults are automatically detected;

(ii) at least 80% of detected faults are located in a single replaceable module;

(iii) at least 10% of detected faults are located in a group of modules;

(iv) 10% of faults may need the attention of a technician for identification and location.

The techniques used for fault detection and location will depend strongly on the design of the equipment. However, some general principles will apply.

*Antenna* The antenna contains no active components and the introduction of devices which measure signals at built-in monitor points is likely to reduce its reliability significantly. Fault detection must rely on the measurement of the far-field beam patterns produced by the antenna and use the r.f. cables normally connected to it. A particular approach to antenna fault detection is described in Section 11.2.3.

*Interrogator* The interrogator uses mainly analog modules and test points will produce analog values. The measured values are compared with high and low limits to determine appropriate operation. This comparison can be performed by using either threshold detectors, producing "go" or "no go" indications, or analog-to-digital converters with the limit determination performed digitally. Both the transmitter and receiver sections of the interrogator are mainly serial in nature so that a series of monitor points can enable a fault to be isolated to a single module.

*Plot extractor* Mainly digital circuits are used in the design of the plot extractor. When integrated logic circuits are used, such as for the reply decoder, techniques such as signature analysis can be effectively applied. Computer-based sections, e.g. the reply correlator and the surveillance processor, can exploit the power of the computer to perform its own rigorous checks. Read–write and read-only memory can be tested by using writing and reading back for the former and check sums for the latter. Data flow can be monitored by routinely inspecting the contents of data buffers and recognizing buffer overflow. It is useful to inject predefined reply and target signals at the input to the plot extractor and to have the computer section check for their existence and correctness at the output of the extractor. This provides a form of overall check of the plot extractor functions. Often a synthetic target report, injected at the input to the plot extractor, will be output as a test report to the ATC center as a means of confirming the correct operation of the land lines used.

The plot extractor can also perform more complete checks on the SSR system by monitoring the correct detection of the fixed transponder used as a site monitor. Since, using monopulse bearing measurement, each aircraft reply can yield the direction of the aircraft, all replies from one aircraft can be compared for consistency of measurement. By these two means the total equipment—antenna, interrogator and plot extractor—can be monitored.

### 11.2.3 Antenna pattern measurement

Typical SSR antennas contain no active components and demonstrate very high reliability. Occasionally an ingress of moisture occurs which affects the distribution system causing a degradation of the antenna beam patterns. Such a degradation can be difficult to identify and to distinguish from faults in the r.f. cables or the transmitter or receiver.

It is usually difficult to determine whether the performance of the antenna has deteriorated without removing the antenna from the tower. Such a process is expensive and time consuming and removes the SSR sensor from operational service until a replacement antenna is installed. It is not unknown for the antenna to exhibit no fault when subsequently measured on a test range, as the fault was elsewhere but was incorrectly attributed to the antenna.

The antenna, r.f. cables and receiver contribute to the performance of the SSR sensor. A malfunction may be manifest as a shift in the azimuth of the site monitor. In the case of a monopulse SSR system the consistency of the bearing measurement of each reply from an aircraft during a single antenna scan can be a more sensitive indication. An alternative technique is possible whereby replies from the site monitor are used and their signal strength is measured. Measurement of the sidelobes of the antenna requires the inhibition of the P2 sidelobe suppression pulse. Interrogation Mode B is used so that ring-around responses are not produced from aircraft and the local fruit rate is not increased.

The special interrogations for measuring the antenna beam patterns can be made during the interrogator dead time, i.e. after the time of the maximum range reply and before the next interrogation. In this way the normal operation of the sensor is not affected. Figure 11.5 shows antenna beam patterns measured in this way and includes examples of specifically induced antenna faults.

## 11.3 Model of the secondary surveillance radar environment

Many different external factors can influence the performance of the SSR system. Some of these, if known, can be accommodated in the design of the equipment. Apart from multipath effects, which are discussed in Chapter 9, the other major consideration is the performance of the airborne transponder in the surrounding SSR environment.

An aircraft can fail to reply to an interrogation if it is in the process of replying to another interrogator or has been sidelobe suppressed by that other interrogator. Onboard DME and u.h.f. communication equipment can suppress the transponder while it is operating. When replying to an interrogator, the replies can be detected by other interrogators where they appear as fruit.

The installation of a new SSR sensor will increase the local interrogation and fruit rate and could degrade system performance for existing sensors. Many countries protect the performance of their SSR system by the careful regulation of ground interrogators and control of their interrogation rates and transmitter powers. The prediction of the performance of new equipment and its effect on the existing system can be determined by the use of an SSR environment model. Such a model will include details of all interrogator sites in the region of interest [11.6]. Important parameters of each site are as follows:

location
radiated power
interrogation rate (pulse repetition frequency)
antenna rotation rate
gain and width of the antenna main lobe
extent and level of antenna sidelobes and backlobe
mode interlace pattern

The model should be capable of predicting the mean interrogation rate and sidelobe suppression rate at any location in the area and from these the transponder reply probability. In order to do this it will determine the range of the main lobes and sidelobes taking into account the curvature of the Earth and atmospheric attenuation.

Peak interrogation rates will depend upon the probability that a number of interrogator antenna beams could illuminate the same aircraft at one time. Particular attention should be given to sensors which operate on similar pulse repetition frequencies since synchronism between them can cause losses of replies which are nonrandom and may persist for many interrogations. Note should also be taken of the range of the antenna control pattern where it exceeds the range of the sidelobes. In these areas only the P2 pulses will be detected by transponders. No suppression will be caused, but chance timing with an interrogation from another site can alter the interrogation mode as seen by the transponder. Modelling the airborne element of the system is particularly difficult because of the wide range of transponder receiver sensitivities and reply powers that are permitted in the system specifications. Airborne antenna patterns can vary widely with aircraft orientation.

Examples of interrogation rates calculated using an SSR model are given in Figs 3.7, 3.8 and 3.9 (Chapter 3). Assumptions about the density and location of aircraft in the environment can permit the fruit rate at any ground location to be calculated.

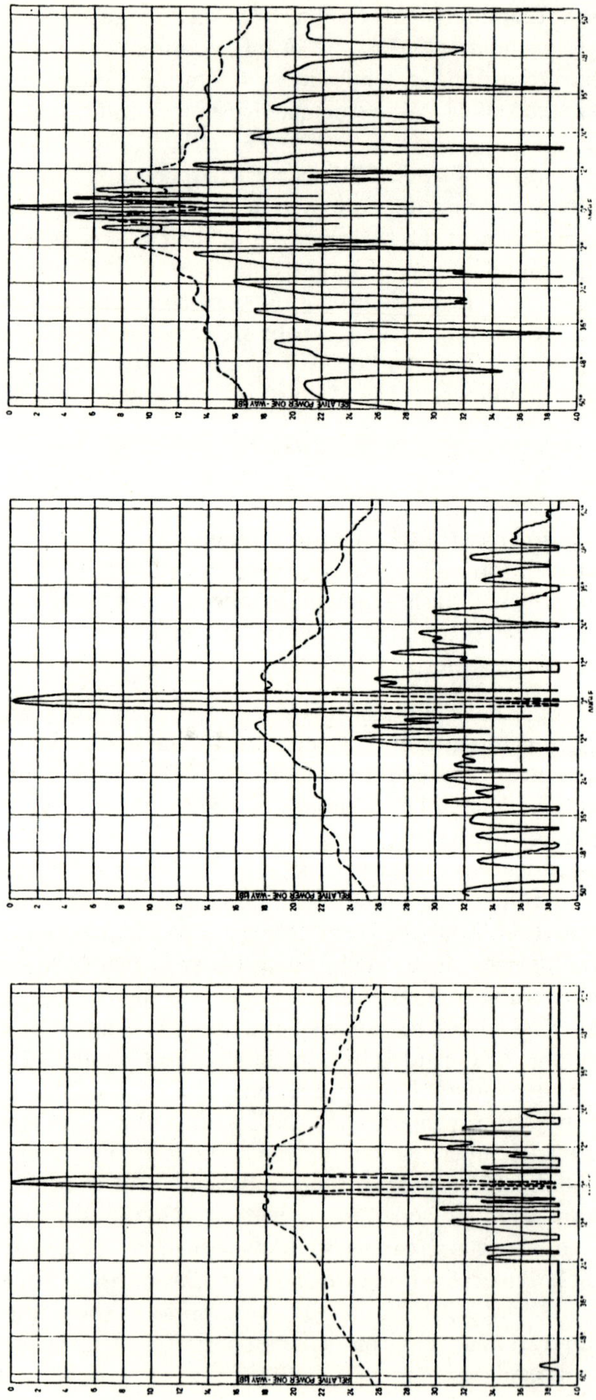

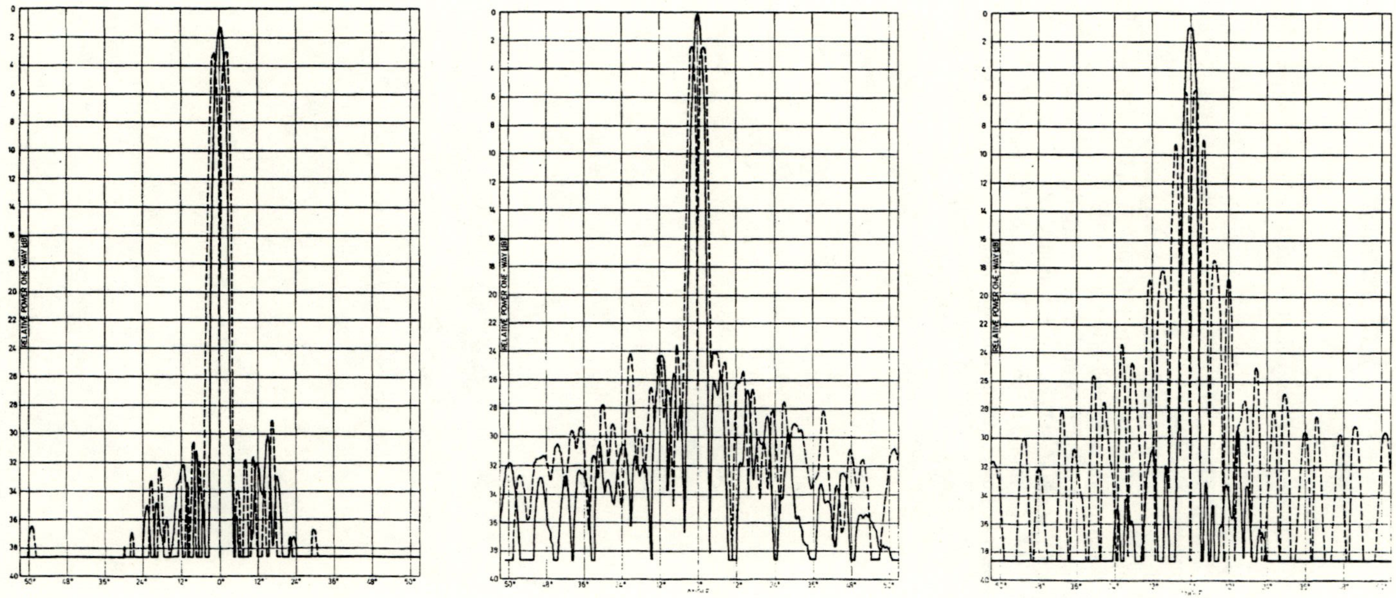

(d) (e) (f)

**Fig. 11.5** Measured antenna beam patterns: (a), (b), (c) sum and control; (d), (e), (f) sum and difference; (a), (d) normal patterns; (b), (e) with single column open-circuit fault; (c), (f) center column group open-circuit fault.

# References


11.1 FAA Specification, Mode Select Beacon Sensor, *FAA-E-2716*, FAA, Washington, DC.

11.2 Renes, J. J. and Best, M. R. Theoretical background to the reconstruction facility MURATREC, *Rep. NRL TR 84008 L*, National Aerospace Laboratory, Amsterdam, 1984.

11.3 Shaw, J. Radar data processing with new generation monopulse SSR radars, *IEE Conf. Proc. RADAR 87*, pp. 195–199, IEE, London, October 1987.

11.4 Shaw, J. Personal communication, July 1987.

11.5 McCarthy, P. Availability and maintainability of the RAMP monopulse SSR systems, *ICAO Bull.*, **42** (1), 9–11, January 1987.

11.6 Stevens, M. C. and Jenkins, V. Model of the secondary surveillance radar environment of the United Kingdom. *Proc. Inst. Electr. Eng.*, **121** (5), 340–344, 1974.

# Chapter 12

# Mode S

## 12.1 Reason for development

Mode S development started independently in the UK and the USA for somewhat different reasons. In the UK the study of the current SSR system [12.1, 12.2] indicated that, with the continuing growth in the number of aircraft, the problems of fruit and garble would become limiting factors in the use of SSR in an automated ATC system. The solution proposed was to interrogate aircraft individually so that their replies could be kept separate from one another. Only one reply per aircraft per scan would be required, and that reply would contain both identity and height data to avoid the need to correlate Mode A and Mode C replies and, more importantly, to reduce the reply rate and thereby the fruit generated for other users. A single reply per aircraft necessitated the use of monopulse to preserve the bearing measurement capability.

A similar concern about the performance limitations of SSR had arisen in the USA. In addition some airports were already experiencing saturation at busy times and this was expected to become worse. To cope with this expansion the number of air traffic controllers required would increase drastically. To improve the productivity of the existing controllers a number of measures were recommended [12.3, 12.4] which included greater automation with intermittent positive control of aircraft from the ground using a data link incorporated in an improved SSR.

The individual addressing of aircraft and the transmission of more data than identity and height information meant that significantly different interrogation and reply signal formats would be needed. At the same time the evolution from the current SSR Modes A and C to the new SSR Mode S system necessitated that the new equipment should be compatible with the old. It was not thought acceptable to require aircraft to carry two sets of equipment, one for use in regions where the new system was operating and the other for regions where it was not. The key to the compatibility requirement was the decision to use the same frequencies for Mode S as for the older modes of SSR, i.e. 1030 MHz for the interrogation and 1090 MHz for the reply. The Mode S interrogation and reply signals must not interfere with the existing equipment or require its modification.

For the interrogation, the prevention of interference with earlier equipment is achieved by transmitting the SSR sidelobe suppression pulses P1 and P2 as the preamble to the interrogation. These pulses suppress a Mode A and C transponder so that it will not attempt to decode the remainder of the signal and possibly generate an unwanted reply. The data part of the Mode S interrogation is completed within the transponder suppression time.

The prevention of interference on the reply path relies on the fact that Mode A and Mode C plot extractors require multiple synchronous replies to recognize the presence of an aircraft. The single Mode S replies of the new system will not produce these multiple synchronous replies.

The Mode S system solves many of the system problems encountered with SSR Modes A and C. All the necessary data are contained in one reply and the accuracy of the reply data is confirmed by the parity. In contrast, the Mode A and Mode C data are sent separately and have to be correctly associated by the ground equipment. There is no guarantee of data accuracy, and validation can only be achieved by repetition and comparison with data received on previous antenna scans.

Figure 12.1 shows a commercially available Mode S transponder and control unit conforming to *ARINC Characteristic 718* [12.5].

## 12.2 Interrogation formats

Figure 12.2 shows the Mode S interrogation format that is used for most of the transactions. The first two pulses, P1 and P2, are 0.8 μs in duration and are spaced 2.0 μs apart to simulate the sidelobe suppression pulses of a Mode A and C interrogation. A non-Mode S transponder would suppress on receipt of these pulses and not attempt to decode the remainder of the interrogation. Following P2 is a long pulse P6, which is either 16.25 or 30.25 μs in duration. Within P6 there are a number of phase reversals which carry the data to be transmitted. The first phase reversal, 1.25 μs after the leading edge of the pulse, is always present and is used by the transponder to synchronize its clock to decode the subsequent data. The synchronizing phase reversal is also used as a time reference for the transmission of the reply in the same way that pulse P3 is used in Modes A and C. The interrogator measures the target range by timing the delay between the transmission of the synchronizing phase reversal and the receipt of the first pulse of the reply.

Data are transmitted by differential phase shift keying (DPSK) whereby a binary 1 is indicated by a phase change of 180° and a binary 0 by no phase change. The spacing between the phase reversal positions is 0.25 μs to give a data bit rate of 4 MHz. The total number of data bits transmitted is either 56 or 112 depending upon requirements.

Sidelobe suppression (SLS) is accomplished in a different way from that for Modes A and C. In Mode S an SLS pulse P5 is transmitted from the antenna control beam (see Chapter 4, Fig. 4.2) to overlap in time the synchronizing phase reversal of pulse P6. If the transmission is being received

(a)

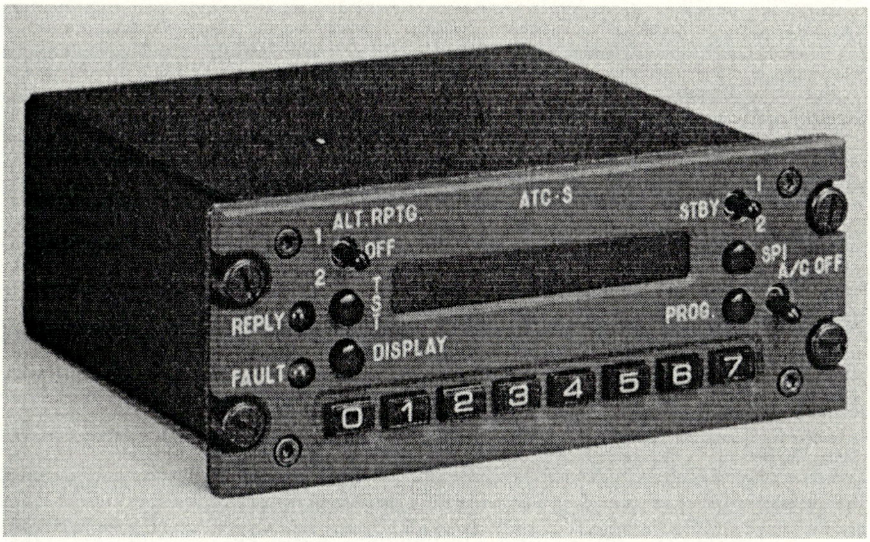


(b)

**Fig. 12.1** Mode S transponder and control unit (TRT type TSR 718 S): (a) transponder showing the internal arrangement of control circuitry and centrally mounted r.f. and power sections; (b) control unit.

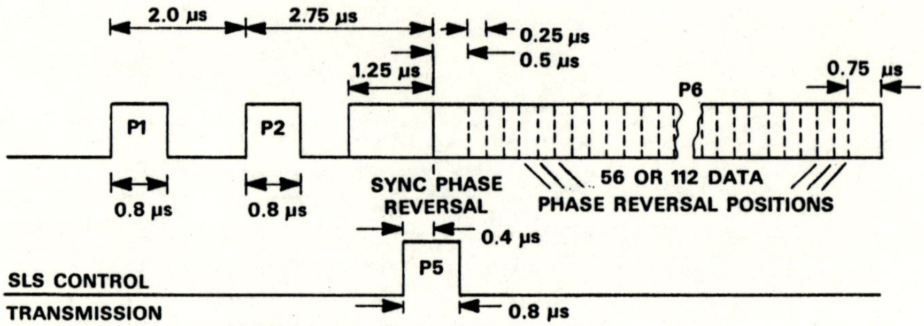


**Fig. 12.2** Mode S interrogation format.

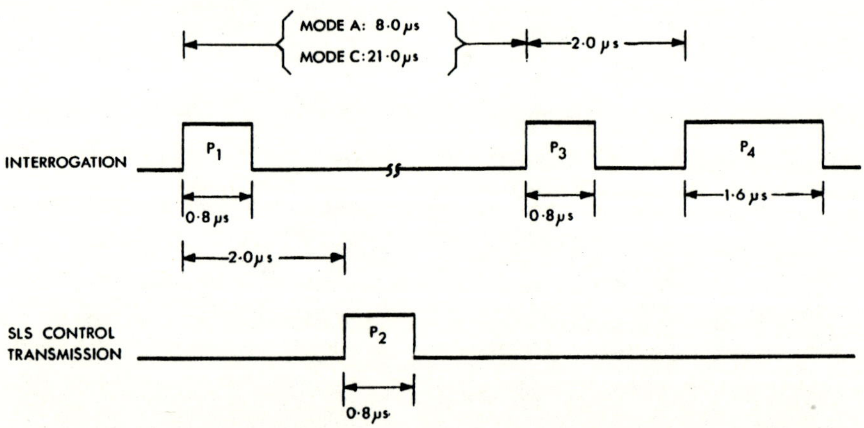


**Fig. 12.3** Mode A/C/S All-call interrogation format.

from an antenna sidelobe, the pulse P5 will be stronger than pulse P6 and the transponder will be unable to detect the synchronizing phase reversal and will not decode or reply to the interrogation. P5 will be weak in the antenna main beam and the synchronizing phase reversal will be detected.

The contents of the data transmitted are described in Sections 12.5–12.7, but the last 24 bits are used to indicate the address of the aircraft for which the interrogation is intended. The transponder will only reply if it finds its own address in an interrogation.

The inclusion of an aircraft address in an interrogation presupposes that

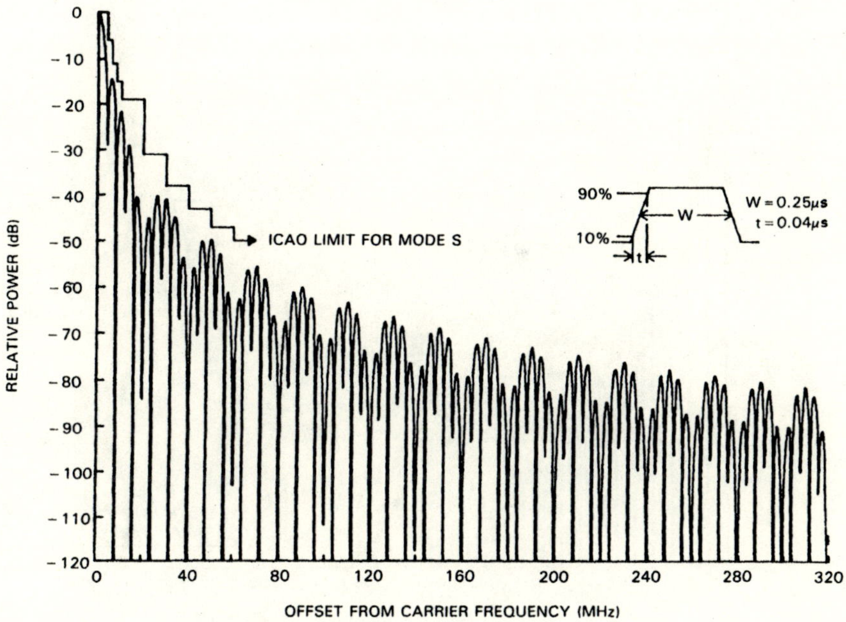


**Fig. 12.4** Mode S interrogation sidebands.

an aircraft's address is known to the interrogator. Aircraft addresses are acquired by means of the All-call interrogation. One form of this uses a Mode S interrogation with 56 data bits and a special address, and this is described in Section 12.5.4. Another form uses a different signal format which is shown in Fig. 12.3. The All-call interrogation of Fig. 12.3 is intentionally similar to the standard Mode A and C interrogation signals. The only difference is the addition of the pulse P4 after P3. A Mode A and C transponder will commence its reply sequence after the receipt of P3 with the first reply pulse F1 being transmitted 3 μs later. The presence of the P4 pulse will have no effect on this process. A Mode S transponder is also capable of replying to a Mode A or C interrogation but will stop this action when it detects P4.

The All-call pulse P4 can have two widths, 0.8 μs or 1.6 μs. A Mode S transponder will transmit a reply announcing its address if the width of P4 is 1.6 μs. A pulse width of 0.8 μs will produce no reply, neither Mode S nor Mode A or C. An interrogation with the shorter P4 pulse is primarily intended for use by an airborne TCAS interrogator (see Chapter 13) as a means of obtaining responses from Mode A and C transponders only.

The interrogation pulses are specified in Table 12.1. Figure 12.4 shows the modulation sidebands generated by the Mode S interrogation r.f. signal, and Fig. 12.5 shows the sidebands after passing through a second-order

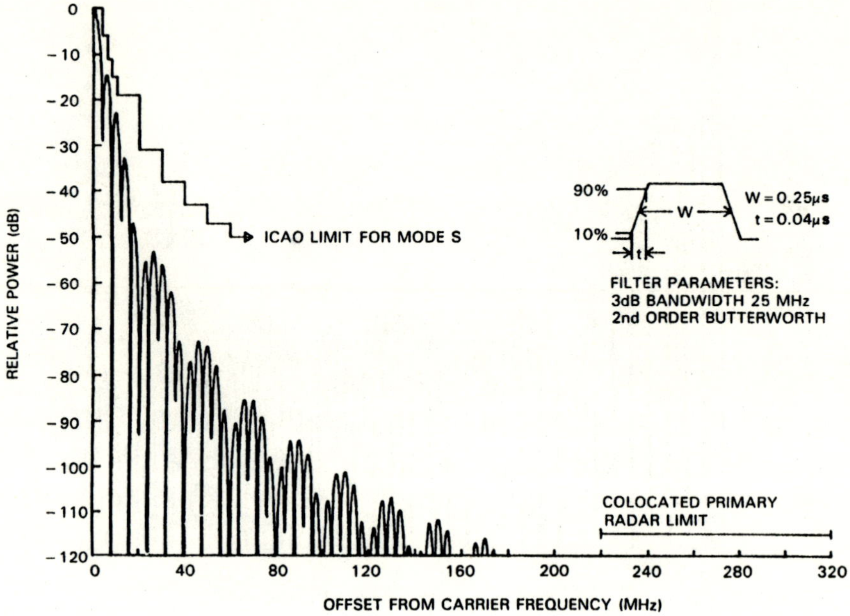


**Fig. 12.5** Mode S interrogation sidebands after filtering.

**Table 12.1** Interrogation pulse specification

| *Pulse designation* | *Pulse duration* (μs) | *Duration tolerance* (μs) | *Rise time* | | *Decay time* | |
|---|---|---|---|---|---|---|
| | | | *Minimum* (μs) | *Maximum* (μs) | *Minimum* (μs) | *Maximum* (μs) |
| P1, P2, P3, P5 | 0.8 | ±0.1 | 0.05 | 0.1 | 0.05 | 0.2 |
| P4 | 1.6 | ±0.1 | 0.05 | 0.1 | 0.05 | 0.2 |
| P6 (short) | 16.25 | ±0.25 | 0.05 | 0.1 | 0.05 | 0.2 |
| P6 (long) | 30.25 | ±0.25 | 0.05 | 0.1 | 0.05 | 0.2 |

Synch phase reversal: $1.25 \pm 0.05$ μs after the rising edge of P6

Further phase reversals: $N \times 0.25 \pm 0.02$ μs after the synch phase reversal, where $N = 2, \ldots, 56$ (short P6) or $n = 2, \ldots, 112$ (long P6)

Phase reversal duration: $<0.08$ μs measured between the 10° and 170° points of the phase transition and between the 80% points of any associated amplitude transient

Phase accuracy: within $\pm 5°$ of 0° or 180°

Butterworth filter of bandwidth 25 MHz in order to limit interference in the primary radar band (see also Chapter 5, Figs 5.9 and 5.10).

## 12.3 Reply format

Figure 12.6 shows the Mode S reply signal format. The reply is recognized by the decoding equipment by detecting the pattern of the first four pulses. These preamble pulses are so spaced that they cannot be replicated by the chance overlap of two Mode A or Mode C replies. Any of the four pulse leading edges can be used for target range measurement if overlap by another reply masks some of the pulses.

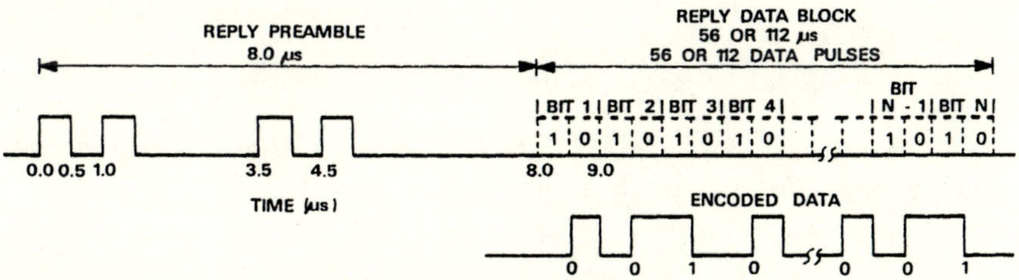


**Fig. 12.6** Mode S reply format.

The reply data are contained in the data block following the four preamble pulses and the data are coded by pulse position modulation. For each bit position, lasting for 1 μs, there is a pulse and a no-pulse period, each of duration 0.5 μs, with a binary 1 being represented by a pulse followed by a no-pulse and a binary 0 by a no-pulse followed by a pulse. In effect the reply data are transmitted twice with each first 0.5 μs indicating a 1 by a pulse and a 0 by the absence of a pulse and also in an inverted form in each second 0.5 μs. This form of coding is very resistant to interference, as it is almost impossible for interference to suppress a pulse in one position and insert a pulse in an adjacent position. As with the interrogation, the reply data can be in the short form of 56 bits or the long form of 112 bits with the last 24 bits forming a combined parity and address field.

Each information bit in the reply message is decoded and assessed for high or low confidence. The simplest decoding procedure is to determine whether the first 0.5 μs period of a reply bit is of greater signal strength than the second 0.5 μs period (binary value 1) or vice versa (binary value 0). The confidence of the decoding decision can be decided by measuring the signal level in the weaker 0.5 μs period: if the signal level is low, no interference is present and the decoding decision is high confidence.

An alternative decoding procedure is to select pulses in the reply preamble to be used as a reference in an analogous way to the selection of an F1 or F2 pulse to aid SSR reply decoding (Chapter 7, Section 7.2.11). Each reply data pulse is compared with the reference for correlation in terms of signal amplitude and/or monopulse value. Correlation indicates a high confidence decision—either binary value 1 or 0 depending upon the position of the pulse in the 1.0 μs bit period. This alternative approach reduces the probability that a strong interfering pulse changes the decoded data or a multipath signal degrades the decision to low confidence.

As an additional check on the quality of the decoded data, their accuracy is also confirmed by parity. In the case of the Mode S reply the parity system adopted allows for error correction to be performed if the reply is corrupted by a single SSR reply of duration 21 μs. The parity system is described in Section 12.4.

## 12.4 Parity and address

The message parity of 24 bits is calculated using a modified cyclic redundancy code based on the following polynomial:

$$G(x) = x^{24}+x^{23}+x^{22}+x^{21}+x^{20}+x^{19}+x^{18}+x^{17}+x^{16}+ \\ +x^{15}+x^{14}+x^{13}+x^{12}+x^{10}+x^{3}+1$$

The parity is calculated at the sending end, combined with the 24-bit aircraft address and added to the message. The parity is recalculated at the receiving end and compared with the combined parity and address attached to the message to confirm, or otherwise, that the message has been transmitted without error and was intended for that aircraft.

The polynomial $G(x)$ has the useful property of supporting both error detection and error correction [12.6, 12.7]. However, only error detection is used on the uplink in order to keep the transponder design simple. The greater processing power available in the ground equipment allows error correction to be employed on the downlink if desired.

Cyclic redundancy codes have the property that if they are cyclically shifted, i.e. if the data are moved sideways and the bits overflowing at one end are added in at the other end, the result is still a valid code. For the polynomial $G(x)$ used for Mode S, the length of the full cyclic code is over two million bits which is clearly both impractical and unnecessary for the application. However, the code can be shortened without losing some of its more important properties.

Another principle is that addition and subtraction is made modulo 2, i.e. without a carry to the adjacent digit. Consequently, addition and subtraction are identical processes. Multiplication is realized by multiple additions and shifts, while division is achieved by multiple subtractions and shifts. Very similar, but not identical, circuits are used for multiplication and division.

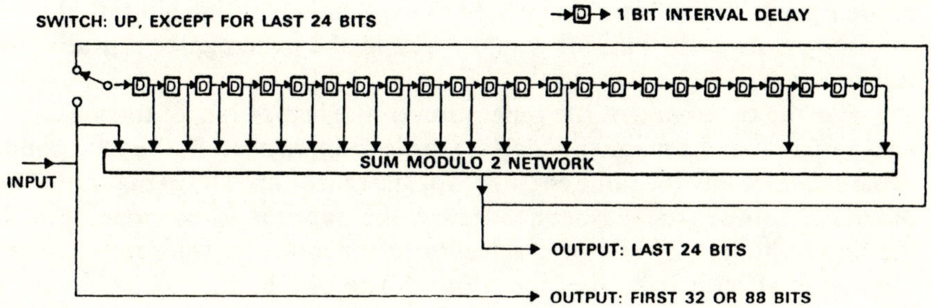


**Fig. 12.7** Sensor encoder.

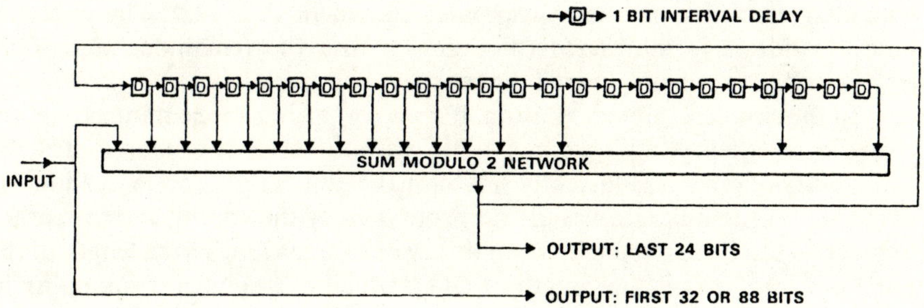


**Fig. 12.8** Transponder decoder.

The circuit shown in Fig. 12.7 is used to generate the sensor message with the taps on the shift register implementing the polynomial $G(x)$. The first 32 bits (or 88 bits for a long message) are transmitted unmodified, but while they are being transmitted they are also being multiplied by $G(x)$ in the feedback shift register. For the final 24 bits of the sensor message the feedback loop is broken and the aircraft 24-bit address is substituted. The result is that the last 24 bits of the message contain a 24-bit parity field added (modulo 2) to the aircraft address multiplied by $G(x)$.

The circuit of Fig. 12.8 is used to decode the message. It is similar to that used to encode the message except that there is no switch at the input to the shift register. The function of the shift register is to divide the message by $G(x)$. The first 32 bits (or 88 bits for a long message) are taken without modification. The division process re-forms the aircraft address as the last 24 bits which are compared with the aircraft's own address for equivalence.

If the wrong aircraft successfully decoded the message, it would discover the wrong address. It would not know whether the difference was due to a transmission error or to the message's being intended for another aircraft. This does not matter since no action is required from the aircraft if the decoding fails. If the transmission had been intended for that aircraft, the lack of response from the aircraft would result in the interrogator repeating the transmission.

For the transponder, the generation of the last 24 bits of the message is also performed in a manner which is simple to implement in the transponder using substantially the same circuit as for the sensor decoding (Fig. 12.9). The difference between the sensor encoder and the transponder encoder is that for the latter the last 24 bits of the feedback input to the shift register are a succession of zeros instead of the address. The 24-bit address is still taken to the modulo 2 circuit so that it is added to the calculated message parity, but this time the address is not multiplied by $G(x)$.

The sensor decoder is an alternative form of division circuit which leaves the remainder in parallel form in the shift register. The circuit used is identical with that employed in the transponder shown in Fig. 12.9. The expected aircraft address is subtracted (modulo 2) from the remainder, and if no transmission error has occurred the result will be zero.

In the simplest implementation, if an error occurs a repeat interrogation is made to attempt to obtain a reply without error. At many sites the probability of error is sufficiently small for the number of repeats to add little overhead to the link. However, the properties of the coding system permit burst errors to be corrected in the majority of cases as long as the length of the burst does not exceed the length of $G(x)$, which is 24 bits corresponding to 24 μs on the downlink. The most likely cause of error is corruption due to a single SSR fruit reply which has a duration of approximately 21 μs. If the remainder in the shift register of Fig. 12.9 is nonzero, its bit pattern is an indicator of the pattern of the burst error. The remainder is an exact map of the error if it occurred in the last 24 bits. Usually this is not the case, and the location of the burst error is not known initially. Of course, it is only sensible to correct errors where an error is suspected, i.e. where the decoding process has indicated that interference is present and the message bits are low confidence.

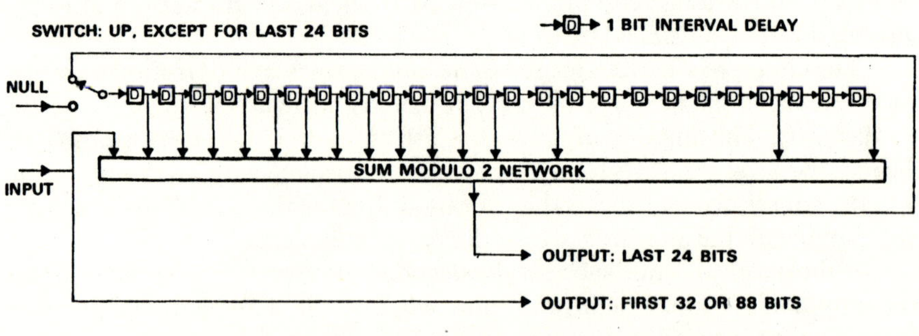


**Fig. 12.9** Transponder encoder and sensor decoder.

As explained above, the full cyclic code remains valid if cyclically shifted, in which case shifting would result in the error residing in the last 24 bits in due course. For each shift the remainder could be recalculated for that "message" and the new remainder could be compared with the pattern of low confidence bits in the last 24 bits. This process would be repeated until a complete correlation occurred when the error would be found and the correct data established.

Fortunately a similar process is available with the truncated version of the cyclic code used for Mode S and, further, the recalculation at each shift can be simplified. The message is subjected to a single left shift and the error pattern is recalculated by either a left shift if its most significant bit is a 0 or a left shift with the addition of $G(x)$ if its most significant bit is a 1. This process is continued until a pattern of low confidence bits which matches the associated remainder error pattern is found in the lowest 24 bits of the message. A fuller description of this procedure is given in ref. 12.6.

## 12.5 Interrogation types

The Mode S interrogation format shown in Fig. 12.2 supports several different forms of interrogation. The interrogation message consists of a number of different data fields, with the last 24 bits always being the combined address–parity field. Interrogation messages of 24 different types are possible, of which eight have been allocated to specific purposes with the remainder currently unassigned [12.8, 12.9]. A summary of the 24 message types is given in Fig. 12.10.

The first few bits of the message identify the message type. If the first two bits are 11 then a Comm-C interrogation is indicated, and this message type is discussed below. In all other cases the first five bits, designated the UF field, define the type. Two of the interrogation types, those with UF fields of UF = 00 and UF = 16, are used for air-to-air surveillance in connection with the TCAS system and are discussed further in Chapter 13.

### 12.5.1 Surveillance interrogation

A main role for a Mode S station is as a surveillance radar to update range, bearing, height and identity data regularly on each aircraft of interest. The principal interrogation types for this purpose are those with UF fields of UF = 04 or UF = 05. The former type, UF = 04, is normally used as this obtains a reply from the aircraft which contains its altitude as would be reported in an SSR Mode C reply. On target acquisition, and at other times as necessary, the message type UF = 05 can be used to obtain a reply containing

```
[UF=00] -3-[RL:1]----4----[AQ:1]-----18-----[AP:24]  Short Air-air Surveillance
[UF=01] --------------27 or 83--------------[AP:24]
[UF=02] --------------27 or 83--------------[AP:24]
[UF=03] --------------27 or 83--------------[AP:24]
[UF=04] [ PC:3 ] [ RR:5 ] [ DI:3 ] [ SD:16 ] [AP:24]  Surveillance, Alt. Request
[UF=05] [ PC:3 ] [ RR:5 ] [ DI:3 ] [ SD:16 ] [AP:24]  Surveillance, Ident Request
[UF=06] --------------27 or 83--------------[AP:24]
[UF=07] --------------27 or 83--------------[AP:24]
[UF=08] --------------27 or 83--------------[AP:24]
[UF=09] --------------27 or 83--------------[AP:24]
[UF=10] --------------27 or 83--------------[AP:24]
[UF=11] [ PR:4 ] [II:4]--------19---------[AP:24]  Mode S Only All-call
[UF=12] --------------27 or 83--------------[AP:24]
[UF=13] --------------27 or 83--------------[AP:24]
[UF=14] --------------27 or 83--------------[AP:24]
[UF=15] --------------27 or 83--------------[AP:24]
[UF=16] -3-[RL:1]--4--[AQ:1]----18---[MU:56] [AP:24]  Long Air-air Surveillance
[UF=17] --------------27 or 83--------------[AP:24]
[UF=18] --------------27 or 83--------------[AP:24]
[UF=19] --------------27 or 83--------------[AP:24]
[UF=20] [PC:3] [RR:5] [ DI:3 ] [SD:16] [MA:56] [AP:24]  Comm-A, Altitude Request
[UF=21] [PC:3] [RR:5] [ DI:3 ] [SD:16] [MA:56] [AP:24]  Comm-A, Identity Request
[UF=22] --------------27 or 83--------------[AP:24]
[UF=23] --------------27 or 83--------------[AP:24]
[11] [RC:2] [NC:4] [          MC:80          ] [AP:24]  Comm-C (ELM)
Note:  XX:M denotes a field designated "XX" which is assigned M bits.
       ---N--- denotes an unallocated field of N bits.
       Unallocated formats are for future expansion.
```

**Fig. 12.10** Summary of Mode S interrogation messages.

the aircraft's Mode A identity code as this information is likely to be of continuing value for ATC purposes. The FS field of a reply (see Fig. 12.11 later) will indicate a recent pilot change of the Mode A code so that the sensor can initiate a readout of the new code.

The PC field in the surveillance interrogation is mainly concerned with acknowledging the conclusion of various data-link transactions, and the details are discussed elsewhere. For completeness the contents of the PC field are listed in Table 12.2.

The RR field (Table 12.3) indicates the type of reply required from the aircraft with the first bit of the field specifying whether the reply is to be short (56 bits) or long (112 bits).

The fields DI and SD are interpreted together with only the values DI = 0, DI = 1 and DI = 7 currently assigned. The meaning of the SD field for the two cases DI = 1 and DI = 7 is shown in Table 12.4. When DI = 0 only the IIS subfield of the SD field has a meaning, and this is discussed below.

**Table 12.2** Contents of the PC field

| *PC code* | *Meaning* |
|---|---|
| 0 | No changes in transponder state |
| 1 | Nonselective All-call lockout |
| 2 | Not assigned |
| 3 | Not assigned |
| 4 | Comm-B close out |
| 5 | Uplink ELM close out |
| 6 | Downlink ELM close out |
| 7 | Not assigned |

**Table 12.3** Contents of the RR field

| *RR code* | *Reply length* | *MB field contents* |
|---|---|---|
| 0–15 | Short | — |
| 16 | Long | Air-initiated Comm-B |
| 17 | Long | Data-link capability |
| 18 | Long | Aircraft Mode A identity |
| 19–31 | Long | Not assigned |

**Table 12.4** Structure of the SD field

| *Position* | *Subfield* | *Name* | *Significance* |
|---|---|---|---|
| *DI = 1* | | | |
| 17–20 | IIS | Interrogator identification | Identification of interrogating site |
| 21–22 | MBS | Multisite Comm-B | Reserve/close out multisite Comm-B |
| 23–25 | MES | Multisite ELM | Reserve/close out multisite ELM |
| 26 | LOS | Lockout | Multisite lockout control |
| 27–28 | RSS | Reservation status | Reservation status request |
| 29–32 | TMS | Tactical message | Comm-A linking |
| *DI = 7* | | | |
| 17–20 | IIS | Interrogator identification | Identification of interrogating site |
| 21–24 | RRS | Reply request | Specifies BDS2 code of MB messages |
| 25 | — | Not assigned | |
| 26 | LOS | Lockout | Multisite lockout control |
| 27–28 | — | Not assigned | |
| 29–32 | TMS | Tactical message | Comm-A linking |

### 12.5.2 Comm-A interrogation

The Comm-A interrogation (UF = 20 and UF = 21) has a similar function to the surveillance interrogation with an additional 56 bits that can be used to pass data to the aircraft. Up to four Comm-A interrogations can be linked to expand the data capacity to 224 bits. The TMS subfield of the SD field is used to indicate the initial component, which must be first, the second and third components (if used), and the final component, which must be last. Each Comm-A interrogation must obtain a reply from the aircraft, otherwise the sensor will repeat the interrogation.

To allow for linked Comm-A messages to be transmitted from several sensors, all of which are within range of the aircraft, a ground site address is included in the IIS subfield of the SD field. The IIS subfield has the same contents as the II field of a Mode S All-call interrogation and is discussed further in Section 12.5.4. The use of linked Comm-A messages for the transmission of data is efficient since the surveillance role is included automatically and the initiation and termination are also accomplished with minimum overhead. Longer data messages require the use of the Comm-C interrogation.

### 12.5.3 Comm-C interrogation

A Comm-C interrogation, which is also known as an uplink extended-length message (ELM), has the first two bits of the message set at 11. The MC field contains 80 bits of data and between two and sixteen Comm-C interrogations can be linked to provide a total message length of up to 1280 bits. The NC field indicates the number of the segment (0, ..., $n-1$) being transmitted. The initial interrogation contains, paradoxically, the last part of the message with the NC field indicating the number of that segment ($n-1$). By this means the initial interrogation informs the transponder of the total number of segments that are to be transmitted.

The transponder does not reply until the final segment (RC = 2) is received. The reply is a Comm-D reply (see Section 12.6.3) which confirms each segment that has been received without error. If necessary the sensor will repeat the missing segments until reception of the complete message is indicated by the transponder. The process is terminated by sending PC = 5 in subsequent surveillance or Comm-A interrogation.

The Comm-C interrogation cannot be used for surveillance purposes, i.e. to update aircraft position, since no new altitude data are available. To accomplish a position update, a surveillance or Comm-A interrogation is also transmitted as a separate activity.

If the aircraft is within the active cover of more than one Mode S sensor, it is possible that more than one sensor may attempt a multiple Comm-C message at the same time. The resultant message at the transponder could be a mix of segments of two unrelated messages. To prevent this occurrence a sensor may first attempt to reserve an uplink ELM activity with the transponder. The reservation is accomplished by including in an earlier surveillance or Comm-A interrogation details of the request and the identity of the sensor making the request. The interrogation would contain the following data:

DI code = 1 (multisite SD)
MES code = 1 (uplink ELM reservation)
RSS code = 2 (report uplink reservation status in UM)
IIS code = sensor site address

The transponder surveillance or Comm-B reply contains, in its UM field, the site address of the sensor for which the ELM activity is currently reserved. Only if this site address is the same as that of the sensor does the sensor proceed with the ELM transaction at that time.

When the uplink ELM message delivery has been completed the sensor must release the ELM activity. This function is performed on transmission of the next surveillance or Comm-A interrogation by including the following data:

DI code = 1 (multisite SD)
MES code = 2 (uplink ELM close out)
IIS code = sensor site address

All data transfers between air and ground are controlled by the ground. To facilitate this control a single-segment Comm-C interrogation is used to control a downlink ELM message which uses multiple Comm-D replies. This aspect is discussed further in Section 12.6.3.

### 12.5.4 All-call interrogation

One form of All-call interrogation uses a modified form of an SSR interrogation which is shown in Fig. 12.3 and discussed in Section 12.2. This form of All-call interrogation has limitations on some occasions, and an alternative form based on the Mode S interrogation format shown in Fig. 12.2 is available.

An All-call interrogation is indicated by UF = 11 (see Fig. 12.10). Apart from the UF and AP fields, only two other fields—PR and II—are defined for the All-call interrogation. The PR field assigns a probability of reply to the transponder controlled by an internal changing random number. This facility is provided to assist sensor recovery if there is a complete failure and knowledge of all aircraft is lost. By reducing the reply probability below unity the self-garbling of aircraft Mode S replies by aircraft in close proximity is reduced to increase the likelihood of receiving some replies clear of interference. These aircraft so acquired can be locked out of replying to further All-call interrogations by data in the PC or SD fields of a surveillance or Comm-A interrogation. The reply probability can then be increased to acquire some of the aircraft which may not have replied to the first All-call interrogation. This process is continued until the reply probability is unity and all aircraft have been acquired. The contents of the PR field are given in Table 12.5.

**Table 12.5** Contents of the PR field

| *PR code* | *Meaning* |
|---|---|
| 0 | Reply with probability = 1 |
| 1 | Reply with probability = 1/2 |
| 2 | Reply with probability = 1/4 |
| 3 | Reply with probability = 1/8 |
| 4 | Reply with probability = 1/16 |
| 5, 6, 7 | Do not reply |
| 8 | Disregard lockout, reply with probability = 1 |
| 9 | Disregard lockout, reply with probability = 1/2 |
| 10 | Disregard lockout, reply with probability = 1/4 |
| 11 | Disregard lockout, reply with probability = 1/8 |
| 12 | Disregard lockout, reply with probability = 1/16 |
| 13, 14, 15 | Do not reply |

The II field can contain a 4-bit sensor site address as in the IIS subfield of the SD and UM fields. This field permits a sensor to disable an acquired transponder from replying to further All-call interrogations from that sensor without inhibiting the transponder from replying to All-call interrogations from other sensors which may yet need to acquire the aircraft.

In the case of the All-call interrogation the address–parity field AP cannot contain the aircraft address since this is not yet known. The "address" used for the All-call interrogation is the same as the broadcast address (see Section 12.5.5) and consists of 24 binary 1s.

### 12.5.5 Broadcast interrogation

Instead of directing a Comm-A interrogation to a specific aircraft, messages of general interest can be broadcast to all aircraft in the antenna beam. Such a broadcast interrogation uses an address consisting of 24 binary 1s instead of the address of a single aircraft. The aircraft receiving such an interrogation are not required to reply since to do so may give rise to garbling of replies from closely spaced aircraft.

The interrogator has no knowledge of whether all aircraft received the broadcast interrogation and cannot repeat the interrogation for those aircraft which unsuccessfully received the data. Consequently the use of the broadcast interrogation is limited to data of noncritical importance or to cases where the same data are repeated at frequent intervals.

## 12.6 Reply types

All Mode S replies are conveyed by the signal format shown in Fig. 12.6. Like the interrogations they can be of short or long form (either 56 bits or 112 bits in length), with the last 24 bits being the parity–address. Figure 12.11 shows the 24 different types of reply message available, of which eight are currently assigned. If the first two bits of a message are 11 then a Comm-D reply is indicated, and this message type is used for the transmission of large amounts of data in an analogous manner to the Comm-C interrogation. Other message types are defined by the first five bits, designated the DF field. The reply types with DF = 00 and DF = 16 are air-to-air surveillance replies used

```
[DF=00] [VS:1] ---7---- [RI:4] ----2---- [AC:13] [AP:24]  Short Air-air Surveillance
[DF=01] --------------27 or 83-------------- [AP:24]
[DF=02] --------------27 or 83-------------- [AP:24]
[DF=03] --------------27 or 83-------------- [AP:24]
[DF=04] [ FS:3 ] [ DR:5 ] [ UM:6 ] [ AC:13 ] [AP:24]  Surveillance, Altitude
[DF=05] [ FS:3 ] [ DR:5 ] [ UM:6 ] [ ID:13 ] [AP:24]  Surveillance, Identity
[DF=06] --------------27 or 83-------------- [AP:24]
[DF=07] --------------27 or 83-------------- [AP:24]
[DF=08] --------------27 or 83-------------- [AP:24]
[DF=09] --------------27 or 83-------------- [AP:24]
[DF=10] --------------27 or 83-------------- [AP:24]
[DF=11] [ CA:3 ] [          AA:24          ] [AP:24]  All-call Reply
[DF=12] --------------27 or 83-------------- [AP:24]
[DF=13] --------------27 or 83-------------- [AP:24]
[DF=14] --------------27 or 83-------------- [AP:24]
[DF=15] --------------27 or 83-------------- [AP:24]
[DF=16] [VS:1] --7-- [RI:4] --2-- [AC:13] [MB:56] [AP:24]  Long Air-air Surveillance
[DF=17] --------------27 or 83-------------- [AP:24]
[DF=18] --------------27 or 83-------------- [AP:24]
[DF=19] --------------27 or 83-------------- [AP:24]
[DF=20] [FS:3] [DR:5] [UM:6] [AC:13] [MB:56] [AP:24]  Comm-B, Altitude
[DF=21] [FS:3] [DR:5] [UM:6] [ID:13] [MB:56] [AP:24]  Comm-B, Identity
[DF=22] --------------27 or 83-------------- [AP:24]
[DF=23] --------------27 or 83-------------- [AP:24]
[11] --1-- [KE:1] [ND:4] [      MD:80      ] [AP:24]  Comm-D (ELM)
Note:  XX:M denotes a field designated "XX" which is assigned M bits.
       ---N--- denotes an unallocated field of N bits.
       Unallocated formats are for future expansion.
```

**Fig. 12.11** Summary of Mode S reply messages.

**Table 12.6** Comm-A interrogations and equivalent Comm-B replies

| *UF field* | *Interrogation* | *DF field* | *Reply* |
|---|---|---|---|
| | (Altitude request) | | (Altitude response) |
| UF = 04 | Surveillance or | DF = 04 | Surveillance or |
| UF = 20 | Comm-A | DF = 20 | Comm-B |
| | (Identity request) | | (Identity response) |
| UF = 05 | Surveillance or | DF = 05 | Surveillance or |
| UF = 21 | Comm-A | DF = 21 | Comm-B |

by the TCAS system and are discussed in Chapter 13. The surveillance or Comm-B reply matches the surveillance or Comm-A interrogation as shown in Table 12.6. The longer form of interrogation or reply is only used when data are to be transferred concurrently with the update of aircraft range, bearing and altitude or identity.

## 12.6.1 Surveillance reply

The most common Mode S reply is the surveillance reply containing altitude data. This form of reply is indicated by a DF field of DF = 04 with the AC field containing the altitude code. This field is allocated 13 bits which is sufficient for all A, B, C and D SSR code pulses plus the X code pulse not currently used in SSR. In the AC field the seventh pulse, corresponding to the SSR X pulse, is renamed the M bit. If the M bit is 0, the altitude code is similar to that used in the SSR Mode C reply—in this case the pulse D1 is zero as it is for SSR. If the M bit is set to 1, the remaining 12 pulses in the AC field give the aircraft altitude in metric units.

The alternative surveillance reply, with DF = 05, replaces the AC field with the ID field which contains the SSR Mode A 4096 identity code. The X pulse in the SSR Mode A code is currently unused as it is in SSR.

The FS field indicates the flight status as shown in Table 12.7.

**Table 12.7** Contents of the FS field

| *FS code* | *Mode A code change* | *SPI pulse present* | *Aircraft* |
|---|---|---|---|
| 0 | No | No | Airborne |
| 1 | No | No | Ground |
| 2 | Yes | No | Airborne |
| 3 | Yes | No | Ground |
| 4 | Yes | Yes | Either |
| 5 | No | Yes | Either |
| 6, 7 | — | Not assigned | — |

**Table 12.8** Contents of the DR field

| *DR code* | *Meaning* |
|---|---|
| 0 | No downlink request |
| 1 | Request to send Comm-B message |
| 2, 3 | Reserved for TCAS |
| 4 | Comm-B broadcast message 1 available |
| 5 | Comm-B broadcast message 2 available |
| 6, 7 | Reserved for TCAS |
| 8–15 | Not assigned |
| 16–31 | Used for downlink ELM (see Section 12.6.3) |

**Table 12.9** Structure of the UM field

| *Position* | *Subfield* | *Name* | *Significance* |
|---|---|---|---|
| 14–17 | IIS | Interrogator identification | Reserved site number for multisite protocols |
| 18–19 | IDS | Interrogator designator | Reservation type: Comm-B, uplink ELM or downlink ELM |

The DR field is a 5-bit field mainly used to indicate to the ground interrogator that a downlink data message is awaiting transmission. All data-link transactions are controlled from the sensor and it must be triggered into requesting the downlink data. The contents of the DR field are shown in Table 12.8.

The UM field contains control and status data as indicated in Table 12.9. The 6-bit UM field is divided into two subfields—the IIS subfield of 4 bits and the IDS subfield of 2 bits. It is used to indicate the current data-link status of the transponder, with the IDS field showing activity with a Comm-B transaction or with an uplink and/or downlink ELM transaction. The interpretation of the IDS subfield is as follows:

0 no activity
1 Comm-B active
2 uplink ELM active
3 downlink ELM active

The IIS subfield indicates the identity of the interrogator involved in the data-link activity.

## 12.6.2 Comm-B reply

The Comm-B reply, DF = 20 or DF = 21, has a similar function to the surveillance reply but with an additional 56 bits that can be used to pass data from the aircraft to the interrogator. All data transfers are controlled by the sensor so that an aircraft can only transmit downlink data when specifically requested by an interrogator. The transfer of the downlink data may be at the instigation of the interrogator, but on occasion the interrogator must be told by the aircraft that downlink data are awaiting transmission.

In the case of sensor-initiated transfer of data, the request is indicated by an RR code greater than 16 (see Section 12.5.1) contained in a surveillance or Comm-A interrogation. The aircraft replies with the data requested. If the data are not received correctly by the interrogator, a repeat request is made. No other acknowledgment action is necessary.

When an air-initiated Comm-B message is available, its existence is indicated to the ground by setting the least significant bit of the DR field in surveillance and other Comm-B replies. The sensor responds with an interrogation with RR = 16 so that the subsequent reply contains the air-initiated message. In this particular case an acknowledgment is necessary since the transponder will not clear the bit in the DR field until it is advised that no repeat of the information is necessary. The next surveillance or Comm-A message includes PC = 4 to indicate the successful reception of the downlink message, and the transaction is completed when the sensor receives a reply to that further interrogation.

Complications can arise when an aircraft is within cover of more than one sensor and more than one air-initiated Comm-B reply is awaiting delivery. The possibility exists that repeats of interrogations or replies, which may be necessary to overcome losses due to interference, could inadvertently terminate a transaction before the data had been correctly transferred. In cases of multiple cover the multisite protocol is used and the initiating surveillance or Comm-A interrogation contains the following:

RR code = 16 (read air-initiated Comm-B)
DI code = 1 (multisite SD field)
MBS code = 1 (Comm-B reservation)
RSS code = 1 (report Comm-B reservation status in UM)
IIS code = sensor site address

The transponder stores the address of the first site requesting the Comm-B data and includes this address in the IIS subfield of the UM field in the reply and other replies. The initiating site and other sensors obtaining replies can determine whether it or another site has responsibility for handling the message.

After the Comm-B message is successfully delivered an acknowledgment is made by including the following in the next surveillance or Comm-A message:

DI code = 1 (multisite SD field)
MBS code = 2 (Comm-B close out)
IIS code = sensor site address

The transponder clears the message (DR = 0) and the transaction is completed when the sensor receives a reply to that further interrogation.

In a similar way as for Comm-A interrogations, up to four Comm-B replies can be linked to increase the data capacity.

### 12.6.3 Comm-D reply

A Comm-D reply, which is also known as a downlink ELM, has the first two bits of the message set at 11. The MD field contains 80 bits of data, and between two and 16 Comm-D replies can be linked to provide a total message length of up to 1280 bits.

All downlink messages are controlled by the ground sensor following an indication of a waiting message in a surveillance or Comm-B reply. In the case of a downlink ELM the surveillance or Comm-B reply has its DR code greater than 15 and the number of Comm-D segments awaiting transfer is determined by subtracting 15 from the DR code. The interrogator transmits a single Comm-C interrogation with field RC = 3. In this interrogation the first 16 bits of the MC field indicate the requested Comm-D segments to be transmitted. The aircraft response to the interrogation is to transmit the requested Comm-D segments. The Comm-D replies are spaced by 136 μs between the start of the replies so that the interrogator knows the duration of the expected response. When this time has elapsed the interrogator determines whether all the expected replies have been correctly received. A further Comm-C interrogation will either request a repeat of selected missing segments or indicate that the complete data have been transferred and terminate the transaction.

If the aircraft is within the active cover of more than one Mode S sensor, it is possible that more than one sensor may attempt a multiple Comm-D message at the same time. The result could be a partial delivery only of the message at each of the two sensors. To prevent this occurrence a sensor will first attempt to reserve a downlink ELM activity with the transponder. The reservation is accomplished by including in an earlier surveillance or Comm-A interrogation details of the request and the identity of the sensor making the request. The interrogation will contain the following data:

DI code = 1 (multisite SD)
MES code = 3 (downlink ELM reservation)
RSS code = 3 (report downlink reservation status in UM)
IIS code = sensor site address

The transponder surveillance or Comm-B reply contains, in its UM field, the site address of the sensor for which the ELM activity is currently reserved.

Only if this site address is the same as that of the sensor does the sensor proceed with the ELM transaction at that time.

When the downlink ELM mesage delivery has been completed the sensor must release the ELM activity. This function is performed on transmission of the next surveillance or Comm-A interrogation by including the following data:

DI code = 1 (multisite SD)
MES code = 4 (downlink ELM close out)
IIS code = sensor site address

## 12.6.4 All-call replies

All-call replies are transmitted in response to one of the two types of All-call interrogations shown in Figs 12.2 and 12.3. It is the main process whereby an interrogator can determine the presence of an aircraft equipped with Mode S and its unique 24-bit address. The All-call reply is indicated by the DF field (Fig. 12.11) with the value DF = 11.

The All-call reply is the one exception where an address is not combined with the calculated 24-bit parity field at the end of the message. The reason for this is that the sensor does not yet know the aircraft address; its acquisition is the purpose of the All-call interrogation and reply, and therefore any bit errors in the reply transmission could not otherwise be recognized. The aircraft address is separately reported in the AA field.

It is also important for the sensor to understand the data-link capability of the airborne equipment. This is reported in the CA field as shown in Table 12.10. Further details of the transponder capability, such as the type of peripherals available on board, are reported in a Comm-B reply upon specific request from the ground.

Once a sensor has acquired an aircraft and determined its Mode S address it no longer needs that aircraft to reply to All-call interrogations. To continue to reply would cause unnecessary interference and could reduce the probability of acquiring other aircraft in a timely manner. The transponder can be inhibited from replying to All-call interrogations by subsequent surveillance or Comm-A interrogations in which the PC field has a value of 1. The All-call

**Table 12.10** Contents of the CA field

| *CA code* | *Capability* |
|---|---|
| 0 | No data-link capability |
| 1 | Comm-A and Comm-B |
| 2 | Comm-A, Comm-B and Comm-C |
| 3 | Comm-A, Comm-B, Comm-C and Comm-D |

lockout is limited to a period of 18 s unless reconfirmed by an instruction from the sensor. By this means a loss of contact with the aircraft would result in the ability to re-acquire the aircraft in due course by using further All-call interrogations.

The All-call lockout is perhaps the weakest part of the Mode S system since its action is to deny detection to other interrogators also within range of the aircraft. To overcome this limitation the interrogator site address is used so that All-call interrogations can be labelled with a specific sensor identification. Alternative All-call lockouts, which specify the site address, inhibit All-call replies only to those All-call interrogations which contain that site address in their II field.

## 12.7 Applications of the data link

The main justification for the adoption of Mode S is the value of the data link in supporting a greater degree of automation in the air traffic management system. The use of the data link would permit the flight of aircraft in increasingly dense airspace with improved safety whilst at the same time allowing aircraft to fly more economical flight paths better suited to the technical capabilities of the airframe.

The increased application of data processing, both on the ground and in the air, is resulting in the origination of more flight data in digital form. The transfer of these data from ground to air and from air to ground can be more effectively achieved by the use of a digital data link. This application reduces the controller and pilot workloads and virtually eliminates misunderstandings.

Automatic collision prediction is a particular example of the use of the data link whereby a sensor determines the occurrence of a future proximity of two aircraft, alerts the pilots and issues advice as to how to avoid a collision. An airborne version of collision avoidance is TCAS which is discussed in Chapter 13.

The Mode S data link is designed for flexibility and expandability and to be suitable for operation in various ATC environments around the world. The applications of the data link can be expected to evolve and expand continually.

Other forms of data link do exist and will exist in the future. Present data links include the analog types using h.f. and v.h.f. voice radio as well as digital types such as the v.h.f. ACARS and SITA Aircom which are mainly used for airline company business. In future data links using satellites can be anticipated.

ATC applications for the Mode S data link can be broadly divided into the following classifications [12.10]:

- ATC instructions
  - Pre-departure clearances
  - In-flight clearances
  - Sector hand-off
  - En route metering to increase system capacity
  - Vector delivery (heading, airspeed, altitude)
  - Altimeter setting
- ATC and flight management support
  - Mode C altitude echo to allow pilot to check for correct operation
  - Position echo to allow pilot to check onboard navigation accuracy
  - Flow management advisories to inform pilot of airport delays etc.
  - Pilot-requested altitude profile
  - Conflict alert
  - Terrain-avoidance advisory
- Weather and environment
  - Airport surface weather conditions
  - Airport forecasts
  - En route forecasts
  - Airborne weather conditions measured by the aircraft and telemetered to the ground
  - Weather boundaries as measured by ground radar
  - Severe weather warnings
  - NOTAM, SIGMET, AIRMET advisories

## 12.8 The open systems interconnection model

SSR started as a means of reinforcing the primary radar echo from an aircraft so as to indicate that it was a "friend". This application required special equipment to be carried by the aircraft, and it was a simple step to code the response to convey additional data such as identity and flight level. Mode S takes this process a step further by increasing the available data capacity. In some eyes SSR Mode S is now a communication system first and a radar system second and should be subject to international communication protocols as far as the data transfer is concerned. The open systems interconnection (OSI) model is such a scheme.

The OSI concept permits dissimilar data communications systems to be interconnected so as to allow the transfer of messages without regard to the media used for the transfer. The essence of OSI is the division of the system into a number of layers, each of which has a functional responsibility with counterparts in the rest of the system. The benefit is that technical differences in the means of transmitting data can be accommodated within the standardized system. The concept allows the data to be passed between the end users without the requirement for the users to define the transmission medium. Mode S is only one such medium, and v.h.f. and satellites may be others.

Seven functional layers are defined which range from the physical means of transferring data to the final processing of the message by the recipient:

*application layer* user control of access to the lower levels
*presentation layer* the general function of supporting the application layer such as encryption or data compression
*session layer* the function of establishing control between the two end-user entities
*transport layer* the end-to-end concept of the communication system as a whole including alternative available networks
*network layer* management of the network connections including routing decisions where internal multiple link connections are available
*data-link layer* control of the access to the transmission media including synchronization, sequencing and error detection
*physical layer* the hardware and transmission medium of the transmission system, such as the transmitter, propagation medium and receiver

The passage of data involves the movement of information starting at the application layer and moving down and up as the network requires. At each layer transition control data, specific to that layer, can be added or removed, with control data appropriate to higher layers being transmitted in a transparent manner.

A data message from ATC to the pilot or flight management computer would start at the application layer and pass through the lower levels until it reached the physical layer, which is the ground lines to the interrogator site. At the site it would rise to the transport layer, which manifests as a gateway to the Mode S system, before descending to another physical layer for the transmission of the signal at radio frequency. The transponder interface is at the transport level as a gateway to the onboard data network at a further physical layer. The data are then passed up to an application layer for the end user.

## References

12.1 Ullyatt, C. Secondary radar in the era of auto-tracking, *IEE Conf. Publ.*, **28**, 140, 1967.
12.2 Ullyatt, C. Sensors for the ATC environment with special reference to SSR, *Electron. Civil Aviat.*, **3**, C1–C3, 1969.

12.3 *Report of the Air Traffic Control Advisory Committee*, Department of Transportation, Washington, DC, December 1969.
12.4 Klass, P. Warning issued on traffic control, *Aviat. Week Space Technol.*, **91**, 37–39, 15 September 1969.
12.5 Mark 3 air traffic control transponder (ATCRBS/Mode S), *ARINC Characteristic 718-3*, ARINC, Annapolis, MD, 10 September 1981.
12.6 Gertz, J. L. Fundamentals of Mode S parity coding, *DOT/FAA/PM-83/6 ATC-177*, Lincoln Laboratory, MIT, Lexington, MA, 2 April 1984.
12.7 Kasami, T. and Matoba, S. Some efficient shortened cyclic codes for burst-error correction, *IEEE Trans. Inf. Theory*, **10**, 252–253, 1964.
12.8 Orlando, V. A. and Drouilhet, P. R. Mode S beacon system: functional description, *DOT/FAA/PM-86/19 ATC-42 (Rev D)*, Lincoln Laboratory, MIT, Lexington, MA, 29 August 1986.
12.9 Report of the SSR improvements and collision avoidance systems panel (SICASP), *ICAO Doc. 9450*, ICAO, Montreal, 1984.
12.10 *Working Paper 49 to SICASP Panel Working Group 3*, ICAO, Montreal, 1984.

## Chapter 13

# Traffic Advisory and Collision Avoidance System

The Traffic Advisory and Collision Avoidance System (TCAS) is an airborne system, developed in the USA, which uses the Mode S system to communicate avoidance decisions between aircraft. TCAS equipment includes a Mode S transponder as part of the system. The system has been given the name Airborne Collision Avoidance System (ACAS) by ICAO, but the American name is used here in recognition of the term in most common usage.

The main purpose of this book is to describe the ground-based civil SSR system and a discussion of TCAS is included to complete the understanding of the SSR Mode S. The reader is directed to specialist publications on TCAS for a detailed description [13.1–13.4].

Apart from its application as an airborne safety system, TCAS is significant as its widespread installation in aircraft is an important stimulus to the deployment of the ground-based Mode S enhancement of SSR. TCAS comprises an interrogator, a transponder and a computer element. Three levels of performance sophistication are accommodated, with corresponding equipment cost implications, to make appropriate models available for general aviation and commercial aircraft. These models are designated TCAS I, TCAS II and TCAS III.

TCAS I, which is intended for small aircraft, is capable of detecting nearby aircraft at a similar flight level and indicating the approximate direction of the other aircraft. The pilot is required to locate the threat aircraft visually and to decide his own avoiding action.

TCAS II is intended for large air carriers. Nearby aircraft are tracked in both the horizontal and the vertical planes, and the time of their closest approach is predicted. The direction of the threat aircraft is indicated and the pilot is advised to avoid certain maneuvers which could make a collision more likely. If the pilot has been unable to identify the other aircraft visually and decide on his own avoiding action, the equipment will give him positive advice on how to avoid a collision. The advice is limited to "climb" or "do not climb", "descend" or "do not descend", as bearing measurement accuracy is insufficient to support horizontal maneuvers. If the threat aircraft is also equipped with TCAS, the escape maneuver is communicated and coordinated with that of the other aircraft using the Mode S data link.

TCAS III (once known as enhanced TCAS II) is also intended for use by larger commercial aircraft and is similar to TCAS II. Its principal difference is the inclusion of an antenna with greater accuracy so that horizontal avoiding maneuvers can be provided.

## 13.1 Identification of intruding aircraft

The first step in the process of avoiding a mid-air collision is the detection of the aircraft that may give rise to a conflict. Until the majority of aircraft are equipped with Mode S transponders, the most common method of target detection is Mode C interrogation. A Mode C interrogation will elicit a Mode C altitude reply from an aircraft fitted with an SSR transponder and an encoding altimeter, which is the minimum fit required. Mode A interrogations are not used because flight level rather than identity is the essential parameter.

Mid-air collisions are more likely in congested airspace. However, garbled replies are also more likely in congested airspace, and steps must be taken to reduce the number of replies received so that those aircraft that are replying have a high probability of detection. Three methods of controlling the number of replies are used; the whisper–shout interrogation, the directional interrogation and the Mode S/A/C All-call interrogation.

### 13.1.1 Whisper–shout interrogation

Figure 13.1 illustrates the transmitter signal format of whisper–shout interrogation. A low level transmission of two 0.8 μs pulses spaced 21 μs apart forms a Mode C interrogation to which close aircraft, or more distant aircraft with more sensitive transponders, will respond. After a delay of about 1 ms, a stronger Mode C interrogation is made which is preceded, 2 μs before the first Mode C pulse, by another 0.8 μs pulse at, or just below, the earlier signal strength. The first two pulses of this new interrogation form an SSR suppression pair which causes the first replying transponders to suppress instead of replying a second time.

Aircraft which did not reply to the first Mode C interrogation may reply

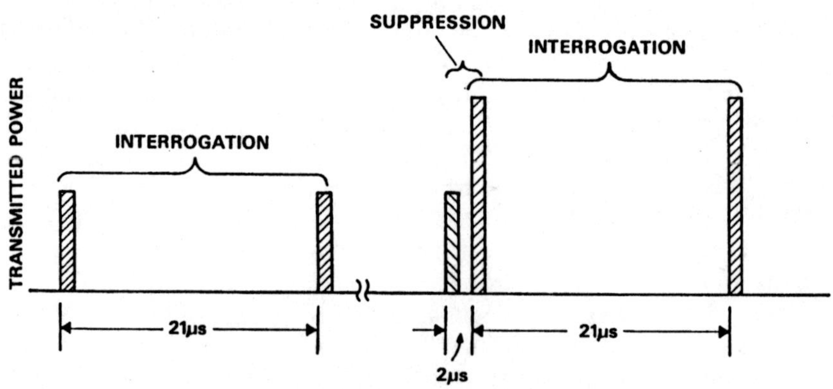


**Fig. 13.1** Whisper–shout interrogation (courtesy of ICAO).

to the second stronger interrogation. The interrogations can be followed by a third and fourth interrogation, each stronger than before. For each interrogation a new set of aircraft will reply with earlier sets being suppressed. In this way the number of aircraft replying at any one time is reduced so that those that do reply have less chance of being garbled by replies from adjacent aircraft.

### 13.1.2 Directional interrogation

Additional subdivision of replying aircraft can be achieved by using a directional antenna analogous to that employed in a ground antenna (see Chapter 4). In the airborne case the antenna is of necessity much smaller and consequently the beam widths are much larger than for the ground antenna. Airborne antenna beam widths are typically 45°–90°.

The directional antenna has a beam with a high gain main lobe, which transmits the P1 and P3 Mode C pulses, and a second beam with approximately omnidirectional characteristics which radiates the P2 sidelobe suppression pulse. Aircraft which receive pulse P2 more strongly than pulse P1 will suppress and not reply.

The antenna must be stepped around 360° so that all directions can be covered. Aircraft ahead could be coming in the opposite direction so that the closing speed is high, and more advanced warning is needed than for aircraft to the rear for which the closing speed is lower. The directional antenna can spend more time in the forward direction, improving the detection of more distant aircraft, than to the rear where the threat is less.

### 13.1.3 Mode C All-call interrogation

A further technique can be used to reduce the number of aircraft replying at a given time so as to reduce the possibility of garble. That is the use of the SSR-only All-call interrogation. This interrogation is shown in Fig. 13.2 and is similar to the Mode A/C/S All-call interrogation shown in Fig. 12.3 (Chapter 12) with the exception that the width of the pulse P4 is only 0.8 μs.

Aircraft equipped with a Mode S transponder will not reply to an interrogation with a short P4 pulse. Such aircraft are acquired by indirect means.

Mode-S-equipped aircraft generate self-triggered "squitter" transmissions at intervals of approximately 1 s. These transmissions are detected by the TCAS-equipped aircraft which keeps a list of the addresses of all Mode-S-equipped aircraft within range and so is able to recognize a new aircraft. Alternatively, All-call replies triggered by a ground Mode S interrogator can be used instead of the squitter transmissions. Both All-call replies and squitter

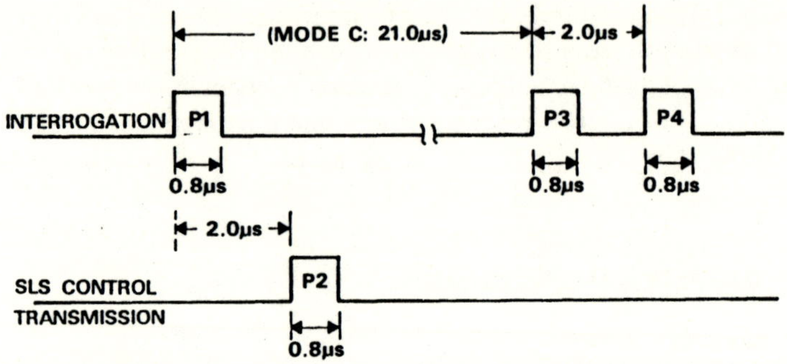


**Fig. 13.2** SSR-only Mode C All-call interrogation.

transmissions are transmitted with a similar Mode S reply format with DF = 11 (Chapter 12, Fig. 12.11).

## 13.2 Track initiation and maintenance

Surveillance is maintained on other aircraft using either the Mode-C-only All-call interrogation in the whisper–shout role, if the aircraft is SSR equipped, or a Mode S interrogation for Mode-S-equipped aircraft. Aircraft within a range of 14 n miles, and at a similar height, are tracked and the tracks are updated about once per second by surveillance interrogations. Following each track update the range, altitude and, in the case of TCAS III aircraft, bearing are updated. Range rate and altitude rate are computed from the track data.

The Mode S interrogations use the short air-to-air form with UF = 00 (Chapter 12, Fig. 12.11). A first Mode S interrogation has the acquisition bit AQ set to a 1 and the reply length bit RL to a 0. This produces a short air-to-air surveillance reply with the reply information field RI indicating the maximum airspeed capability of the replying aircraft as shown in Table 13.1. The VS bit, if set to a 1, indicates that the aircraft is on the ground, in which case it is not tracked.

## 13.3 Threat traffic and resolution advisories

Following the updating of the aircraft tracks, threat detection is performed at two levels. The first level determines that the traffic may approach sufficiently close to cause future concern. A traffic advisory is presented to the pilot so that he can attempt a visual sighting and, if he thinks fit, take corrective action to increase the separation.

**Table 13.1** Contents of the RI field

| *RI code* | *Meaning* |
|---|---|
| 0 | Reply to interrogation with AQ = 0 |
| 1–7 | Reserved |
| 8 | No airspeed data available |
| 9 | Maximum airspeed ⩽ 75 knots |
| 10 | Maximum airspeed > 75 knots and ⩽ 150 knots |
| 11 | Maximum airspeed > 150 knots and ⩽ 300 knots |
| 12 | Maximum airspeed > 300 knots and ⩽ 600 knots |
| 13 | Maximum airspeed > 600 knots and ⩽ 1200 knots |
| 14 | Maximum airspeed > 1200 knots |
| 15 | Not assigned |

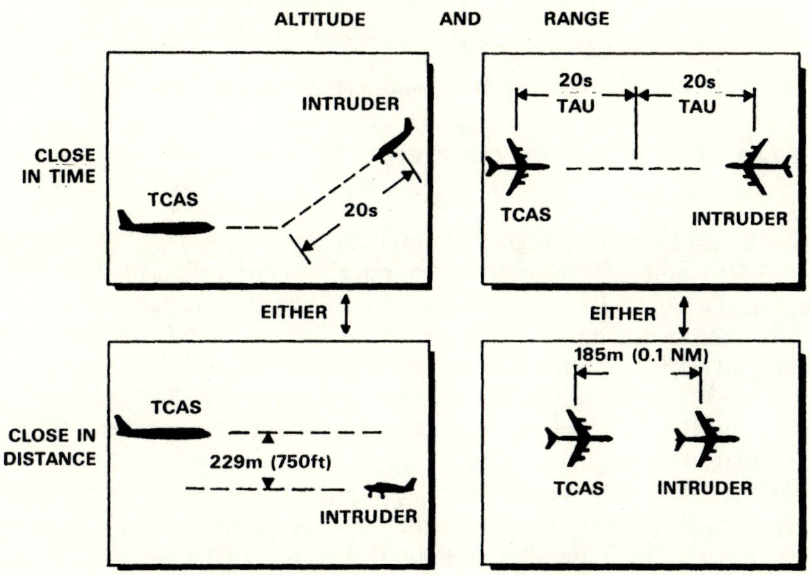


**Fig. 13.3** Threat proximity criteria (courtesy of ICAO).

If the two aircraft continue to approach, typically such that within a minimum time of about 20 s they will be within a range of 0.1 n mile and within 750 ft altitude, a resolution advisory will be issued. Figure 13.3 shows the proximity limits diagrammatically. A resolution advisory will be displayed to the pilot showing a maneuver which will increase the separation of the aircraft (Fig. 13.4). The suggested maneuvers are "climb", "do not climb", "descend" or "do not descend". The advisories may be complemented by limits to the rate of climb or descent. Figure 13.5 shows an instantaneous vertical speed indicator (IVSI) modified with up and down red arrows and

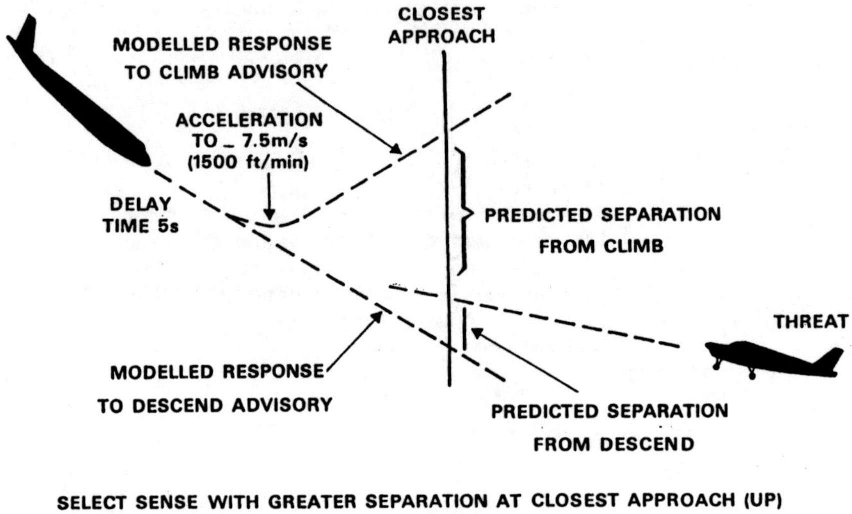


**Fig. 13.4** Maneuver to increase separation at closest approach (courtesy of ICAO).

amber height band bars. Figure 13.5(a) indicates the advisory "descend" and Fig. 13.5(b) indicates the advisory "do not climb faster than 1000 ft/min". In the case of a TCAS-III-equipped aircraft, which has the capability of bearing measurement with an accuracy to about 10°, the advisory can include horizontal maneuver information.

If the threat aircraft is only replying with empty Mode C brackets, such that no altitude data are available, then only traffic advisories can be issued since the relative height of the other aircraft is not known.

If the threat aircraft is also TCAS equipped, the advisory information can be exchanged between them to ensure that they do not both select the same avoidance maneuver, thereby continuing the conflict. The exchange of data is achieved by the use of the Mode S long air-to-air surveillance interrogation and reply (UF = 16 and DF = 16 respectively) with the MU and MV fields carrying the data.

## 13.4 Coordination with ground air traffic control

If the TCAS-equipped aircraft is within range of a ground Mode S station, ground ATC can be informed of the resolution advisories determined using a Mode S Comm-B reply. All Mode S data-link transactions are controlled by the sensor, in this case the ground Mode S interrogator, so that the TCAS aircraft must first inform the sensor that a Comm-B message is

(a) (b)

**Fig. 13.5** Modified instantaneous vertical speed indicator: (a) descend; (b) limit climb to 1000 ft/min.

waiting. When the sensor next interrogates the TCAS aircraft the reply contains data in the DR field (Chapter 12, Fig. 12.11) indicating the existence of an air-derived Comm-B message. The next ground interrogation will then request the Comm-B reply.

## 13.5 Sensitivity level control

While the purpose of the TCAS equipment is to give timely warning of a possible mid-air collision, it must perform this function without causing an unacceptable number of false alarms. Failure to control the number of false alarms would bring the system into disrepute such that valid warnings would not be believed. In addition, unnecessary maneuvers could themselves generate risks.

The balance between high detection efficiency and low false alarm rates requires different detection thresholds for different traffic situations. Aircraft flying at high altitude tend to meet other fast-moving aircraft which are maintaining relatively steady courses. In contrast, aircraft at low altitudes can expect dense traffic in a terminal area which is slow moving and maneuvering.

TCAS threat detection criteria can be changed in one of three ways. The criteria can be set automatically according to aircraft altitude or by a data-link message from a ground Mode S interrogator or manually by pilot action. Sensitivity changes are implemented by changing the minimum altitude

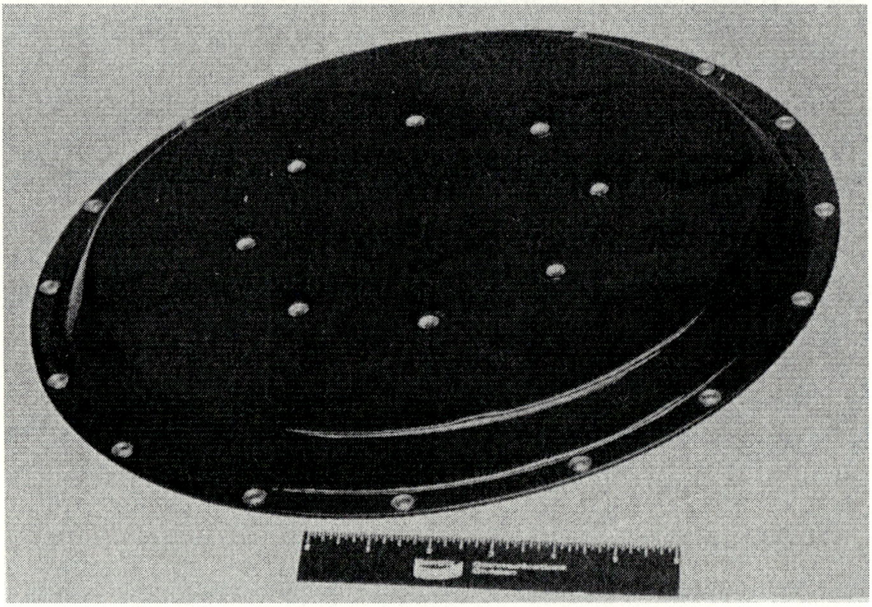

Fig. 13.6 TCAS III directional antenna (Allied Bendix).

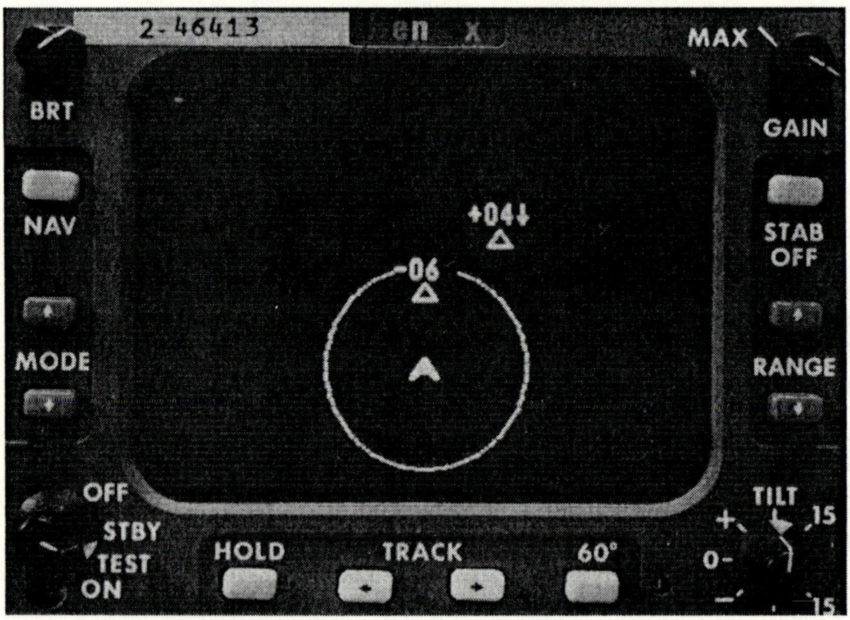


Fig. 13.7 TCAS III pilot display indicating the direction of nearby aircraft (Allied Bendix).

separation (between 950 and 750 ft), the minimum horizontal separation (between 1 and 0.1 n mile) and the time to closest approach (between 30 and 20 s).

## 13.6 TCAS III

TCAS I and TCAS II do not determine the horizontal direction in which the threat may lie and are limited to avoidance advisories in the vertical plane. In some cases an easier maneuver would be a turn, if the necessary direction of the turn were known. TCAS III represents an enhancement in which the direction of the threat is measured and used in the avoidance determination.

An essential element of TCAS III is the directional antenna (Fig. 13.6) which consists of eight top-loaded monopoles on a circle of diameter 10.4 in [13.4]. Electronic steering provides 64 different sum and difference beams. On transmit the difference pattern is used as a control pattern to limit the effective beam width to about 22.5°. On receive the antenna feeds a monopulse receiver which provides a bearing measurement accuracy of $\pm 4°$ bias with 1° standard deviation. Figure 13.7 shows a pilot display with three aircraft in proximity with their relative flight levels. One aircraft, at a relative bearing of 10°, is indicated as a threat.

## References

13.1 Report of the second meeting of the SSR improvements and collision avoidance systems panel (SICASP), *ICAO Doc. 9450*, ICAO, Montreal, 1984.

13.2 Minimum operational performance standards for an active traffic alert and collision avoidance system I (Active TCAS I), *RTCA DO-147*, RTCA, Washington, DC, 1987.

13.3 Airborne collision avoidance systems, *ICAO Circ. 195-AN/118*, ICAO, Montreal, 1985.

13.4 Sinsky, A. I., Reed, J. E. and Fee, J. Enhanced TCAS-II tracking accuracy, *AIAA/IEEE 6th Digital Avionics Systems Conf.*, pp. 577–585, IEEE, New York, December 1984.

# Glossary of Terms

| | |
|---|---|
| ACARS | Aircraft Communications Addressing and Reporting System. A v.h.f. data link used for airline business in the USA and operated by ARINC |
| ACAS | Airborne Collision Avoidance System. The ICAO name for TCAS |
| ADF | Automatic Direction Finder |
| Adsel | Former UK name for Mode S. Acronym for ADdress SELective |
| AFTN | Aeronautical Fixed Telecommunication Network |
| Aircom | Alternative name for ACARS as operated by SITA outside the USA |
| Aircraft address | The 24-bit unique address assigned to Mode-S-equipped aircraft |
| All-call interrogation | A Mode S interrogation which elicits a reply from all transponders in a group |
| ANC | ICAO Air Navigation Commission |
| *Annex 10* | Aeronautical Telecommunications, *Annex 10 to the Convention on International Civil Aviation*, Vol. 1, ICAO, Montreal, April 1985. The principal international document defining SSR |
| Antenna diversity | The use of dual SSR transponder antennas, one mounted on top of the fuselage and one underneath |
| ARINC | Aeronautical Radio INCorporated |
| ASDE or ASMI | Airport Surface Detection Equipment. Also known as ASMI for Airport Surface Movement Indicator |
| ASR | Airport Surveillance (primary) Radar |
| ATC | Air Traffic Control |
| ATCRBS | Alternative name for SSR used in the USA. Acronym for Air Traffic Control Radar Beacon System (pronounced at-crabs) |
| AWOP | ICAO All Weather Operations Panel |
| Beacon | An emitter for purposes of identification. US term for SSR |
| Beacon digitizer | An alternative name for plot extractor used in the USA |
| Broadcast interrogation | A Mode S interrogation directed at all Mode-S-equipped aircraft within cover but for which no reply is required |
| CAA | Civil Aviation Authority, UK |
| Capability report | Mode S reply data indicating the capability of the aircraft Mode S installation |
| Capture | The inability of a transponder to reply to an interrogation |

| | |
|---|---|
| | because it is replying or has just replied to another interrogation or has been suppressed by a sidelobe interrogation |
| Central limit theorem | The postulation that a large combination of random events combine to form a normal, or Gaussian, distribution regardless of the distribution of the individual events |
| DABS | Former US name for Mode S. Acronym for Discrete Address Beacon System |
| dB | Decibel |
| Decay time | The time taken by a pulse to terminate. Usually measured between the 90% point and the 10% point referenced to pulse amplitude. Alternative to fall time |
| DME | Distance-Measuring Equipment |
| DOT | Department of Transportation, USA |
| Downlink | The signal path from aircraft to ground |
| DVOR | Doppler Very high frequency Omnidirectional radio Range |
| ECAC | European Civil Aviation Conference |
| ELM | Extended Length Message. A group of up to 16 individual Mode S message segments, each of 80 bits, which are linked to provide a single interrogation or reply message |
| Emergency codes | Special Mode A reply codes used to indicate certain distress situations, i.e. code 7700—emergency, code 7600—radio communications failure, code 7500—hijack |
| Emergency reply | Special Mode 1, 2 and 3 reply sequence to indicate an emergency. The sequence is a reply, which may or may not be 7700, followed by three replies spaced 4.35 μs each with the code 0000 |
| e.r.p. | Abbreviation for Effective Radiated Power. Transmitter power as enhanced by the gain of the antenna |
| Eurocae | European Organization of Civil Aviation Electronics |
| Eurocontrol | European Organization for the Safety of Air Navigation |
| FAA | Federal Aviation Administration, USA |
| Falling edge | Terminating edge of a pulse |
| Fall time | The time taken by a pulse to terminate. Usually measured between the 90% point and the 10% point referenced to pulse amplitude. Alternative to decay time |
| FIR | Flight Information Region |
| FIS | Flight Information Service |
| Fruit | Unwanted SSR replies, received by an interrogator, which have been triggered by other interrogators. Acronym of False Replies Unsynchronized to Interrogator Transmission—alternatively False Replies Unsynchronized In Time or Friendly Replies Unsynchronized In Time (IFF) |
| Garble | Term applied to chance overlapping of two replies so that the pulse positions of one reply fall close to the pulse positions of |

| | |
|---|---|
| | the other reply, thereby making the decoding of reply data prone to error. See also Synchronous garble |
| IATA | International Air Transport Association |
| ICAO | International Civil Aviation Organization |
| IFF | Identification Friend or Foe. A military identification system operating on the same principles as SSR and on the same frequencies. IFF interrogation modes are 1, 2, 3 and 4. Mode 3 is common to civil Mode A |
| IFR | Induced Field Ratio (in radomes) |
| IFR | Instrument Flight Rules |
| IISLS | "Improved" Interrogator Side Lobe Suppression. A technique whereby interrogation pulse P1 is transmitted via both the main beam and the control beam of an antenna so that a transponder, in a sidelobe direction, more reliably receives a P1–P2 pair |
| ILS | Instrument Landing System |
| Interrogation | Transmission of pulses P1 and P3, and usually P2, with the purpose of obtaining transponder replies within the main lobe of the associated antenna |
| Interrogator | SSR system element which transmits interrogations to transponders fitted on board aircraft which then respond with a reply. Originally the term referred to the transmitter function only, but now it generally includes the receiver function which detects the transponder reply |
| Interrogator identifier | A 4-bit address assigned to a Mode S ground interrogator |
| ISLS | Interrogator Side Lobe Suppression. See Sidelobe suppression |
| JTIDS | Joint Tactical Information Distribution System. A high capacity jamming-resistant military data link using randomly hopped frequencies over several hundred megahertz in L band |
| Label | Block of data accompanying aircraft position plots on an air traffic controller's display. The data normally include SSR Mode A identity (often converted to flight number) and SSR Mode C flight level |
| Leading edge | Has two uses. One is to describe the initiating edge of a pulse (see Rising edge). The other is to describe the detection of the commencement of synchronous replies in a sliding window detector |
| Lockout | Means by which a transponder can be inhibited from replying to specified All-call interrogations |
| LVA | Large Vertical Aperture (antenna) |
| Mk X | IFF identification system similar to SSR |

| | |
|---|---|
| Mk XII | IFF secure identification system. Uses Modes 1, 2, 3, 4 and C |
| Mk XV | IFF secure identification system to replace Mk XII. Uses spread spectrum techniques including frequencies used by Mk X and SSR. Uses multiple interrogation and reply formats |
| Mode 1 | Mk X IFF identity mode. P1–P3 spacing is 3 μs |
| Mode 2 | Mk X IFF identity mode. P1–P3 spacing is 5 μs |
| Mode 3 | Mk X IFF identity mode. Common to civil Mode A. P1–P3 spacing is 8 μs |
| Mode 4 | Mk XII IFF secure identity mode. Uses different interrogation and reply formats to the other modes and encryption techniques to prevent copying by hostile aircraft |
| Mode A | SSR identity mode. Common to military Mode 3. P1–P3 spacing is 8 μs |
| Mode B | Once used as a second SSR identity mode. Not currently used. P1–P3 spacing is 17 μs |
| Mode C | SSR flight level reporting mode. P1–P3 spacing is 21 μs |
| Mode D | SSR mode reserved for possible future use. P1–P3 spacing is 25 μs |
| Mode S | Extension of SSR previously known as Adsel in the UK and DABS in the USA. Each aircraft has a unique 24-bit address which is included in the interrogation. Only the addressed aircraft replies, thereby avoiding the problems of synchronous garble. Both interrogations and replies can contain additional data |
| Monopulse | Radar-receiving processing technique used to provide a precise bearing measurement. Originally developed for primary radar. Its application to SSR is a spin-off from early Mode S development. Monopulse SSR provides greatly improved target detection and reply decoding performance compared with earlier versions of SSR |
| *MOPS* | *Minimum Operational Performance Standards* |
| MOTNE | Meteorological Operational Telecommunication Network Europe |
| MTBF | Mean Time Between Failures |
| MTD | Moving Target Detection (primary radar) |
| MTI | Moving Target Indication (primary radar) |
| MTL | Minimum Trigger Level. Transponder parameter which defines the minimum interrogation signal level which produces a transponder reply for 90% of the time |
| MTTR | Mean Time To Repair |
| Multipath | Interference and distortion effects due to the presence of more than one path between transmitter and receiver. See also Reflections |

| | |
|---|---|
| MURATREC | MUlti RAdar Trajectory REConstitution. A technique for the accurate determination of an aircraft track using simultaneous observations from a number of radars |
| NATO | North Atlantic Treaty Organization |
| Nautical mile (n mile) | A measure used in navigation. There are various definitions but the international unit is equal to 1.151 statute miles, 6076.1033 ft and 1.852 km |
| NDB | Non-Directional radio Beacon |
| NOTAM | NOtices To AirMen |
| OSI | Open Systems Interconnection |
| PANS | Procedures for Air Navigation Services |
| PAR | Precision Approach Radar |
| PARROT | Acronym for Position Adjustable Range Reference Orientation Transponder. A ground-based transponder sited near an SSR interrogator station and used to provide a calibrated response to demonstrate the correct functioning of the station |
| Plot extractor | Signal-processing equipment which converts receiver video into digital target reports suitable for transmission by land lines |
| PPI | Plan Position Indicator. A display which presents radar data in map form |
| PRF | Abbreviation for Pulse Repetition Frequency. Often used to mean interrogation rate (per second). The usage ignores the fact that the interrogation itself includes several pulses |
| Primary radar | A radar system which relies on the detection of reflected energy from the skin of an aircraft in order to locate that aircraft |
| Pulse width | The duration of a pulse usually measured at the 50% voltage points on the rising and falling edges |
| Radar | RAdio Detection And Ranging. A means of locating objects by emitting radio waves and detecting reply signals. The use of a directional antenna enables the object's bearing to be determined. Range is determined by measuring the time delay to the reception of the reply |
| Radome | Radio-transparent window used to protect an antenna principally against the effects of wind and ice |
| Reflections | False responses produced in SSR systems by interrogation via reflecting objects acting as mirrors. The objects are usually man-made such as hangars, office buildings and chain-link fencing, but occasionally can be terrain |
| Reply rate limiting | Used to protect the transponder transmitter from overdissipation in areas of high interrogation rate. The action is to reduce the sensitivity of the transponder receiver so as not to |

reply to the weakest, and hence the most distant, interrogators

Rise time — The time taken by a pulse to commence. Usually measured between the 10% point and the 90% point referenced to pulse amplitude

Rising edge — The initiating edge of a pulse

RSLS — Receiver Side Lobe Suppression. Use of a receiver connected to the antenna control beam and thereby providing means to detect whether a reply is being received via an antenna sidelobe (the signal is then stronger in the control beam than in the main beam)

RSRE — Royal Signal and Radar Establishment. A UK government research establishment that undertakes research into SSR and IFF

RTCA — Radio Technical Commission for Aeronautics

Scan — The rotation period of the ground interrogator antenna. The update period of an ATC display

Sidelobe suppression — A technique to avoid eliciting transponder replies in response to interrogations transmitted via antenna sidelobes

SIF — Selective Identification Facility. An enhancement now a standard part of Mk X IFF

SIGMET — SIGnificant METeorological information

SITA — Société International de Télécommunications Aéronautiques

Sliding window — A method of detecting synchronous replies on successive antenna scans. A minimum number of replies are required for a target to be recognized

SLS — Side Lobe Suppression. A technique to avoid eliciting transponder replies in response to interrogations transmitted via antenna sidelobes

SPI — Special Position Identification pulse. An extra pulse that can be added to an SSR Mode A reply, 4.35 μs after F2, to resolve identity ambiguities. The SPI pulse is activated by pilot action on request from the ground ATC authority. Now little used for its original purpose but sometimes used to assist acquisition by ground computer systems

Squawk — ATC terminology for activating an airborne transponder and setting its identity code on its control panel switches

Squitter — Random reply by a transponder not triggered by an interrogation. A Mode S transponder can deliberately produce a squitter Mode S transmission to announce its presence

SSR — Secondary Surveillance Radar. A radar system which depends upon the cooperation of aircraft by carrying transponders. Measures aircraft range and bearing, as does

| | |
|---|---|
| | primary radar, but additionally provides identification and height. Uses transmitter powers three orders of magnitude less than primary radar for the same range |
| *STANAG* | (NATO) *STANdardization AGreement. STANAG 4193* replaces the earlier *STANAG 5017* |
| Super Beacon | An earlier name for DABS |
| Sweep | The time between the interrogation and that allowed for the reception of replies. Originally the time that the electron beam on a display swept between minimum and maximum range |
| Synchro-DABS | A proposal whereby Mode S interrogations were time marked so that other aircraft could determine the time of transmission and hence measure the range of the responding aircraft. Proposal not pursued |
| Synchronous garble | Garble situation arising and persisting when two aircraft are within 3 km slant range of an interrogator and at a similar bearing. The replies overlap in the receiver |
| Target report | A digital message produced by a plot extractor following receipt of synchronous SSR replies. The target report typically contains data on aircraft range, bearing, identity and flight level. The target report is normally transmitted via modem and land line to an ATC center |
| TCAS | Traffic advisory and Collision Avoidance System; alternatively Threat Conflict and Alert System. A collision detection and warning system based on Mode S and developed in the USA. The US name for ACAS |
| TMA | Terminal Movement Area |
| Track jitter | Radar measurement error which causes an aircraft track to deviate randomly from its true course on an ATC display. Also called track wander |
| Track wander | See Track jitter |
| Trailing edge | Has two uses. One is to describe the terminating edge of a pulse (see Falling edge). The other is to describe the detection of the cessation of synchronous replies in a sliding window detector |
| Transponder | Airborne unit of the SSR system. Detects an interrogator's transmission and responds with a coded reply stating either the aircraft's identity or its flight level |
| Uplink | Ground-to-air signal path |
| VFR | Visual Flight Rules |
| VMC | Visual Meteorological Conditions |
| VOR | Very high frequency Omnidirectional Range |
| X pulse | The center pulse in a transponder reply. Not currently used and therefore always set to zero |

# INDEX

Principal references are shown in bold type.

*The Artech House Radar Library*

Adamy, David L., **Preparing and Delivering Effective Technical Presentations**

Alison, W.B.W., **A Handbook for the Mechanical Tolerancing of Waveguide Components**

Arams, Frank R., ed., **Infrared-to-Millimeter Wavelength Detectors**

Arbenz, Kurt, and Alfred Wohlhauser, **Advanced Mathematics for Practicing Engineers**

Banakh, V.A., and V.L. Mironov, **Lidar in a Turbulent Atmosphere**

Barton, David K., **Modern Radar System Analysis**

Barton, David K. and Harold R. Ward, **Handbook of Radar Measurement**

Beckmann, Peter, and Andre Spizzichino, **The Scattering of Electromagnetic Waves from Rough Surfaces**

Blackman, Samuel S., **Multiple-Target Tracking with Radar Applications**

Blake, Lamont V., **Radar Range-Performance Analysis**

Brookner, Eli, ed., **Radar Technology**

Brookner, Eli, **Phased Array Radars**

Currie, Nicholas C., and Charles E. Brown, eds., **Principles and Applications of Millimeter-Wave Radar**

DiFranco, J.V., and W.L. Rubin, **Radar Detection**

Erst, Stephen J., **Receiving Systems Design**

Fielding, John E., and Gary D. Reynolds, **RGCALC: Radar Range Detection Software and User's Manual**

Hansen, R.C., ed., **Significant Phased Array Papers**

Hovanessian, S.A., **Radar System Design and Analysis**

Hughes, Richard Smith, **Logarithmic Amplification**

Knott, Eugene F., John F. Shaeffer, and Michael T. Tuley, **Radar Cross Section**

Kolosov, A.A., et al., **Over-The-Horizon Radar**

Leonov, A.I., and K.I. Fomichev, **Monopulse Radar**

Lewis, Bernard, Frank Kretschmer, and Wesley Shelton, **Aspects of Radar Signal Processing**

Maksimov, M.V., et al., **Radar Anti-Jamming Techniques**

Meeks, M.L., **Radar Propagation at Low Altitudes**

Mensa, Dean L., **High Resolution Radar Imaging**

Ostroff, Edward D., et al., **Solid-State Radar Transmitters**

Ostrovityanov, R.V., and F.A. Basalov, **Statistical Theory of Extended Radar Targets**

Schleher, D. Curtis, **Introduction to Electronic Warfare**

Sherman, Samuel M., **Monopulse Principles and Techniques**

Skillman, William A., **SIGCLUT: Surface and Volumetric Clutter-to-Noise, Jammer, and Target Signal-to-Noise Radar Calculation Software and User's Manual**

Stevens, Michael C., **Secondary Surveillance Radar**

Torrieri, Don J., **Principles of Secure Communication Systems**

Wehner, Donald R., **High Resolution Radar**

Whalen, Timothy, **Writing and Managing Winning Technical Proposals**

Wiley, Richard G., **Electronic Intelligence: The Analysis of Radar Signals**

Wiley, Richard G., **Electronic Intelligence: The Interception of Radar Signals**

Wiley, Richard G., and Michael B. Szymanski, **Pulse Train Analysis Using Personal Computers**

9 780890 062920